An Introduction to
Quarks and Partons

An Introduction to Quarks and Partons

F. E. Close

Rutherford Laboratory
Didcot, Oxfordshire

1979

ACADEMIC PRESS
LONDON   NEW YORK   SAN FRANCISCO
A Subsidiary of Harcourt Brace Jovanovich, Publishers

Academic Press Inc. (London) Ltd
24–28 Oval Road
London NW1

US edition published by
Academic Press Inc.
111 Fifth Avenue,
New York, New York 10003

Library of Congress Catalog Card Number: 78-54530
ISBN: 0-12-175150-3

Printed in Great Britain by
J. W. ARROWSMITH LIMITED
Winterstoke Road
Bristol BS3 2NT

Preface

Around 1968, electron scattering experiments at Stanford, California (SLAC), gave the first clear hints that pointlike particles existed inside the proton. These were named "partons". Earlier, in 1964, Gell-Mann and Zweig had proposed that the proton and the other elementary particles known at that time were in fact built from more basic entities named "quarks".

The quark model had some success in the 1960s; in particular it enabled sense to be made out of the multitude of meson and baryon resonances then being found. When interpreted as excited states of multi-quark and quark–antiquark systems, these resonances and their properties were understood. Even so, there was much argument at the time as to whether these quarks were really physical entities or just an artefact that was a useful mnemonic and aid when calculating with unitary symmetry groups.

When the Stanford electron scattering data were combined with subsequent neutrino data from CERN, Geneva, it began to look as if the "partons" and "quarks" were one and the same thing. Hence early in the 1970s physicists began to take more seriously the idea that quarks are the fundamental building blocks of the world about us.

Since those days a renaissance has taken place in high-energy physics. A charmed quark was predicted to exist in order that rather elegant ideas on unifying weak and electromagnetic interactions might be consistent with data. Evidence for this charmed quark being present in Nature was found in 1974 and by 1976 its properties were being studied and were being found to be exactly as had been predicted. Quarks were postulated to have a new sort of charge called "colour". This gave predictions that appear to be manifested experimentally. A theory of strong interactions, "quantum chromodynamics", has been developed exploiting this. Although it is still early days, great optimism abounds that we at last have a theory of strong interactions as well as weak-electromagnetic. All of these ideas have grown out of the realization that quarks underwrite the phenomena of the universe, and today practically all high-energy physicists accept this (indeed, very complicated experiments are designed, and their results analysed, with the quark–parton model being used as an axiom).

Given all this interest in quarks, and the extensive use professionals are making of quark model ideas, it seemed that a basic text was called for spelling out in some detail the *basic* ideas. In 1973, I gave a series of lectures on quarks and partons at Daresbury Laboratory, England, which were produced as a report and proved very popular. After the discovery of the J/Ψ meson in November 1974 and the subsequent realisation that charm had been found, I was asked to give lectures on quarks, symmetries and the new physics. Again there was much demand for these. It seems sensible therefore to put all these together and expand them, hereby producing a useful text on the techniques required when working with quarks and partons. This is what I set out to do and I must thank Drs D. M. Scott and C. H. Llewellyn Smith for having encouraged me to begin this venture.

The book is in three parts. Part I describes the basic quark model ideas (if one wishes to think chronologically, this part of the book is roughly "pre-1970"). I have gone into some aspects of this in great detail: building up the baryon wavefunctions is one example, the hope being that one can thereby understand the physics behind the various wavefunction symmetries. In Part 2 the parton model ("post-1970") ideas are presented. I have also dealt with electron scattering kinematics in some detail. Again, here I have treated some topics more than once and from different approaches. I wanted thereby to get the important physics across, in particular the relation between scaling, pointlike particles, and absence of length scales.

During the writing of the book another quark was discovered and a veritable explosion of ideas was taking place. In Part 3 I have given a superficial survey of some exciting recent developments.

In much of the book my intention has been that a first-year experimental graduate student, or good undergraduate theorist, would be able to follow it. In some places there are kinematic details or theoretical ideas which will be of more interest to professionals. My hope is that the book will stimulate and excite the newcomer sufficiently to go on and read the specialist reviews listed in the bibliography.

This book could not have been written were it not for the great benefit I have gained over the years learning from many colleagues in Oxford, SLAC, CERN and at Daresbury and Rutherford Laboratories. To them all I give my thanks. I would also like to thank Drs D. M. Scott, D. Sivers and A. J. G. Hey for their comments on parts of the early manuscript, Dr G. Karl for acting on my behalf in preparing the proofs during my absence in the USA, Mrs J. F. Ling for her excellent typing and finally my wife for her patience and stimulation throughout this project.

CERN, Geneva   F. E. CLOSE
October 1978

Contents

Der kleine Gott . . . In jeden Quark begräbt er seine Nase.

Goethe, *Faust*

Three quarks for Muster Mark.

J. Joyce, *Finnegan's Wake*

INTRODUCTION

1 Why Quarks?

At the beginning of this century Thomson suggested that atoms consisted of electrons embedded in a ball of positive electrical charge. The recently discovered α and β particles had been observed to be deflected in magnetic fields and so were natural weapons with which one could hope to study the atomic substructure, in particular its charge distribution. Beams of β particles were found to pass through atoms with little or no deflection causing Lenard to observe that atoms have vast empty spaces within. The massive positively charged α particles were seen to be deflected by the atom, sometimes undergoing violent collisions which scattered them through large angles. This discovery of significant large angle scattering led Rutherford to suggest that the atom contained a localised massive nucleus of positive charge, the collisions of the α particles with these compact heavy nuclei giving rise to the large angle deflections observed. Furthermore, to support this intuitive picture, Rutherford was able to show by explicit calculation that the angular distribution of the scattered α particles agreed with that expected if they indeed interacted with a massive scattering centre of positive charge Ze.

Thus was the existence of the nucleus and the proton inferred and later the proton was isolated in the laboratory. The 90 or so elements were found not to be "elements" at all but instead composites of Z ($=1 \ldots 90 \ldots$) electrons and a nucleus containing Z protons with accompanying neutrons. These became the new "elements" or elementary particles.

In the 1960s very high energy electron beams were utilised at the Stanford Linear Accelerator Center (SLAC) in an experiment that was analogous to the old α particle one in which the atomic structure was revealed. The electron beam was fired at protons and it was found that the electrons were scattered with large transfers of momentum more frequently than had been anticipated (Panofsky, 1968). The observation of these violent collisions suggested that the proton contained discrete scattering centres within. Furthermore the distribution of the scattered electrons in energy and angle exhibited a phenomenon called scale invariance which suggested that the scattering centres had no internal structure of their own, and hence were "pointlike" (Bjorken, 1967; Feynman, 1969; Bjorken and Paschos, 1969). More recently, collisions of two protons head on at the CERN ISR and elsewhere have shown that debris emerge with large momenta transverse to the collision axis (large p_T) and with a probability much larger than one would expect if the proton were a diffuse distribution of matter. Hence one is again seeing evidence for discrete scattering centres within the proton.

Comparison of the data on electron scattering with the analogous probing by neutrino beams has enabled us to learn about the nature, or quantum numbers, of the constituents of the proton. The large p_T phenomena in proton–proton collisions are hypothesised to arise from direct collisions between these pointlike constituents and from these data we are beginning to learn the rules governing their basic interactions.

The reference to the constituents as pointlike means that they have no internal structure or, more probably, that we have not yet resolved any that they may have. To Rutherford the nucleus appeared pointlike; to uncover the substructure of the proton required higher energy beams or equivalently shorter wavelength probes. In turn, to uncover any substructure to these constituents of the proton will require yet higher energy beams. The arrival of much higher energy lepton beams at Fermilab and the CERN SPS might enable us to uncover such a substructure.

As a result of the above experiments, we have learned that the proton and neutron are therefore not elementary but are made instead of "partons" (which is a generic name for the nucleon constituents). There appear to be two types of parton:

  i. electrically neutral particles that are called gluons. Theoretical prejudice suggests that these may be massless vector particles.

ii. Spin $\frac{1}{2}$ fermion fields called quarks and carrying electrical charges which are fractions $\frac{2}{3}$ or $-\frac{1}{3}$ of a proton's charge. For several years it was believed that three varieties, or "flavours", of quark existed and these were named the up, down and strange quarks. Recently it has become apparent that a fourth flavour exists, known as a charmed quark. It is possible that there are further flavours awaiting discovery. (A fifth flavour was confirmed while this book was in production. A sixth is now confidently predicted.)

1.1 Hadron spectroscopy

It was not a complete surprise to discover that the proton was not elementary. The continuing discovery throughout the 1950s and 1960s of more and more "elementary" particles showed no sign of abating and today we know of several hundreds. Naturally one supposes that they are composites of a few more elementary entities rather like the many atoms (elements!) which were discovered to be composites of electrons, protons and neutrons.

It was found that if three flavours of quark were hypothesised then all particles discovered before November 1974 could be understood as composites of three such quarks (baryons) or of one quark and an antiquark (mesons).

1.1.1 MESONS (QUARKONIUM)

Quarks have spin $\frac{1}{2}$, like the electron, so the mesons which are bound states of quark and antiquark can be thought of as "quarkonium" in analogy to positronium, the bound state of e^- and e^+. To highlight this analogy and see how the three flavours of quark are required we will first recapitulate the energy level structure of positronium.

The e^- and e^+ can couple their spins to 1 (triplet state) or 0 (singlet state). In addition they can have relative orbital angular momentum $0, 1, 2 \ldots$ called $S, P, D \ldots$ states. Coupling the spin ($\mathscr{S}$) with the orbital angular momentum ($\mathbf{L}$) yields the total angular momentum of the system $\mathbf{J} = \mathbf{L} + \mathscr{S}$. The resulting energy levels in atomic physics are labelled $^{2\mathscr{S}+1}L_J$ and so the lowest levels are $^1S_0$ and $^3S_1$, then $^1P_1\,^3P_0\,^3P_1\,^3P_2$ and so on.

If there is only one flavour of quark then the quarkonium system will have a series of levels like those of positronium and each level of quarkonium will correspond to one meson. Indeed there are mesons observed corresponding to each of these levels, namely

$$\left.\begin{matrix} ^1S_0 \\ \pi \end{matrix}\right\}\left.\begin{matrix} ^3S_1 \\ \rho \end{matrix}\right\}; \qquad \left.\begin{matrix} ^1P_1 \\ B \end{matrix}\right\}\left.\begin{matrix} ^3P_0 \\ \delta \end{matrix}\right\}\left.\begin{matrix} ^3P_1 \\ A_1 \end{matrix}\right\}\left.\begin{matrix} ^3P_2 \\ A_2 \end{matrix}\right\}$$

Actually the meson spectroscopy is not quite as simple as this. The π comes in three forms distinguished by their electrical charge, namely $\pi^+\pi^0\pi^-$. There is also another neutral meson (η^0) which has $^1S_0$ quantum numbers. Precisely the same pattern is found at the $^3S_1$ level where we have $\rho^+\rho^0\rho^-$ and another neutral meson, the ω^0. This pattern of three charge states for each of the cited mesons plus the brother neutral appears to run throughout the quarkonium system and would emerge naturally if there were two flavours of quark, up (u) and down (d), since at each level $^1S_0$, $^3S_1$, etc. four possibilities can arise ($u\bar{d}$, $u\bar{u}$, $d\bar{d}$, $d\bar{u}$) and hence four mesons.

Since the discovery of strange particles it has become clear that there are in fact nine mesons at each level. This is certainly the case for $^1S_0$, $^3S_1$, $^3P_0$, $^3P_2$ and will probably prove to be true for the other levels also. Nine is naturally the number of mesons expected at each level of quarkonium if three flavours of quark exist (up, down and strange).

The current situation with regard to the quarkonium or meson spectroscopy is summarised in Table 1.1.

The analogy between positronium and quarkonium is not perfect. If we ignore the fine and hyperfine structure, the energy levels of positronium have the familiar Coulomb structure like hydrogen (Fig. 1.1).

TABLE 1.1
Quarkonium meson spectroscopy

$^1S_0$	$^3S_1$	$^1P_1$	$^3P_0$	$^3P_1$	$^3P_2$
π^+	ρ^+	B^+	δ^+	A_1^+?	A_2^+
π^0	ρ^0	B^0	δ^0	A_1^0?	A_2^0
π^-	ρ^-	B^-	δ^-	A_1^-?	A_2^-
η^0	ω^0	?	ε	D?	f^0
η'^0	ϕ^0	?	S^*	E?	f'^0
K^+	K^{+*}	Q_B^+?	κ^+	Q_A^+?	K^{**+}
K^0	K^{0*}	Q_B^0?	κ^0	Q_A^0?	K^{**0}
$\bar{K}^0$	$\bar{K}^{0*}$	$\bar{Q}_B^0$?	$\bar{\kappa}^0$	$\bar{Q}_A^0$?	$\bar{K}^{**0}$
K^-	K^{-*}	Q_B^-?	κ^-	Q_A^-?	K^{**-}

This has the consequence that if the P-wave excitation is an amount Δ of energy above the S-wave, then a quantity 2Δ of energy injected into a positronium system will excite the constituents into the continuum where they are liberated as free particles.

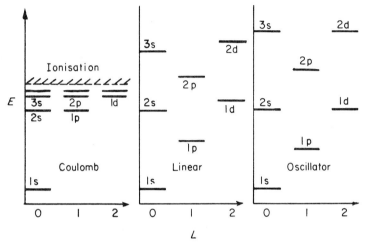

FIG. 1.1. Comparative spectroscopies of Coulomb, linear and oscillator potentials.

In quarkonium the separation of S- and P-wave states is of the order of 500 MeV. If the quarkonium system were Coulombic (like positronium) we would infer that 1 GeV of energy impinging on a pion would liberate free quarks from within. This does not happen in practice. The energy levels of quarkonium are roughly equally spaced in energy (like a harmonic oscillator potential for instance) and pumping in energy just continues to excite higher states which decay and produce many pions, but never free quarks.

1.1.2 BARYONS (3 QUARK NUCLEI)

The other way that quarks bind to form the observed particles is by forming groups of three. The resulting particles are called baryons (of which the proton and neutron are examples) and the resulting system is analogous to the nuclei made from three baryons. As in the quarkonium–positronium analogy we again find that while baryons can be liberated from nuclei, quarks seem not to be liberated from baryons. In this "quark nuclei"–"baryon nuclei" example there is a second place in which the analogy fails and this provides a clue that may explain the different potential structure in quarkonium as against positronium, and hence in turn the origin of quark confinement.

First we will study the similarity with the H^3 and He^3 nuclei made from three nucleons, pnn and npp respectively. Analogously for quarks u and d, the ground state quark nuclei udd and duu are the neutron and proton. The three spin $\frac{1}{2}$ quarks in S-wave couple to give $J = \frac{1}{2}$ or $\frac{3}{2}$, hence the nucleon and Δ. As one excites a quark into P, D, F . . . states so the parity of the state will alternate S(+), P(−), D(+) and so on. This pattern is indeed seen in the observed spectroscopy of baryons.

In the quark world there is a third flavour, s, needed in order to understand the strange particles, in particular the Λ which is an sud quark system. In the nuclear world one can regard these as analogues of hypernuclei where an n is replaced by a Λ, e.g. $^\Lambda He^3$.

This far it looks as if the quark picture of baryons faithfully imitates the nuclear example and so all that we need to do is to go to a nuclear physics text and make trivial modifications throughout. However, there are two crucial differences.

The quark model picture of baryon spectroscopy is analogous to the H^3, He^3, $^\Lambda He^3$ nuclear system but no analogues of He^4, deuterium or U^{235} appear to exist, i.e. if we denote a quark by the symbol q then qq, $qqqq$, q^{235} are not seen while qqq states are seen. Even more dramatic is that the basic nuclear elements—the nucleons (which come in two "flavours", namely the proton and neutron)—can be removed from a nucleus and isolated in the laboratory whereas the basic element of the baryons and mesons, namely the quark, has never been removed from a proton nor does it appear to be freely available in the Universe in isolation.

This nonobservation of quarks raises the question as to whether quarks are real or merely artefacts. The concensus of opinion today is that they are more than a mnemonic and have a genuine dynamical role to play in Nature. Predictions based on quark model ideas have been made that a fourth flavour of quark should exist and that hadrons containing this quark should have rather distinctive properties. Such "charmed" hadrons have recently been discovered giving dramatic support to the quark hypothesis.[1] If quarks indeed have a "reality" then

---

[1] Hence we expect there to be $4 \times 4 = 16$ mesons for each level of quarkonium. Charmed baryons should also exist. See Chapter 16 for more details.

their nonobservation in isolation is a central problem. This question of quark confinement is at the centre of much current research.

The second important difference between the quark and conventional nuclear cases is that in the nuclear case there is no state $ppp(Li^3)$ with the three protons each in S-wave since the Pauli principle excludes two protons having the same quantum numbers. The proton has spin $\frac{1}{2}$ and hence two can exist in S-wave with total spin of 0 but a third cannot then be accommodated in S-wave. In the three-quark system on the other hand, the Ω^- requires three strange quarks each in S-wave in order to make $\Omega^-(sss)$. One can either suppose that the quarks do not obey Pauli statistics individually but instead obey parastatistics of rank 3, i.e. are like bosons individually but take on fermion characteristics when in groups of three, or alternatively one can hypothesise that a further hidden degree of freedom exists for the quarks so the sss can exist with each quark having the same L and $\mathscr{S}$ quantum numbers so long as each one is in a different state of the additional degree of freedom. It is therefore hypothesised that the quarks have a further degree of freedom called colour. Each flavour of quark comes in three colours. By requiring each s quark in the Ω^- to have a different colour then each one can have the same orbital and spin angular momentum quantum numbers without violating the Pauli principle.

1.1.3 COLOURED QUARKS AND QUANTUM CHROMODYNAMICS

Currently theoretical research is investigating whether this colour degree of freedom, which quarks possess but which nucleons do not, is the crucial feature that differentiates the systematics of quark nuclei and conventional nuclei. Although it has yet to be proven, it does appear possible that whereas colourless nucleons can form nuclei with arbitrary numbers of nucleons, the colour degree of "freedom" actually constrains the possible number of quark systems to only a few, e.g. qqq, $q\bar{q}$ but not qq, $qq\bar{q}$, q^{235}, etc. It is suspected that the colour freedom causes an isolated coloured quark to combine with an antiquark of the same colour or with a pair of quarks of different colours so that "white" mesons or baryons are respectively formed. In the meson case the colours of quark and antiquark have "annihilated" whereas in the baryon case the three primary colours, red, yellow and green, have yielded white. As a general rule it is believed that Nature forbids all but "white" hadrons to exist as isolated systems. The background to this belief, and a description of some of the research into it, will be described in Chapter 15.

This colour degree of freedom might consequently also generate the difference between the potentials in positronium (e^+e^-) and quark-onium ($q\bar{q}$ mesons). We stated that the partons come in two forms: quarks and gluons. The quarks have flavour and come in three colours whereas the gluons are flavourless but are hypothesised to come in eight colours. This enables a quark (in any of 3 colours) to couple to another quark (many of 3 colours) and a gluon, the latter having 9 possible primary colour mixtures. One of these is a colour singlet (like the photon which couples electromagnetically to the quarks); the remaining 8 gluons are hypothesised to couple with some strength α_s to the quarks. One can then build up a field theory of strong interactions where quarks carrying colour charge exchange massless coloured vector gluons analogous to QED where fermions with electrical charge couple to photons. This theory of strong interactions is called QCD or quantum chromodynamics.

In QED an isolated electrical charge polarises the vacuum and surrounds itself with a virtual cloud of electron–positron pairs (Fig. 1.2(a)). A test charge placed some distance away will feel a small charge density due to this cloud. Bringing the test particle nearer will cause it to

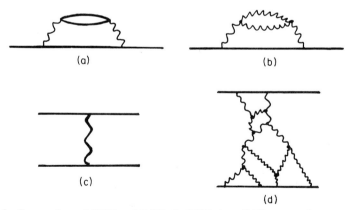

FIG. 1.2. Comparison of QED and QCD. In QED the solid lines are electrons and the wiggly lines photons; only topologies (a) and (c) exist. In QCD the solid lines are quarks and the wiggly lines gluons. Topologies like (b) and (d) can now also exist due to the existence of triple gluon vertices.

probe inside the cloud and so the effective charge density that it feels will rise. Hence α_{eff} increases in QED at short distances. This can be shown rigorously.

In QCD an isolated colour charge (quark) will analogously surround itself with a virtual cloud of coloured quark–antiquark pairs, and these will cause α_s to increase at short distances (Fig. 1.2(a)). However, the gluons themselves carry colour charge (contrast the electrically neutral photon in QED). Hence a gluon can couple to a pair of gluons and so the quark can also surround itself with gluon pairs (Fig. 1.2(b)). This tends to cause α_s to decrease at short distances. In a world of eight gluon colours and less than 16 quark flavours the net effect is that α_s decreases at short distances, tending asymptotically to zero. This implies that in experiments where the proton is probed at short distances (e.g. high momentum transfer processes) the quarks will appear to be quasi-free. This indeed appears to be the case experimentally. At moderately short distances α_s will be small and single gluon exchange will occur between the quarks in the hadron (Fig. 1.2(c)). This will yield phenomenology similar to QED; in particular a hyperfine splitting will arise between the singlet and triplet spin states of $q\bar{q}$ analogous to that in e^+e^- or in hydrogen (the latter being the source of the 21 cm wavelength photons which yield the "autograph" of hydrogen). This is hypothesised to be the source of the $\pi-\rho$, $K-K^*$ and $N-\Delta$ mass differences.

In QED the Coulomb potential persists to large separations giving rise to the Coulomb–Darwin–Breit potential ($1/r$ to leading order) with its associated energy level spectrum with rapid ionisation and free e^- or e^+ production. In QCD at large separation α_s increases and diagrams like Fig. 1.2(d) (where gluons fragment and recombine) can contribute. These topologies have no counterpart in QED where only one photon exists as against eight colours of gluon in QCD. It is hypothesised that this generates a non-Coulomb potential (linear, Harmonic oscillator . . . ?) with energy levels as in Fig. 1.1 and no quark ionisation (quark confinement).

The basic ideas as to how quarks generate the observed hadron spectroscopy are described in Part 1 (in particular Chapters 3, 4, 5) and some more recent ideas on the hyperfine splittings are found in Chapter 17. The new spectroscopy associated with the existence of a fourth flavour of quark (charm) is described in Chapter 16. Some attempts to understand the effects that quark confinement might have in spectroscopic and other properties of hadrons are described in Chapter 18.

An introduction to the upsurge of interest in field theories and quantum chromodynamics is given in Chapter 15.

1.2   Hadron substructure and scaling

When a lepton scatters from a target it does so by transferring energy and momentum to the target via a virtual current (photon or W boson for electromagnetic or weak interactions respectively). When large energy and momentum are transferred then, by the uncertainty principle, the current can resolve very small space–time distances and hence reveal the granular ("parton") substructure of the target.

If the partons are truly elementary then the interaction of the current with a parton in the target is a function only of the *ratio*, x, of the parton's energy to the current's energy in the laboratory or target rest frame (in a general frame x is the ratio of the parton and target Mandelstam invariants $s = (p_1 + p_2)^2$ where p_1 is the current and p_2 the parton or target four momenta). In particular the scattering is independent of Q^2, the invariant squared mass of the virtual current probe. The dependence on only the *dimensionless* ratio x is known as scale invariance or scaling since no energy or length scale governs the interaction (the partons are "pointlike").

From the energy and angle of scatter of the lepton one can infer the value of x of the parton with which it has interacted and by accumulating events at a given x value one can verify the Q^2 independence (scaling) to good approximation in the data.

If the partons (quarks) have a structure themselves then this may be resolved if a current with better resolution (shorter wavelength, larger Q^2) probes the system. The parton carrying momentum fraction x will be seen to be a system of "prepartons" each carrying some fraction y of the momentum x $(0 \leqslant y \leqslant 1)$. As Q^2 is increased and this new structure is resolved there will be a violation of scaling (the size of the parton clouds sets an intrinsic scale of length which is resolved when $Q^2 \gtrsim Q^2_{\text{critical}}$). Hence at a given x there will now be a Q^2 dependence in the scattering.

Eventually at very high Q^2 the prepartons will be seen. If their internal structure is not resolved then a new scaling regime will emerge in the data. The average momentum of a preparton will be smaller than the average momentum of a parton. However, by momentum conservation the total momentum of all the prepartons will be the same as that of the partons. Hence if the distribution of partons' momenta is like Fig. 1.3 (dotted line) the prepartons' momentum distribution might be like the solid line. Since the partons are resolved at moderate Q^2 and the

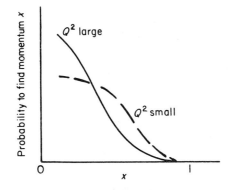

FIG. 1.3. Parton and preparton momentum distributions (dotted and solid lines respectively) resolved at small and large Q^2 respectively (large and short distances).

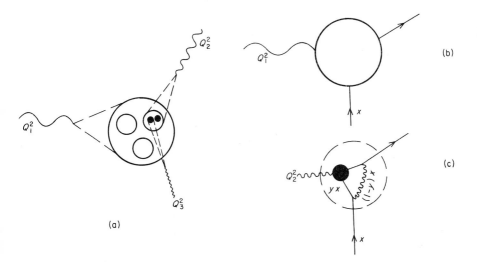

FIG. 1.4. Increasing resolution with increasing Q^2 yields a different picture of the proton. (a) Prequarks: $Q_1^2 < Q_2^2 < Q_3^2$ yields coherent proton, quark structure and finally prequark structure as successive granular layers are revealed. (b) and (c) Field theory: (b) small Q_1^2 sees a parton of momentum x; (c) at larger Q^2 we see that the quark has radiated a gluon and lowered its momentum to xy.

prepartons at much larger Q^2 then the data for the momentum distribution of the scattering centres as a function of Q^2 will show a transition from the dotted to the solid curve (scaling violation).

There is some indication in the data that such a pattern of scaling violation may be present. The data at moderate Q^2 values are consistent with quarks being the granular scattering centres in the proton. At larger Q^2 we may be seeing the first revelation of a deeper layer of matter (prequarks ...) (Fig. 1.4(a)) or evidence for a field theory structure based on quarks and gluons where quarks are dressed with gluon and quark–antiquark clouds. As Q^2 increases this cloud is probed (Fig. 1.4(b), (c)). These ideas are developed in more detail in Chapters 9 and 11.

The general phenomenology of inelastic lepton scattering and large p_T hadron collisions is described in Part 2 with particular reference to the possibility that these phenomena indeed reveal evidence for a quark substructure in the proton.

PART 1
SYMMETRIES, HADRON SPECTROSCOPY
AND QUARKS

2 SU(N) Symmetries

2.1 Multiplets and quarks

On studying the spectrum of the known particles one finds a large
number of cases where two or three particles exist with the same spin
and parity and mass (to within a per cent or so), the only distinction
among them being the different magnitudes of their electrical charge.
Obvious examples include the neutron and proton

$$n(939 \cdot 5 \text{ MeV}), \ p(938 \cdot 3 \text{ MeV})$$

or the three pions $\pi^-(139 \cdot 6 \text{ MeV})\pi^0(135 \text{ MeV})\pi^+(139 \cdot 6 \text{ MeV})$.

These families are called isospin multiplets and one can imagine that
if electromagnetism did not exist, or could be "turned off", then the
neutron and proton would become a single entity—the nucleon;
similarly the π^-, π^0, π^+ would become a single entity—the pion. This
suggests that the eigenstates of the strong interactions exhibit a
degeneracy which is lifted when electromagnetic interactions are
turned on.

Degenerate energy levels or eigenstates are traditionally a
consequence of an underlying symmetry of the Hamiltonian. A familiar
example in atomic physics is the Zeeman effect where a degeneracy of
energy levels is lifted when a magnetic field is applied. A rotational
symmetry of the Hamiltonian has been broken by the magnetic field
which has defined a direction ($\hat{z}$) in space. The $2J_z + 1$ orientations
along $\hat{z}$ of the state's angular momentum J give rise to different energy
levels.

The interrelation between symmetry and degeneracy of energy levels carries over to the particle case. The Hamiltonian of the strong interactions has an SU(2) symmetry structure, isospin symmetry. Electromagnetism (charge) defines a direction in isospin-space (I_3), breaks the isospin symmetry and hence lifts the degeneracy. An example is the nucleon which shows two faces—the neutron and proton.

The simplest example of a multiplet of SU(2) is the fundamental or two-dimensional representation which is an object that can occur in two states labelled up and down or (u, d). The proton and neutron form an example of such a representation of the SU(2) isospin group with the proton being the up ($I_3 = +\frac{1}{2}$) and neutron the down ($I_3 = -\frac{1}{2}$) versions of the nucleon. Higher dimensional representations can exist within the SU(2) symmetry and in the particle spectrum the three-dimensional ("regular representation") is seen (e.g. $\pi^- \pi^0 \pi^+$) and also the four-dimensional ($\Delta^- \Delta^0 \Delta^+ \Delta^{++}$). Even higher dimensional representations of SU(2) are allowed mathematically but do not appear to be employed in the particle spectrum.

The strange particles $K^+ K^0$; $\Sigma^- \Sigma^0 \Sigma^+$ etc. also form isospin multiplets. One can find several examples of isospin multiplets having the same spin and parity but with different magnitudes of strangeness. The masses of these multiplets are of the same order of magnitude, increasing by about 150 MeV for each unit increase in the magnitude of the strangeness. An illustrative example of this is the system $\Delta(1235)$, $\Sigma^*(1385)$, $\Xi^*(1530)$, $\Omega(1670)$ having spin $\frac{3}{2}$ and strangeness $0, -1, -2, -3$ respectively.

Taking this and other similar examples one can form families of particles characterised by isospin and strangeness which are in

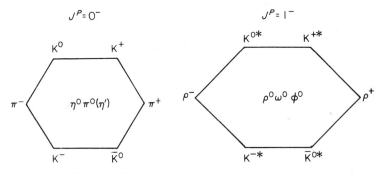

FIG. 2.1. Pseudoscalar and vector meson nonets.

representations of SU(3) (the SU(2) of isospin having become SU(3) due to the incorporation of strangeness). This symmetry is clearly broken by about 20 per cent or more but the underlying multiplet structures are still apparent in the spectrum (e.g. Figs 2.1 and 2.2).

The fundamental representation of SU(3) is three dimensional. Higher dimensional representations are allowed, e.g. **6**, **8**, **10**, **27**, etc.

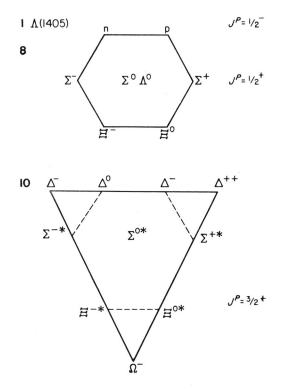

FIG. 2.2. Octet $\frac{1}{2}^{+}$ and decuplet $\frac{3}{2}^{+}$ baryons.

Are these multiplets realised in Nature? Some of them are but not all. The data can be summarised as follows.

Mesons are found in singlets and octets (nonets). Two familiar examples of such nonets are shown in Fig. 2.1 for pseudoscalar (0^{-}) and vector (1^{-}) mesons. We refer to this mesonic structure as nonet since for every octet of particles with a given spin and parity there appears to be a singlet nearby in mass with the same spin-parity. The physical states are often mixtures of **1** and **8** (e.g. ω and ϕ in the vector mesons).

Baryons are found in singlets, octets *and* decuplets. The octet with spin-parity $\frac{1}{2}^+$ and decuplet with $\frac{3}{2}^+$ are familiar and illustrated in Fig. 2.2. The lowest mass singlet state known to exist is the $\Lambda(1405)$ with $J^P = \frac{1}{2}^-$, there being no singlet $\frac{1}{2}^+$ partner to the **8**, $\frac{1}{2}^+$ containing the proton.

Some questions that immediately come to mind are: (i) Why do mesons show nonet structure whereas baryon octets do not appear to have nearby singlet states? (ii) Why are there decuplets of baryons but not of mesons?

Another noticeable feature of the spectroscopy is that while the particles appear to be in **1**, **8** and **10** dimensional SU(3) multiplets, the basic triplet representation is not utilised by Nature.

Gell-Mann (1964) and Zweig (1964) suggested that the fundamental triplet does exist and that it contains three *quarks*. The jargon refers to these as three flavours of quark. The quark was hypothesised to be a building block in the sense that individual quarks do not exist in isolation but bind with antiquarks or pairs of quarks to form $q\bar{q}$ states (mesons) or qqq (baryons). From the rules for combining representations of SU(3) one then finds that

$$q\bar{q} \equiv \mathbf{3} \otimes \mathbf{\bar{3}} = \mathbf{1} \oplus \mathbf{8}$$

$$qqq \equiv \mathbf{3} \otimes \mathbf{3} \otimes \mathbf{3} = \mathbf{1} \oplus \mathbf{8} \oplus \mathbf{8} \oplus \mathbf{10}$$

The absence of decuplet mesons and the **1**, **8**, **10** pattern of baryons therefore emerges naturally from the $q\bar{q}$ and qqq structure. Before showing how these multiplet dimensionalities actually come about we will survey some of the essential features of unitary symmetry. This can be discussed mathematically without reference to quarks. The dimensions of the representations of the group can be obtained and the particles assigned to them. The transitions among the particles and other of their static properties can be related by the machinery of SU(3) Clebsch–Gordan coefficients (e.g. Wigner, 1959; Rose, 1957; Particle Data Group, 1976; de Swart, 1963). The philosophy would be that one accepts that an SU(3) symmetry occurs in the underlying strong interaction Hamiltonian and is manifested in the particle spectrum but the origin of this symmetry is not discussed.

To motivate the source of this strong interaction SU(3) symmetry one supposes that quarks fill the fundamental representation. Higher dimensional representations can then be formed by combining fundamental representations qq, qqq, $q\bar{q}$ etc. One can regard the quarks merely as a mnemonic and a useful tool which aid one in calculating SU(3) matrix elements or in building up the wavefunctions for various representations of the group. This is illustrated in Chapter 3 and the reader whose interest is primarily in quarks can profitably proceed there directly and bypass the formal discussion of unitary symmetry.

Of course one can believe that the quarks are more than just a mnemonic and instead hypothesise that they are the entities out of which the mesons and baryons are physically constructed. In the mid-1960s the feeling was that quarks were probably just a mnemonic and had no physical meaning. Today physicists tend very much towards the belief that quarks are physical entities that are dynamically confined within mesons and baryons. The observed multiplet structure and approximate SU(*N*) symmetry of particle physics then has its origin in *N* flavours of quark filling the fundamental representation of SU(*N*) and the composite quark systems yield the hadron spectroscopy. In this picture the discussion of Chapter 3 has immediate physical relevance and the following description of unitary symmetry is primarily of use for mathematical completeness and for defining concepts that will pervade subsequent chapters.

2.2   Rudiments of unitary symmetry: SU(*N*)

2.2.1   SU(2)

2.2.1.1   *The fundamental representation and general idea*

The spin independence of nuclear forces and isospin symmetry in particle physics are two familiar examples of phenomena associated with an underlying SU(2) symmetry group structure.

The fundamental representation of SU(2) is a doublet

$$\chi \equiv \begin{pmatrix} u \\ d \end{pmatrix} \tag{2.1}$$

As an illustrative example we can consider a particle with spin $\frac{1}{2}$. This particle can have its spin projected up or down along the z-axis; we shall refer to these two states as u, d respectively. Apart from a phase factor these two states will remain invariant under a rotation about the z-axis.

Rotation[1] about some other axis (e.g. the y-axis) by an angle θ will transform the states as follows (e.g. Gasiorowicz, 1966, p. 38):

$$\begin{pmatrix} u' \\ d' \end{pmatrix} = \begin{pmatrix} \cos\dfrac{\theta}{2} & \sin\dfrac{\theta}{2} \\ -\sin\dfrac{\theta}{2} & \cos\dfrac{\theta}{2} \end{pmatrix} \begin{pmatrix} u \\ d \end{pmatrix} \tag{2.2}$$

i.e.

$$u \equiv \begin{pmatrix} 1 \\ 0 \end{pmatrix}, \qquad d \equiv \begin{pmatrix} 0 \\ 1 \end{pmatrix} \tag{2.3}$$

$$u' = \begin{pmatrix} \cos\dfrac{\theta}{2} \\ -\sin\dfrac{\theta}{2} \end{pmatrix}, \qquad d' = \begin{pmatrix} \sin\dfrac{\theta}{2} \\ \cos\dfrac{\theta}{2} \end{pmatrix} \tag{2.4}$$

Notice that the 2×2 matrix is unitary and hence the norm

$$\chi'^{+}\chi' \equiv \chi^{+}U^{+}U\chi = \chi^{+}\chi \tag{2.5}$$

is preserved $\left(\text{where } \chi \equiv \begin{pmatrix} u \\ d \end{pmatrix} \text{ etc.}\right)$.

In general we can consider the transformations

$$\chi' = U\chi \tag{2.6}$$

where U is a 2×2 unitary matrix. The group of such transformations is known as the SU(2) group. The general form for U is conventionally written as[1]

$$U \equiv \exp\left(\tfrac{1}{2}i\theta\hat{\mathbf{n}} \cdot \boldsymbol{\sigma}\right) \tag{2.7}$$

where θ is a measure of the rotation about the axis $\hat{\mathbf{n}}$ and $\tfrac{1}{2}\boldsymbol{\sigma}$ are 2×2 matrices. These matrices, $\tfrac{1}{2}\boldsymbol{\sigma}$, are called the generators of the infinitesimal transformations since for infinitesimal θ

$$\chi' \to \chi + \delta\chi$$
$$\delta\chi \equiv i\theta\hat{\mathbf{n}} \cdot \left(\tfrac{1}{2}\boldsymbol{\sigma}\chi\right) \tag{2.8}$$

Since U is unitary then det $U = 1$. Having written $U \equiv \mathrm{e}^{\frac{1}{2}i\hat{\mathbf{n}}.\boldsymbol{\sigma}\theta}$ then utilising the fact that $\det(\mathrm{e}^{A}) \equiv \mathrm{e}^{\text{trace }A}$ for matrices A, it follows that the

[1] We rotate the axes by θ, the states by $-\theta$.

2×2 matrices $\boldsymbol{\sigma}$ have

$$\text{Trace } \boldsymbol{\sigma} = 0 \tag{2.9}$$

Furthermore, $U^{-1} \equiv e^{-\frac{1}{2}i\hat{n}.\boldsymbol{\sigma}\theta}$ while $U^{+} \equiv e^{-\frac{1}{2}i\hat{n}.\boldsymbol{\sigma}^{+}\theta}$ and these two quantities are identical for unitary matrices. Hence

$$\boldsymbol{\sigma}^{+} = \boldsymbol{\sigma} \tag{2.10}$$

Therefore the matrices $\boldsymbol{\sigma}$ are the set of 2×2 traceless Hermitian matrices, viz.

$$\boldsymbol{\sigma} \equiv \begin{pmatrix} a & b \\ b^* & -a \end{pmatrix} \tag{2.11}$$

with a real and normalised so that $a^2 + b^2 = 1$. There are only *three* independent matrices of this type and these are conventionally chosen to be the Pauli matrices

$$\sigma_1 = \begin{pmatrix} 0 & 1 \\ 1 & 0 \end{pmatrix}, \quad \sigma_2 = \begin{pmatrix} 0 & -i \\ i & 0 \end{pmatrix}, \quad \sigma_3 = \begin{pmatrix} 1 & 0 \\ 0 & -1 \end{pmatrix} \tag{2.12}$$

Notice that these matrices do not commute; instead they satisfy the following commutation relations:

$$[\tfrac{1}{2}\sigma_i, \tfrac{1}{2}\sigma_j] = i\varepsilon_{ijk}(\tfrac{1}{2}\sigma_k) \tag{2.13}$$

(where i, j, k = 1, 2, 3; $\varepsilon_{123,231,312} = +1$; $\varepsilon_{213,132,321} = -1$). This is known as the algebra of the generators of $SU(2)$ and the ε_{ijk} are the structure constants of the group.

We can generalise from here and define generators abstractly by the algebra

$$[S_i, S_j] = i\varepsilon_{ijk}S_k \tag{2.14}$$

We have found particular 2×2 matrices which satisfy this algebra and act on the (fundamental) two-dimensional representation of $SU(2)$. For general N-dimensional representations of $SU(2)$ one can find $N \times N$ matrices satisfying this algebra, and the multiplets which are infinitesimally transformed by these matrices are N-dimensional representations of $SU(2)$. A particular example, known as the regular representation, will be illustrated in section 2.2.1.4.

Returning to the two-dimensional fundamental representation, notice that the matrix $\tfrac{1}{2}\sigma_3$ is diagonal and hence that (u, d) are eigenstates of it with eigenvalues $(+\tfrac{1}{2}, -\tfrac{1}{2})$ respectively.

$$(\tfrac{1}{2}\sigma_3)u \equiv \tfrac{1}{2}\begin{pmatrix} 1 & 0 \\ 0 & -1 \end{pmatrix}\begin{pmatrix} 1 \\ 0 \end{pmatrix} = \tfrac{1}{2}\begin{pmatrix} 1 \\ 0 \end{pmatrix} \equiv \tfrac{1}{2}u$$

$$(\tfrac{1}{2}\sigma_3)d \equiv \tfrac{1}{2}\begin{pmatrix} 1 & 0 \\ 0 & -1 \end{pmatrix}\begin{pmatrix} 0 \\ 1 \end{pmatrix} = -\tfrac{1}{2}\begin{pmatrix} 0 \\ 1 \end{pmatrix} \equiv -\tfrac{1}{2}d$$

(2.15)

Hence the various states in an SU(2) multiplet are characterised by their values of $\langle\tfrac{1}{2}\sigma_3\rangle$. The matrices $\sigma_{\pm} \equiv \tfrac{1}{2}(\sigma_1 \pm i\sigma_2)$ acting on the two-dimensional multiplet generate transformations between states differing by one unit of $\langle\tfrac{1}{2}\sigma_3\rangle$. This can be seen explicitly since, e.g.

$\sigma_{+} \equiv \begin{pmatrix} 0 & 1 \\ 0 & 0 \end{pmatrix}$, and hence

$$\sigma_{+}u \equiv \begin{pmatrix} 0 & 1 \\ 0 & 0 \end{pmatrix}\begin{pmatrix} 1 \\ 0 \end{pmatrix} = 0$$

$$\sigma_{+}d \equiv \begin{pmatrix} 0 & 1 \\ 0 & 0 \end{pmatrix}\begin{pmatrix} 0 \\ 1 \end{pmatrix} = \begin{pmatrix} 1 \\ 0 \end{pmatrix} = u$$

(2.16)

The $\sigma_{\pm}$ are known as raising and lowering operators and satisfy the commutation relations

$$[\tfrac{1}{2}\sigma_3, \sigma_{\pm}] = \pm\sigma_{\pm}$$

$$[\sigma_{+}, \sigma_{-}] = 2(\tfrac{1}{2}\sigma_3)$$

(2.17)

As a result we can form a combination of generators that commutes with all the generators of the group. Such an operator is called a "Casimir operator" and here is

$$C = \tfrac{1}{2}(\sigma_{+}\sigma_{-} + \sigma_{-}\sigma_{+}) + \tfrac{1}{4}\sigma_3^2$$

$$\equiv \tfrac{1}{4}(\sigma_1^2 + \sigma_2^2 + \sigma_3^2)$$

$$\equiv (\tfrac{1}{2}\boldsymbol{\sigma})^2$$

(2.18)

Abstracting from the above we can immediately generalise from the 2×2 to the $N \times N$ dimensional case. Replace $\tfrac{1}{2}\sigma_{1,2,3}$ by $S_{1,2,3}$ and $\sigma_{\pm}$ by $S_{\pm}$ in all of the above. Hence states are labelled by the eigenvalues of S_3 and the Casimir operator is $\mathbf{S}^2$. Since $\mathbf{S}^2 S_{\pm} = S_{\pm}\mathbf{S}^2$ then application of raising or lowering operators $S_{\pm}$ generates new states differing by one unit of $\langle S_3 \rangle$ but having the same value of $\langle \mathbf{S}^2 \rangle$. Hence different representations can be specified by the eigenvalues of S^2 while states within that representation are characterised by the eigenvalue of S_3. For an $N = 2S + 1$ dimensional representation of SU(2) (where S is the maximum eigenvalue of S_3), the eigenvalue of $\mathbf{S}^2$ is $S(S + 1)$. This is easily seen by acting on the maximally stretched member of the multiplet (i.e. the one with maximum $S_3 = S$). Then

$$C = \tfrac{1}{2}(S_+S_- + S_-S_+) + S_3^2$$
$$\equiv S_-S_+ + S_3 + S_3^2 \qquad (2.19)$$

Acting on the maximally stretched state then

$$S_+\chi_{\text{max}} = 0$$

and hence

$$C\chi_{\text{max}} = S(S+1)\chi_{\text{max}} \qquad (2.20)$$

A particular example of the above is the two-dimensional representation in which case $S = \tfrac{1}{2}$ (the maximum eigenvalue of S_3) and the Casimir operator is $S(S+1) = \tfrac{3}{4}$.

2.2.1.2 SU(2) *breaking*

As a particular example illustrating our abstract discussion of SU(2) we referred to spin and visualised rotations taking place in real space. Another realisation would be isospin and the analogous rotations would take place in an internal "isospace". Just as angular momentum (spin) is conserved if a theory is invariant under rotations in real space, so will isospin be conserved if there is invariance under rotations in isospace.

Examples of isospin multiplets include

$$I = \tfrac{1}{2}: \quad \text{neutron}(I_3 = -\tfrac{1}{2}), \text{proton}(I_3 = +\tfrac{1}{2})$$

$$I = 1: \quad \pi^-(I_3 = -1), \pi^0(I_3 = 0), \pi^+(I_3 = +1)$$

The proton and the pion have the same electrical charge but different magnitudes of I_3. Therefore define the charge operator by

$$Q = \tfrac{1}{2}B + I_3 \qquad (2.21)$$

where B is the baryon number equalling one for the proton and zero for the pion. From the isospin commutation relations (equation (2.14), namely

$$[I_i, I_j] = i\varepsilon_{ijk}I_k$$

it follows that

$$[Q, I_3] = 0; \qquad [Q, I_{1,2}] \neq 0 \qquad (2.22)$$

Therefore the charge operator is not invariant under isorotations (it has selected the 3-axis as special) and hence the charge violates isospin conservation (more generally electromagnetic interactions violate isospin conservation).

If the Hamiltonian of the world commutes with the generators G_i of some symmetry group

$$[H, G_i] = 0 \quad \text{for all } i \tag{2.23}$$

then that symmetry will be an exact property of Nature. Let us consider the case $G_i = I_i$, the generators of isospin SU(2). Write $H = H_{str} + H_{em}$. It is plausible that isospin is an exact symmetry of the strong interactions and that

$$[H_{str}, I_i] = 0$$
$$[H_{em}, I_i] \neq 0 \tag{2.24}$$

The electromagnetic contributions to the Hamiltonian are small compared to the strong contributions (about 1 per cent or so) and hence the isospin symmetry of the strong interactions can be regarded as an approximate symmetry of Nature. The existence of the underlying symmetry can be identified by the existence of *multiplets* (like proton and neutron or the three pions) and also by various *intensity rules* (various $N^* \to \pi N$ amplitudes are related by isospin symmetry relations known as Clebsch–Gordan coefficients, e.g. Wigner, 1959; Rose, 1957; Particle Data Group, 1976).

If isospin were an exact symmetry of Nature then the proton and neutron masses would be identical. In practice they are slightly different and in light of the above discussion it is natural to blame this on electromagnetic effects which break the isospin symmetry. This will be discussed in detail in section 17.5 using the framework of the quark model.

2.2.1.3   *Conjugate representation of* SU(2)

We have seen how the proton and neutron can be regarded as forming the up and down states in the fundamental two-dimensional representation of SU(2) of isospin. If we label the antinucleons as $\bar{u}$ and $\bar{d}$ respectively then the states $(\bar{d}, \bar{u})$ have $I_3 = (+\tfrac{1}{2}, -\tfrac{1}{2})$ just like the doublet

(u, d). This new representation is known as the conjugate representation and it is conventional to denote the dimensionality of these
representations by

$$2 \equiv (u, d), \qquad 2^* \equiv (\bar{d}, \bar{u}) \tag{2.25}$$

How do the 2^* states transform under isorotations?

The fundamental doublet $\phi \equiv \begin{pmatrix} u \\ d \end{pmatrix}$ transformed under isorotations as

$$\phi' = U\phi$$

$$U \equiv \exp\left(\tfrac{1}{2}i\theta\hat{\mathbf{n}} \cdot \boldsymbol{\tau}\right) \equiv \cos\frac{\theta}{2} + i\hat{\mathbf{n}} \cdot \boldsymbol{\tau} \sin\frac{\theta}{2} \tag{2.26}$$

($\boldsymbol{\tau}$ playing the role of the Pauli spin matrices $\boldsymbol{\sigma}$ in equation 2.7). In
particular a rotation about the 2-axis yields

$$u' = \cos\frac{\theta}{2}u + \sin\frac{\theta}{2}d \tag{2.27}$$

$$d' = -\sin\frac{\theta}{2}u + \cos\frac{\theta}{2}d \tag{2.28}$$

if we use the standard representation of the 2×2 $\boldsymbol{\tau}$ matrices as at
equation (2.12).

Now act on both sides with the charge conjugation[1] so that $u \to \bar{u}$, $d \to \bar{d}$ etc. Equation (2.28) then becomes

$$\bar{d}' = -\sin\frac{\theta}{2}\bar{u} + \cos\frac{\theta}{2}\bar{d} \tag{2.28a}$$

and equation (2.27) becomes

$$\bar{u}' = \cos\frac{\theta}{2}\bar{u} + \sin\frac{\theta}{2}\bar{d} \tag{2.27a}$$

We have ordered the equations this way so that we immediately see the
effect on the $(I_3 = +\tfrac{1}{2}, I_3 = -\tfrac{1}{2})$ doublet

$$\begin{pmatrix} \bar{d}' \\ \bar{u}' \end{pmatrix} = \begin{pmatrix} \cos\dfrac{\theta}{2} & -\sin\dfrac{\theta}{2} \\ \sin\dfrac{\theta}{2} & \cos\dfrac{\theta}{2} \end{pmatrix} \begin{pmatrix} \bar{d} \\ \bar{u} \end{pmatrix} \tag{2.29}$$

[1] The particular case of rotating about the 2-axis through an angle $\theta \equiv \pi$, followed by charge conjugation, yields the operation called *G*-parity conjugation or just *G*-parity. See section 4.1.1.

Note that in equation (2.29) if the doublet is defined to be $\begin{pmatrix} \bar{d} \\ -\bar{u} \end{pmatrix}$ then

$$\bar{d}' = \cos\frac{\theta}{2}\bar{d} + \sin\frac{\theta}{2}(-\bar{u}) \tag{2.30}$$

$$(-\bar{u})' = -\sin\frac{\theta}{2}\bar{d} + \cos\frac{\theta}{2}(-\bar{u}) \tag{2.31}$$

which is the standard form for the rotation of the doublet (equations 2.27 and 2.28). In general the doublet

$$\tilde{\phi} = \begin{pmatrix} \bar{d} \\ -\bar{u} \end{pmatrix} \tag{2.32}$$

transforms as

$$\tilde{\phi}' = U\tilde{\phi}, \qquad U \equiv \exp\left(\tfrac{1}{2}i\theta\hat{\mathbf{n}} \cdot \boldsymbol{\tau}\right) \tag{2.33}$$

Hence the antiparticles, with the above phases, also transform as **2** under SU(2).

In general SU(N) where $N = 2, 3, 4 \ldots$ there will be basic representations of dimension **N** and **N***. In the SU(2) case we found that **2** and **2*** were equivalent representations, transforming in the same way under rotations. For $N = 3, 4 \ldots$ the **3, 3*** etc. representations are *not* equivalent. For the particular example of SU(3) this will be illustrated in section 2.2.2.

2.2.1.4   *Regular representation*

The simplest representation of the generators of SU(N) are the $N^2 - 1$ Hermitian traceless $N \times N$ matrices, the three Pauli matrices being the example for SU(2). Using these matrices one can define an $N^2 - 1$ dimensional representation of SU(N) which is known as the regular representation. In the case of SU(2) this will be the three-dimensional vector (isovector) representation (e.g. the pions).

To illustrate the regular representation recall the algebra of the generators

$$[S_i, S_j] = i\varepsilon_{ijk}S_k \tag{2.34}$$

If we choose S_3 diagonal then we find a 3×3 matrix representation $(3 \equiv 2^2 - 1)$

$$S_3 = \begin{pmatrix} 1 & 0 & 0 \\ 0 & 0 & 0 \\ 0 & 0 & -1 \end{pmatrix} \qquad S_1 = \frac{1}{\sqrt{2}} \begin{pmatrix} 0 & -1 & 0 \\ -1 & 0 & 1 \\ 0 & 1 & 0 \end{pmatrix}$$

$$S_2 = \frac{i}{\sqrt{2}} \begin{pmatrix} 0 & 1 & 0 \\ -1 & 0 & -1 \\ 0 & 1 & 0 \end{pmatrix} \quad (2.35)$$

which satisfies the SU(2) algebra. The basis on which these act consists of the eigenstates with eigenvalues $S_3 = +1, 0, -1$, e.g. if we are considering isospin then the basis could be the three charge states of the π meson, π^+, π^0, π^-.

Instead of these charge states one often meets the alternative basis

$$|\pi_1\rangle \equiv \frac{1}{\sqrt{2}} (-|\pi^+\rangle + |\pi^-\rangle)$$

$$|\pi_2\rangle \equiv \frac{i}{\sqrt{2}} (|\pi^+\rangle + |\pi^-\rangle) \qquad (2.36)$$

$$|\pi_3\rangle \equiv |\pi^0\rangle$$

which behave like the components of a vector under rotations in isospin space. The matrix elements of the S_i taken between these states becomes

$$\langle \pi_j | S_i | \pi_k \rangle = -i\varepsilon_{ijk} \qquad (2.37)$$

e.g. $\qquad \langle \pi_1 | S_3 | \pi_2 \rangle \equiv -\frac{i}{2} (\langle \pi^+ | S_3 | \pi^+ \rangle - \langle \pi^- | S_3 | \pi^- \rangle)$

$$= -i \equiv -i\varepsilon_{312} \qquad (2.38)$$

This is a particular example that is true for any SU(N). If the algebra of the group is defined by

$$[G_i, G_j] = ig_{ijk} G_k \qquad (2.39)$$

with M generators $G_i (i = 1 \ldots M)$ and g_{ijk} the structure constants of the group, then a representation can always be obtained with dimension equal to the number of generators. This is the regular representation and is given by

$$(S_k)_{ij} \equiv -ig_{ijk} \qquad (2.40)$$

For SU(2) with 3 generators we saw by explicit construction

$$(S_k)_{ij} \equiv -i\varepsilon_{ijk}$$

The general proof follows by use of Jacobi's identity. Denoting the generators by λ, then $[[\lambda_i, \lambda_j], \lambda_k] + [[\lambda_j, \lambda_k], \lambda_i] + [[\lambda_k, \lambda_i], \lambda_j] = 0$. Multiply it by a λ and take the traces.

2.2.2   SU(3)

2.2.2.1   *Fundamental representation and general idea*

The extension from SU(2) to SU(3) is immediate if we extend the basic u, d doublet to a triplet u, d, s and investigate transformations of

$$\phi \equiv \begin{pmatrix} u \\ d \\ s \end{pmatrix}$$

of the form

$$\phi' = U\phi$$

where U is now a 3×3 unitary unimodular matrix. Following the SU(2) case write

$$U \equiv \exp\left(\tfrac{1}{2}i\theta\hat{\mathbf{n}} \cdot \boldsymbol{\lambda}\right)$$

where the λ_i are eight independent Hermitian traceless 3×3 matrices analogous to the σ_i of SU(2). Canonically these are chosen to be (Gell-Mann, 1962)

$$\lambda_1 = \begin{pmatrix} 0 & 1 & \cdot \\ 1 & 0 & \cdot \\ \cdot & \cdot & \cdot \end{pmatrix} \quad \lambda_2 = \begin{pmatrix} 0 & -i & \cdot \\ i & 0 & \cdot \\ \cdot & \cdot & \cdot \end{pmatrix} \quad \lambda_3 = \begin{pmatrix} 1 & 0 & \cdot \\ 0 & -1 & \cdot \\ \cdot & \cdot & \cdot \end{pmatrix}$$

$$\lambda_4 = \begin{pmatrix} 0 & \cdot & 1 \\ \cdot & \cdot & \cdot \\ 1 & \cdot & 0 \end{pmatrix} \quad \lambda_5 = \begin{pmatrix} 0 & \cdot & -i \\ \cdot & \cdot & \cdot \\ i & \cdot & 0 \end{pmatrix} \quad \lambda_6 = \begin{pmatrix} \cdot & \cdot & \cdot \\ \cdot & 0 & 1 \\ \cdot & 1 & 0 \end{pmatrix}$$

$$\lambda_7 = \begin{pmatrix} \cdot & \cdot & \cdot \\ \cdot & 0 & -i \\ \cdot & i & 0 \end{pmatrix} \quad \lambda_8 = \frac{1}{\sqrt{3}}\begin{pmatrix} 1 & 0 & 0 \\ 0 & 1 & 0 \\ 0 & 0 & -2 \end{pmatrix} \qquad (2.41)$$

where the dots are zeros and we have written them in this fashion to highlight the SU(2) subgroups contained within SU(3). The $\lambda_{1,2}$ have the structure

$$\begin{pmatrix} \sigma_{1,2} & \vdots & 0 \\ \text{-----} & \vdots & 0 \\ 0 & 0 & 0 \end{pmatrix} \tag{2.42}$$

and hence exhibit the SU(2) isospin subgroup. The $\lambda_{6,7}$ are

$$\begin{pmatrix} 0 & 0 & 0 \\ 0 & \ulcorner\text{---} \\ 0 & \vdots & \sigma_{1,2} \end{pmatrix} \tag{2.43}$$

and exhibit an SU(2) subgroup called U-spin while the $\lambda_{4,5}$ are related to a third subgroup V-spin. In terms of the basic triplet of Fig. 2.3 these

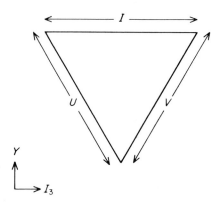

FIG. 2.3. SU(3) triplet with I, U, V SU(2) doublets.

SU(2) doublets are

$$\text{u, d }(I); \qquad \text{d, s }(U); \qquad \text{u, s }(V)$$

The operator $F_3 \equiv \tfrac{1}{2}\lambda_3$ is the isospin operator since acting on u, d, s it has eigenvalues $\pm\tfrac{1}{2}$, 0 respectively. The hypercharge operator is

$$Y = \frac{2}{\sqrt{3}}F_8 \equiv \frac{2}{\sqrt{3}} \cdot \tfrac{1}{2}\lambda_8 \tag{2.44}$$

TABLE 2.1
Structure constants of SU(3)

$$f_{123} = 1$$

$$f_{147} = f_{246} = f_{257} = f_{345} = f_{516} = f_{637} = \tfrac{1}{2}$$

$$f_{458} = f_{678} = \sqrt{3}/2$$

$$d_{118} = d_{228} = d_{338} = -d_{888} = 1/\sqrt{3}$$

$$d_{146} = d_{157} = d_{256} = d_{344} = d_{355} = \tfrac{1}{2}$$

$$d_{247} = d_{366} = d_{377} = -\tfrac{1}{2}$$

$$d_{448} = d_{558} = d_{668} = d_{778} = -\frac{1}{2\sqrt{3}}$$

The commutation relations of the matrices $\tfrac{1}{2}\lambda_i$ can be obtained by explicit calculation. Perform this for yourself and verify that

$$[\tfrac{1}{2}\lambda_i, \tfrac{1}{2}\lambda_j] = if_{ijk}(\tfrac{1}{2}\lambda_k) \qquad (2.45)$$

with the structure constants f_{ijk} having the values in Table 2.1 and being antisymmetric under interchange of any pair of indices. The matrices also satisfy anticommutation relations

$$\{\tfrac{1}{2}\lambda_i, \tfrac{1}{2}\lambda_j\} = \tfrac{1}{3}\delta_{ij} + d_{ijk}(\tfrac{1}{2}\lambda_k) \qquad (2.46)$$

where the d_{ijk} are symmetric under interchange of indices.

As in the SU(2) case we can generalise these results by defining $F_i \equiv \tfrac{1}{2}\lambda_i$ satisfying commutation relations

$$[F_i, F_j] = if_{ijk}F_k \qquad (2.47)$$

$(i = 1 \ldots 8)$. A full study of SU(3) then consists of finding $N \times N$ matrices F_i which transform N-dimensional states by

$$\phi \to \phi' = (1 + i\theta\hat{\mathbf{n}} \cdot \mathbf{F})\phi \qquad (2.48)$$

These states form N-dimensional multiplets of SU(3).

2.2.2.2   SU(3) *Casimir operators*

In SU(2) we found a combination of generators that commuted with all the generators of the group, viz.

$$C = \mathbf{I}^2 = \tfrac{1}{2}(I_+I_- + I_-I_+) + I_3^2$$

$$\equiv \tfrac{1}{2}\{I_+, I_-\} + I_3^2 \qquad (2.19)$$

whose eigenvalue was shown to be $I(I+1)$. In SU(3) the analogous invariant operator is

$$\mathbf{F}^2 \equiv \sum_{i=1}^{8} F_i F_i = \tfrac{1}{2}\{I_+, I_-\} + I_3^2 + \tfrac{1}{2}\{U_+, U_-\} + \tfrac{1}{2}\{V_+, V_-\} + F_8^2$$

(2.49)

where

$$I_\pm \equiv F_1 \pm iF_2, \qquad I_3 \equiv F_3$$

$$U_\pm \equiv F_6 \pm iF_7, \qquad Y \equiv \frac{2}{\sqrt{3}} F_8 \qquad (2.50)$$

$$V_\pm \equiv F_4 \pm iF_5$$

The operators I_+, V_+ and U_- all increase the magnitude of I_3 and so we can define a maximally stretched state such that

$$I_+\phi_{\max} = V_+\phi_{\max} = U_-\phi_{\max} = 0 \qquad (2.51)$$

We will now compute the magnitude of the Casimir operator for any SU(3) representation by acting with F^2 on the maximally stretched state of that representation.

First of all use the values of f_{ijk} (Table 2.1) to verify that

$$[I_+, I_-] = 2I_3$$

$$[U_+, U_-] = \tfrac{3}{2}Y - I_3 \equiv 2U_3 \qquad (2.52)$$

$$[V_+, V_-] = \tfrac{3}{2}Y + I_3 \equiv 2V_3$$

(compare equation 2.17 in the SU(2) case). Now use these to rewrite F^2 with I_+, U_-, V_+ on the right-hand side of any pair of operators. Acting with $\mathbf{F}^2$ on the maximally stretched state $\phi_{\max}$ defined above immediately yields the result

$$\langle \mathbf{F}^2 \rangle = \langle I_3 \rangle^2 + 2\langle I_3 \rangle + \tfrac{3}{4}Y^2 \qquad (2.53)$$

For example the triplet has u as the maximal state for which $I = +\tfrac{1}{2}$, $Y = \tfrac{1}{3}$, hence $\langle F^2 \rangle = \tfrac{4}{3}$.

In order to facilitate computation of $\langle \mathbf{F}^2 \rangle$ for any arbitrary dimensional representation of SU(3) it is useful to do a small calculation to manoeuvre equation (2.53) into a more immediately tractable format.

We assert without proof that any SU(3) representation has a convex boundary in $I_3 - Y$ space (see Gasiorowicz, op. cit., p. 264). Hence we can act on the state ϕ_{max} with the operator V_- repeatedly (p times) until another corner of the boundary is reached, i.e.

$$(V_-)^{p+1}\phi_{max} = 0 \tag{2.54}$$

Then act on this corner state with I_- repeatedly (q times) until yet another corner is reached, i.e.

$$(I_-)^{q+1}(V_-)^p\phi_{max} = 0 \tag{2.55}$$

The SU(3) representation is now completely specified by (p, q). For example the octet in Fig. 2.4 has the point A the state ϕ_{max} (since I_+, V_+, U_- would generate states $A_{1,2,3}$ respectively which are not in the multiplet). Acting with V_- once brings us to the corner B, hence $p = 1$; and I_- once brings us to C, hence $q = 1$. Therefore the octet is $(1, 1)$.

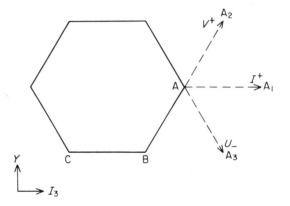

FIG. 2.4. The position ϕ_{max} in the octet.

The maximum state has

$$I_3 = \tfrac{1}{2}(p + q)$$
$$Y = \tfrac{1}{3}(p - q) \tag{2.56}$$

and so the expression for F^2 (equation 2.53) becomes

$$F^2 \equiv \tfrac{1}{3}(p^2 + pq + q^2) + (p + q) \tag{2.57}$$

We illustrate this formula by applying it to various representations and derive the results exhibited in Table 2.2. For a triplet (Fig. 2.5) $p = 1$, $q = 0$. Hence $F^2 = \tfrac{4}{3}$. The same result holds for an antitriplet for which $p = 0$, $q = 1$. For an octet $p = 1$, $q = 1$ and so $F^2 = 3$.

TABLE 2.2

Magnitudes of Casimir operators, F^2, for some common SU(3) representations

Dimension	(p, q)	F^2
1	$(0, 0)$	0
3	$(1, 0)$	$\frac{4}{3}$
3̄	$(0, 1)$	$\frac{4}{3}$
8	$(1, 1)$	3
6	$(2, 0)$	$\frac{10}{3}$
10	$(3, 0)$	6

As an exercise calculate the p and q for a six- and ten-dimensional representation. Compare the (p, q) with that of the fundamental triplet and recall that **6** and **10** are symmetric representations (this is shown in section 3.3). Compare (p, q) for the "mixed" octet. Finally look at (p, q) for the **6** and **10** and compare with **3**. An obvious pattern in p, q will be seen to emerge. Compare also with the explicit quark model constructions of higher dimensional representations in section 3.3 and the discussion of Young tableaux in section 3.4. As a check verify that $F^2 = 6$ for a decuplet and is $\frac{10}{3}$ for a **6**. Finally ask yourself what are p, q for a singlet? Show that F^2 for a singlet is zero and prove that only for a singlet can it be zero. These results will be important in the discussion of quarks with three colours (SU(3) of colour) that appears in section 15.2.

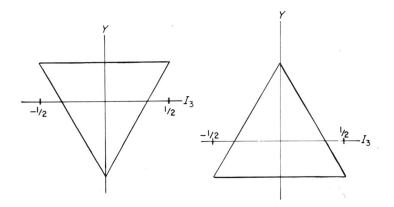

FIG. 2.5. Weight diagrams for **3** and **3***.

Notice that **3** has $(p, q) = (1, 0)$ whereas $\bar{\mathbf{3}}$ is $(0, 1)$. These are there-fore two inequivalent representations of SU(3) unlike that encountered in SU(2) where the **2** and **2*** were found to be equivalent. The diagrammatic representations of **3** and **3*** are shown in Fig. 2.5 and it is clearly seen that it is the hypercharge, which is different in the two cases, that prevents the equivalence. In SU(2) the (u, d) and $(\bar{d}, \bar{u})$ are the same (if we ignore baryon number) since the hypercharge degree of freedom has not entered the picture.

2.2.2   SU(4)

The fundamental representation of SU(4) is (Fig. 2.6)

$$\phi \equiv \begin{pmatrix} c \\ u \\ d \\ s \end{pmatrix}$$

and the transformations $\phi' = U\phi$ now involve 4×4 unitary unimodular

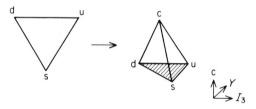

FIG. 2.6. SU(4) quartet.

matrices. As in SU(3) write

$$U \equiv \exp\left(\tfrac{1}{2}i\theta\hat{\mathbf{n}} \cdot \boldsymbol{\lambda}\right)$$

where now λ_i are 15 independent Hermitian traceless 4×4 matrices. We will choose $\lambda_{1\ldots8}$ to be identical to those of SU(3); hence

$$\lambda_i = \begin{pmatrix} 0 & 0 & 0 & 0 \\ 0 & & & \\ 0 & & \lambda_i & \\ 0 & & & \end{pmatrix} \qquad i = 1\ldots8 \qquad (2.58)$$

There are six nondiagonal matrices which are chosen (Amati *et al.*, 1964) analogous to the Pauli matrices

$$
\lambda_9 = \begin{pmatrix} 0 & 1 & & 0 \\ 1 & 0 & & \\ \hline & & & 0 \\ & 0 & & 0 \end{pmatrix}
\qquad
\lambda_{10} = \begin{pmatrix} 0 & -i & & 0 \\ i & 0 & & \\ \hline & & 0 & 0 \end{pmatrix}
$$

$$
\lambda_{11} = \begin{pmatrix} 0 & & 1 & 0 \\ & & 0 & 0 \\ \hline 1 & 0 & & \\ 0 & 0 & & 0 \end{pmatrix}
\qquad
\lambda_{12} = \begin{pmatrix} 0 & & -i & 0 \\ & & 0 & 0 \\ \hline i & 0 & & \\ 0 & 0 & & 0 \end{pmatrix}
\qquad (2.59)
$$

$$
\lambda_{13} = \begin{pmatrix} 0 & & 0 & 1 \\ & & 0 & 0 \\ \hline 0 & 0 & & \\ 1 & 0 & & 0 \end{pmatrix}
\qquad
\lambda_{14} = \begin{pmatrix} 0 & & 0 & -i \\ & & 0 & 0 \\ \hline 0 & 0 & & \\ i & 0 & & 0 \end{pmatrix}
$$

Finally there is a traceless diagonal λ_{15} which is chosen so that it distinguishes the charmed "quark" c from uncharmed uds and is an SU(3) singlet

$$
\lambda_{15} = \frac{1}{\sqrt{6}} \begin{pmatrix} -3 & 0 & & 0 \\ 0 & 1 & & \\ \hline & & 1 & 0 \\ 0 & & 0 & 1 \end{pmatrix}
\qquad (2.60)
$$

Consequently we can define "charm" to be the eigenvalues of

$$
C = \tfrac{1}{4}(1 - \sqrt{6}\,\lambda_{15}) \equiv \begin{pmatrix} 1 & & \\ & 0 & \\ & & 0 \\ & & & 0 \end{pmatrix}
\qquad (2.61)
$$

and hence charm is not a generator of SU(4).

3   Quarks and SU(*N*) Representations

If baryons are to be built from three quarks (qqq) then a quark must have baryon number $\frac{1}{3}$. The antiquark will have baryon number $-\frac{1}{3}$ and so mesons $q\bar{q}$ have baryon number zero.

The quark must occur in two flavours which form an isospin doublet in order that the neutron and proton can be distinguished. This pair of quarks form the basic representation of isospin SU(2). In order to distinguish the proton and Σ^+ a third flavour of quark is needed. This quark must carry strangeness and in company with the isospin doublet will form the basic representation of SU(3).

With the canonical relation between charge, baryon number and strangeness (Gell-Mann, 1963; Nishijima and Nakano, 1953)

$$Q = I_3 + \frac{B+S}{2} \equiv I_3 + \frac{Y}{2} \; (B + S \equiv Y = \text{hypercharge}) \qquad (3.1)$$

then the quarks have quantum numbers exhibited in Table 3.1. The u, d, s flavours of quark therefore form an inverted triangle in $I_3 - Y$ space.

In order to create charmed hadrons a fourth flavour of quark (charmed quark) is required and the four flavours form the basic representation of SU(4). We will concern ourselves with just the three flavours u, d, s in the present discussion. The extension to charm and SU(4) will be deferred until Chapter 16.

TABLE 3.1
Quarks and their quantum numbers

Flavour	u	d	s	c
Charge	$\frac{2}{3}$	$-\frac{1}{3}$	$-\frac{1}{3}$	$\frac{2}{3}$
Isospin	$\frac{1}{2}$	$\frac{1}{2}$	0	0
I_3	$+\frac{1}{2}$	$-\frac{1}{2}$	0	0
Strangeness	0	0	-1	0
Charm	0	0	0	1
Baryon numbers	$\frac{1}{3}$	$\frac{1}{3}$	$\frac{1}{3}$	$\frac{1}{3}$

$$\underbrace{}_{SU(2)}$$

SU(N) group $\qquad \underbrace{ \atop SU(3)}$

$$\underbrace{}_{SU(4)}$$

3.1 Baryons

Baryons are *qqq* states. If we ignore charm then each of the quarks can have any of three flavours u, d, s. In Table 3.2 I list the possible combinations of the quarks uds that can arise after three selections have been made (for the present we will take no account of the *order* in which they were selected, hence uud, udu and duu are equivalent). In column 3 the charge and in column 4 the strangeness of the resulting baryon is

TABLE 3.2
Systems of three quarks with three flavours

Quarks	Symmetry	Charge	Strangeness	Examples
uuu	S	2		Δ^{++}
uud	S M	1		$\Delta^+ P$
udd	S M	0	0	$\Delta^0 N$
ddd	S	-1		Δ^-
uus	S M	1		$\Sigma^{+*}\Sigma^+$
uds	S M M A	0	-1	$\Sigma^{0*}\Sigma^0\Lambda^0\Lambda(1405)$
dds	S M	-1		$\Sigma^{-*}\Sigma^-$
uss	S M	0		$\Xi^{0*}\Xi^0$
dss	S M	-1	-2	$\Xi^{-*}\Xi^-$
sss	S	-1	-3	Ω^-

shown. In column 5 examples of baryons with these quantum numbers are given—the decuplet, octet and singlet $\Lambda(1405)$.

If we have chosen uuu then we know that we have a Δ^{++}, but if we choose uud then how do we distinguish Δ^+ from proton (other than by its spin)? For uds there are four possibilities.

The way we distinguish between them is related to the symbols S, M, A of column 2 which have to do with the symmetry properties of the states. Specifically—if we now worry about the order in which the quarks were selected—then what happens if we change the quarks selected 1st and 2nd around to 2nd and 1st? Clearly if all were the same—uuu for instance—then you get the same state and this is called symmetric, labelled S. In fact you can always define an S combination whatever the three quark content may be, hence **10** such states. If at least one quark differs from the rest you can write a "mixed symmetric state" (M) and there are **8** of these (uds comes in two ways since there are two choices for the "different quark"). Finally if all three are distinct one can form a *single* state antisymmetric (A) under interchange of any pair of quarks.

This brings us to the ideas of *symmetry properties* of states under interchange of their labels. We shall discuss this in detail, and begin with a simple example in SU(2) as a result of which we will be able to write down spin wavefunctions for systems of three spin $\frac{1}{2}$ quarks. Then we extend to SU(3) and make contact with the **10**(S), **8**(M) and **1**(A) structure of Table 3.2. The SU(3) wavefunctions of the three-quark system will be explicitly constructed. We shall then combine these SU(3) wavefunctions with the SU(2) quark spin wavefunctions to formulate the SU(6) structure of the spectrum of qqq states.

3.2　SU(2) representations

The simplest example of this group arises if we have an object that can exist in two "types"—labelled "up and down" or $\begin{pmatrix} u \\ d \end{pmatrix}$. Examples are isospin where the nucleon system has $I = \frac{1}{2}$, the "up" state $(I_z = +\frac{1}{2})$ being the proton and the "down" $(I_z = -\frac{1}{2})$ the neutron. Another familiar example is an object with spin $\frac{1}{2}$: turn on a magnetic field which defines a z direction and its spin points up $\uparrow(S_z = +\frac{1}{2})$ or down $\downarrow(S_z = -\frac{1}{2})$ in the z direction.

3.2.1 SYSTEM OF TWO SUCH OBJECTS

There are clearly four possibilities, given in the first column of Table 3.3. Under interchange of the labels 1 and 2 three states are symmetric and one is antisymmetric.

In the group theory notation we write (2 states)$_1$ × (2 states)$_2$ = **2** ⊗ **2** → **3** ⊕ **1**, showing that there are 3 symmetric and 1 antisymmetric combinations.

A familiar example of this is combining two states with spin $\frac{1}{2}$ to form a state of spin 1 or 0:

$$\tfrac{1}{2} \otimes \tfrac{1}{2} = 1 \oplus 0 \tag{3.2}$$

Rewriting this in terms of the $(2S + 1)$ states one would have

$$\mathbf{2} \otimes \mathbf{2} = \mathbf{3} \oplus \mathbf{1} \tag{3.3}$$

as in our group theory notation above. In fact we can obtain the Table above by referring to the Clebsch–Gordan coefficients for combining the two spin $\frac{1}{2}$ states.

With notation $|S^1 S_z^1; S^2 S_z^2\rangle \rightarrow |SS_z\rangle$ we have from the Clebsch–Gordan Tables (Particle Data Group, 1976).

$$\mathrm{uu} \equiv |\tfrac{1}{2}\tfrac{1}{2}; \tfrac{1}{2}\tfrac{1}{2}\rangle \quad = |11\rangle$$

$$\mathrm{ud} \equiv |\tfrac{1}{2}\tfrac{1}{2}; \tfrac{1}{2}-\tfrac{1}{2}\rangle \quad = \frac{1}{\sqrt{2}}\{|10\rangle + |00\rangle\}$$

$$\mathrm{du} \equiv |\tfrac{1}{2}-\tfrac{1}{2}; \tfrac{1}{2}\tfrac{1}{2}\rangle \quad = \frac{1}{\sqrt{2}}\{|10\rangle - |00\rangle\} \tag{3.4}$$

$$\mathrm{dd} \equiv |\tfrac{1}{2}-\tfrac{1}{2}; \tfrac{1}{2}-\tfrac{1}{2}\rangle = |1-1\rangle$$

TABLE 3.3
Symmetry states for two objects in SU(2)

1st	2nd	1 ↔ 2 interchange	
u	u	uu	
u	d	$\frac{1}{\sqrt{2}}$(ud + du)	$\frac{1}{\sqrt{2}}$(ud − du)
d	u		
d	d	dd	
		symmetric	antisymmetric

and this is identical to the separation made above into three states $(S = 1)$ and one state $(S = 0)$.

3.2.2 SYSTEM OF THREE SUCH OBJECTS

There are now eight combinations. These are shown in Table 3.4 with their value of S_z. One can form the four symmetric combinations and two types of mixed symmetry states, one antisymmetric and the other symmetric under interchange of the first two labels but having no simple symmetry under (13) or (23) interchange. (Actually interchange either of these latter ways and you will obtain a state which is a linear combination of the various quoted states—hence "mixed" symmetry.) The various $\frac{1}{3}, \frac{2}{3}$ factors ensure the orthonormality of these states.

TABLE 3.4
Symmetry states for three objects in SU(2)

1 u	u d u	d u d	d
2 u	u u d	d d u	d
3 u	d u u	u d d	d
uuu	$\dfrac{1}{\sqrt{3}}(uud + udu + duu)$	$\dfrac{1}{\sqrt{3}}(ddu + dud + udd)$	ddd
	$\dfrac{1}{\sqrt{2}}(ud - du)u$	$\dfrac{1}{\sqrt{2}}(ud - du)d$	
	$\dfrac{1}{\sqrt{3}}\left[\dfrac{(ud+du)u}{\sqrt{2}} - uud\sqrt{2}\right]$	$-\dfrac{1}{\sqrt{3}}\left[\dfrac{(ud+du)d}{\sqrt{2}} - ddu\sqrt{2}\right]$	
$S_z = \frac{3}{2}$	$S_z = \frac{1}{2}$	$S_z = -\frac{1}{2}$	$S_z = -\frac{3}{2}$

The physics of the above is clear if we consider the coupling of three particles each with spin $\frac{1}{2}$. The coupling of two such objects was discussed in the previous example and the $S = 1(0)$ states with symmetric (antisymmetric) properties exhibited. This is recapitulated in the first column below. Coupling a third particle is shown in the second column.

$$
\begin{array}{ll}
\textit{From two particles} & \textit{Add a third} \\
(12)_{\text{anti}} \rightarrow S_{12} = 0 & \otimes\frac{1}{2} \rightarrow S = \frac{1}{2} \text{ only} \\
(12)_{\text{sym}} \rightarrow S_{12} = 1 & \otimes\frac{1}{2} \rightarrow S = \frac{1}{2} \text{ and } S = \frac{3}{2}
\end{array}
\tag{3.5}
$$

Hence the $S = \frac{1}{2}$ state can be formed in two ways, one with the $S_{12} = 0$ (antisymmetric) and one with $S_{12} = 1$ (symmetric). This corresponds to the two mixed symmetry types in rows 3 and 4 of Table 3.4.

We can illustrate this by coupling three spin $\frac{1}{2}$ states using the familiar Clebsch–Gordan coefficients. For the $S_z = +\frac{1}{2}$ system one has the possibilities listed below (Table 3.5).

TABLE 3.5
Coupling three spin $\frac{1}{2}$ states

$S = 1 \otimes S = \frac{1}{2}$	$S = \frac{3}{2} \otimes S = \frac{1}{2}$
uu d	$\left\lvert 11, \frac{1}{2} - \frac{1}{2} \right\rangle \rightarrow \frac{1}{\sqrt{3}} \left\lvert \frac{3}{2} \frac{1}{2} \right\rangle - \sqrt{\frac{2}{3}} \left\lvert \frac{1}{2} \frac{1}{2} \right\rangle_s$
$\frac{1}{\sqrt{2}}(ud + du)u$	$\left\lvert 10, \frac{1}{2} \frac{1}{2} \right\rangle \rightarrow \sqrt{\frac{2}{3}} \left\lvert \frac{3}{2} \frac{1}{2} \right\rangle + \sqrt{\frac{1}{3}} \left\lvert \frac{1}{2} \frac{1}{2} \right\rangle_s$
$S = 0 \otimes S = \dfrac{1}{2}$	$S = \dfrac{1}{2}$
$\frac{1}{\sqrt{2}}(ud - du)u$	$\left\lvert 00, \frac{1}{2} \frac{1}{2} \right\rangle \rightarrow \left\lvert \frac{1}{2} \frac{1}{2} \right\rangle_A$

Solving these yields our previous results, e.g.

$$\left\lvert \frac{3}{2} \frac{1}{2} \right\rangle = \frac{1}{\sqrt{3}} \left\lvert 11, \frac{1}{2} - \frac{1}{2} \right\rangle + \sqrt{\frac{2}{3}} \left\lvert 10, \frac{1}{2} \frac{1}{2} \right\rangle \tag{3.6}$$

$$= \frac{1}{\sqrt{3}} uud + \sqrt{\frac{2}{3}} \frac{1}{\sqrt{2}} (ud + du)u \tag{3.7}$$

$$= \frac{1}{\sqrt{3}} (uud + udu + duu) \tag{3.8}$$

and similarly for the other states.

3.2.3 SYSTEM OF 3 SPIN $\frac{1}{2}$ QUARKS

The previous discussion will be useful later when we build the baryons out of three quarks each with spin $\frac{1}{2}$. The total spin of the three-quark system will be described by three-quark spin states as above. We close this section by summarising the essential features of SU(2):

$$\tfrac{1}{2} \times \tfrac{1}{2} = 1_S + 0_A \quad \text{or in } (2S+1) \text{ notation } \mathbf{2} \times \mathbf{2} = \mathbf{3} + \mathbf{1} \tag{3.9}$$

$$(\tfrac{1}{2} \times \tfrac{1}{2}) \times \tfrac{1}{2} = (1 \times \tfrac{1}{2}) + (0 \times \tfrac{1}{2}) \text{ or } (\mathbf{2} \times \mathbf{2}) \times \mathbf{2} = (\mathbf{3} \times \mathbf{2}) + (\mathbf{1} \times \mathbf{2})$$

$$= (\tfrac{3}{2}_S + \tfrac{1}{2}_{M.S}) + \tfrac{1}{2}_{M.A} \qquad\qquad = (4+2) + 2 \tag{3.10}$$

3.3 SU(3) representations

The simplest example arises if we have one object that can exist in three *types* which we shall label (uds). This is a straightforward extension of the (ud) example of SU(2) and so we shall only sketch the steps. The interesting physical application will be when strange and nonstrange uncharmed hadrons are described as systems of quarks, each of which can occur in three types of "flavours" (up, down and strange).

3.3.1 SYSTEM OF TWO SUCH OBJECTS

There are nine possible combinations which can be separated into six symmetric and three antisymmetric states (analogous to Table 3.3 in SU(2)). These are shown in Table 3.6 and form a $\mathbf{6}$ and $\bar{\mathbf{3}}$ of SU(3). That the antisymmetric set is $\bar{\mathbf{3}}$ and not $\mathbf{3}$ can be seen as follows. Note that ud are an $I = \tfrac{1}{2}$ doublet of the SU(2) subgroup with $I_z = +\tfrac{1}{2}, -\tfrac{1}{2}$ respectively. The degree of freedom (s) that extends the group to SU(3) will be labelled by having (minus) one unit of "hypercharge" (Y). Then the uds states fall in an inverted triangle on a $I_3 - Y$ plot (Fig. 3.1). This is called the "weight diagram" for a triplet. The weight diagram for an antitriplet is an inversion of this, i.e. a triangle (which is easily seen by considering $\bar{\mathrm{u}}\bar{\mathrm{d}}\bar{\mathrm{s}}$). The combinations ud, us, ds also form a triangle (Fig. 3.1) and hence are $\bar{\mathbf{3}}$.

3.3.2 SYSTEM OF THREE OBJECTS

Exactly as in the SU(2) example add a third u, d or s to the two body states above. There will be a total of 27 combinations of which 18 come

TABLE 3.6
Symmetry states for two objects in SU(3)

1st	2nd	1 ↔ 2 interchange	
u	u	uu	
u	d } $\frac{1}{\sqrt{2}}$(ud + du)	$\frac{1}{\sqrt{2}}$(ud − du)	
d	u		
d	d	dd	
u	s } $\frac{1}{\sqrt{2}}$(us + su)	$\frac{1}{\sqrt{2}}$(us − su)	
s	u		
d	s } $\frac{1}{\sqrt{2}}$(ds + sd)	$\frac{1}{\sqrt{2}}$(ds − sd)	
s	d		
s	s	ss	
		symmetric	antisymmetric

from $\mathbf{3} \otimes \mathbf{6}$ and 9 from $\mathbf{3} \otimes \bar{\mathbf{3}}$. By explicit construction verify

$$\mathbf{3} \otimes (\mathbf{3} \otimes \mathbf{3}) \rightarrow \mathbf{3} \otimes (\mathbf{6} \oplus \bar{\mathbf{3}})$$

$$\rightarrow (\mathbf{10}_S \oplus \mathbf{8}_{M,S}) \oplus (\mathbf{8}_{M,A} \oplus \mathbf{1}) \qquad (3.11)$$

which is the SU(3) analogue of the $\mathbf{2} \otimes (\mathbf{2} \otimes \mathbf{2})$ in SU(2) in equation (3.10). The $\mathbf{10}$(sym) states are obvious. The $\mathbf{8}_{M,S}$, $\mathbf{8}_{M,A}$ and $\mathbf{1}$ states are given in Table 3.7.

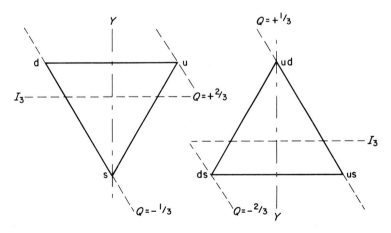

FIG. 3.1. u, d, s triplet and ud, us, ds antitriplets.

Notice that the antisymmetric state exists because the three objects each have three labels available (uds). In the SU(2) example the restriction to two labels meant that only symmetric and (two) mixed symmetry states could be found.

TABLE 3.7

$\phi_{M,S}\phi_{M,A}$: Mixed symmetry representations for the octet states of three quarks

	$\phi_{M,S}$	$\phi_{M,A}$
P	$\dfrac{1}{\sqrt{6}}[(ud+du)u-2uud]$	$\dfrac{1}{\sqrt{2}}(ud-du)u$
N	$-\dfrac{1}{\sqrt{6}}[(ud+du)d-2ddu]$	$\dfrac{1}{\sqrt{2}}(ud-du)d$
Σ^{+}	$\dfrac{1}{\sqrt{6}}[(us+su)u-2uus]$	$\dfrac{1}{\sqrt{2}}(us-su)u$
Σ^{0}	$\dfrac{1}{\sqrt{6}}\left[s\left(\dfrac{du+ud}{\sqrt{2}}\right)+\left(\dfrac{dsu+usd}{\sqrt{2}}\right)\right.$ $\left.-2\left(\dfrac{du+ud}{\sqrt{2}}\right)s\right]$	$\dfrac{1}{\sqrt{2}}\left[\left(\dfrac{dsu+usd}{\sqrt{2}}\right)-s\left(\dfrac{ud+du}{\sqrt{2}}\right)\right]$
Σ^{-}	$\dfrac{1}{\sqrt{6}}[(ds+sd)d-2dds]$	$\dfrac{1}{\sqrt{2}}(ds-sd)d$
Λ^{0}	$\dfrac{1}{\sqrt{2}}\left[\dfrac{dsu-usd}{\sqrt{2}}+\dfrac{s(du-ud)}{\sqrt{2}}\right]$	$\dfrac{1}{\sqrt{6}}\left[\dfrac{s(du-ud)}{\sqrt{2}}+\dfrac{usd-dsu}{\sqrt{2}}-\dfrac{2(du-ud)s}{\sqrt{2}}\right]$
Ξ^{-}	$-\dfrac{1}{\sqrt{6}}[(ds+sd)s-2ssd]$	$\dfrac{1}{\sqrt{2}}[(ds-sd)s]$
Ξ^{0}	$-\dfrac{1}{\sqrt{6}}[(us+su)s-2ssu]$	$\dfrac{1}{\sqrt{2}}[(us-su)s]$

	ϕ_{A}
Λ_{1}^{0}	$\dfrac{1}{\sqrt{6}}[s(du-ud)+(usd-dsu)+(du-ud)s]$

Note: $\tau^{-}u \rightarrow d$ relates P and N; $d \leftrightarrow s$ relates $P \leftrightarrow \Sigma^{+}$; $u \leftrightarrow d$ and $d \leftrightarrow s$ relates $N \leftrightarrow \Xi^{-}$. Note that in the Σ^{0} the ud quarks have $I=1$ while in the Λ^{0} they have $I=0$. The locations in the octet hexagon are shown in Fig. 2.2. The antisymmetric singlet state, ϕ_{A}, is also shown.

Identifying uds with the three flavours of quark, the resulting states are identified with the baryons. You can refer back to Table 3.2 showing baryon states and the S, M, A, notation should now be clear.

3.4   SU(*N*) and Young tableaux

By working through the example of SU(2) in detail and then generalising to SU(3) we have explicitly seen how to combine the fundamental representations and have constructed the SU(3) **1, 8** and **10** representations for the *qqq* system. Now that a fourth flavour of quark is believed to exist one might consider a fundamental quartet (c, u, d, s) and by proceeding through the analogous steps to those just discussed one could obtain the SU(4) representations of the *qqq* system. Alternatively one could construct the SU(6) representations of *qqq* that arise when the uds quarks with spin $\frac{1}{2}$ form the fundamental representation u↑, d↑, s↑, u↓, d↓, s↓, of SU(6). By explicit constructions you would find that

$$6 \otimes \bar{6} = 1 \oplus 35 \tag{3.12}$$

$$6 \otimes 6 \otimes 6 = 56 \oplus 70 \oplus 70 \oplus 20 \tag{3.13}$$

In turn you could calculate the representations of an SU(8) group generated by the fundamental representation c↑, u↑, d↑, s↑, c↓, u↓, d↓, s↓.

Instead of proceeding in this tedious fashion afresh for each SU(*N*) it would be much more elegant if there were some general techniques applicable for arbitrary SU(*N*) which would enable us to easily deduce the dimensions of the irreducible representations arising from products of other representations of the group. Elegant and extremely rapid for calculation are the techniques of Young tableaux. They also have the merit of being fun to play with. Suppose that we are interested in SU(*N*). The fundamental representation will be denoted by a box

$$\square \quad = \text{dimension } N, \tag{3.14}$$

while a column of $N - 1$ boxes denotes the conjugate representation N^*

$$\left.\begin{array}{c}\square \\ \square \\ \square \\ \vdots \\ \square \\ \square\end{array}\right\} N^* \tag{3.15}$$

Hence in SU(2) $\square$ = **2** or **2*** but in SU(3) we find

$$\square = \mathbf{3}, \qquad \boxed{\square\atop\square} = \mathbf{3^*} \qquad\qquad (3.16)$$

and so the fundamental and conjugate representations are the same in SU(2) but differ in SU(3) and higher groups SU($N \geqslant 3$).

Young tableaux with only one row are associated with a totally symmetric representation. A column is associated with an antisymmetric representation.

Now imagine that we wish to obtain the product of two such representations. We can add the second box to the first either to make a line or column of two boxes (the general set of rules is given in Hammermesh, 1963). The dimension of the representations corresponding to the line or column is given underneath both for SU(2) and SU(3). We have already seen how to calculate these dimensions explicitly.

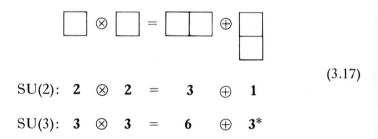

$$(3.17)$$

$$\text{SU(2):} \quad \mathbf{2} \ \otimes \ \mathbf{2} \ = \ \mathbf{3} \ \oplus \ \mathbf{1}$$

$$\text{SU(3):} \quad \mathbf{3} \ \otimes \ \mathbf{3} \ = \ \mathbf{6} \ \oplus \ \mathbf{3^*}$$

In general the product of two fundamental representations has the above row and column box structure. By referring to Tables 3.3 and 3.6 we indeed verify the symmetric–antisymmetric row and column structure.

How do we calculate the resulting dimensions for arbitrary SU(N)? To calculate the dimension of any array of boxes there is a recipe which involves forming the ratio of two numbers. We assert that the way to calculate the numerator and denominator is as follows.

The calculation of the numerator: For any given diagram representing a product of representations of SU(N) insert N in each of the diagonal boxes starting from the top left-hand corner. Along the diagonals immediately above and below insert $N+1$ and $N-1$ respectively. In the next diagonals insert $N+2$ and so forth. A particular example should clarify this, e.g.

N	$N+1$	$N+2$
$N-1$	N	$N+1$
$N-2$	$N-1$	N
$N-3$		

The numerator of our expression will be the product of all these numbers.

The calculation of the denominator: This will be the product of the "hooks". Each box has a value of the hook associated with it and to find this value do the following. Draw a line entering the right-hand end of the row in which the box lies. On entering the box, this line turns downwards through 90° and then proceeds down the column until it leaves the diagram. The total number of boxes that the line has passed through, including the box in question, is the value of the hook associated with that box. The product of all the hooks is the denominator.

We will apply these rules to calculate the dimensions of our particular example met previously. It is trivial to verify that $\square$ indeed has dimension N (as it should!) and that

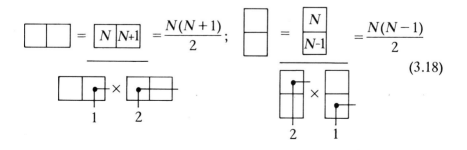

$$(3.18)$$

so that finally

$$\square \otimes \square = \square\square \oplus \begin{array}{c}\square\\\square\end{array}$$

$$= \frac{N(N+1)}{2} \oplus \frac{N(N-1)}{2} \qquad (3.19)$$

Hence for $N = 2$ we have that $\mathbf{2} \otimes \mathbf{2} = \mathbf{3} \oplus \mathbf{1}$ which we found explicitly in section 3.2. For SU(3) we see that $\mathbf{3} \otimes \mathbf{3} = \mathbf{6} \oplus \mathbf{3}^*$ as in section 3.3. In SU(3) the column of two boxes has dimension three by the hook rule and we recognise it as being the conjugate representation (a column of $N-1$ boxes in SU(N)) since it is a two-box column in SU(3).

As an exercise verify that a column of N boxes in SU(N) is always the singlet representation.

Now let's combine three objects in SU(N). We have already seen that

$$\square \otimes \square = \square\square \oplus \begin{array}{c}\square\\\square\end{array} \qquad (3.17)$$

and so we will now add another box to the two-box row and column. There is a detailed list of rules for forming allowed diagrams which are set out in Hammermesh (1963). For our present purpose we need only note that diagrams must not be concave upwards nor concave towards the lower left. Hence

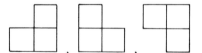

are all forbidden whereas

is allowed. One combines the third box to the original two in all possible ways subject to this constraint. This yields the following topologies.

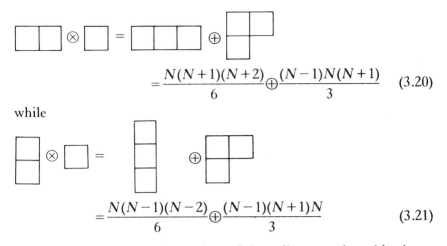

$$= \frac{N(N+1)(N+2)}{6} \oplus \frac{(N-1)N(N+1)}{3} \qquad (3.20)$$

while

$$= \frac{N(N-1)(N-2)}{6} \oplus \frac{(N-1)(N+1)N}{3} \qquad (3.21)$$

and we have written the dimensions of these diagrams alongside. As an exercise use the hook rule already described and verify these dimensions. To save effort in constructing diagrams note that in SU(N) no diagram need be drawn which has more than N boxes in any column since such a diagram trivially has dimension of zero. Hence for $N = 2$ the dimensionality of the tableaux are

$$(2 \otimes 2) \otimes 2 = (4 \oplus 2) \oplus 2 \qquad (3.22)$$

and the column is excluded as it has three boxes and we are discussing $N = 2$. For $N = 3$ we obtain

$$(3 \otimes 3) \otimes 3 = (10 \oplus 8) \oplus (8 \oplus 1) \qquad (3.23)$$

Other examples that are of physical interest include SU(4) and SU(6) where

$$(4 \otimes 4) \otimes 4 = (20 \oplus 20) \oplus (20 \oplus 4) \qquad (3.24)$$

and

$$(6 \otimes 6) \otimes 6 = (56 \oplus 70) \oplus (70 \oplus 20) \qquad (3.25)$$

respectively.

If your appetite is now whetted so that you are stimulated to go and combine more complicated diagrams, e.g. $8 \otimes 8$ in SU(3), then you will need to know the set of rules that are to be obeyed in order to form legal diagrams. Once the legal diagrams have been formed, the dimensions can be calculated by the standard procedure involving the hook rule. The rules for legal diagrams can be found in Hammermesh (1963).

As a final illustration we can combine **3** and **3*** in SU(3) to obtain the representations of the $q\bar{q}$ mesons. We find

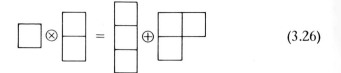

$$(3.26)$$

since all other topologies of connected boxes that contain the original single and two box columns will be concave upwards

or towards the lower left

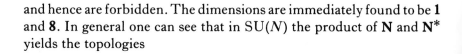

and hence are forbidden. The dimensions are immediately found to be **1** and **8**. In general one can see that in SU(N) the product of **N** and **N*** yields the topologies

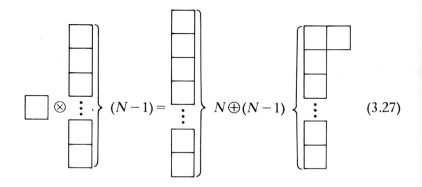

$$(3.27)$$

and so $\mathbf{N} \otimes \mathbf{N}^* = \mathbf{1} \oplus (\mathbf{N}^2 - \mathbf{1})$.

Notice that if ψ is the N-dimensional fundamental representation of SU(N) then $\bar{\psi}\lambda_i\psi$ transforms as the $N^2 - 1$ regular representation and

$\bar{\psi}\mathbf{1}\psi$ as the singlet (see section 2.2.1.4). For example in SU(2) where $\lambda_i \equiv \sigma_i$ then

$$\bar{\psi}\sigma_+\psi = u\bar{d}; \qquad \bar{\psi}\sigma_3\psi = u\bar{u} - d\bar{d}; \qquad \bar{\psi}\sigma_-\psi = d\bar{u} \qquad (3.28)$$

transform as π^+, π^0, π^-. The singlet is formed by the trace

$$\bar{\psi}\mathbf{1}\psi = u\bar{u} + d\bar{d} \qquad (3.29)$$

This leads naturally into the representations of meson ($q\bar{q}$) states.

4   SU(6): Quarks with Spin

4.1 Mesons

Three flavours of quark (uds) combined with any of three antiquark flavours ($\bar{u}\bar{d}\bar{s}$) yield nine meson $q\bar{q}$ combinations. Nine pseudoscalar and nine vector mesons exist corresponding to the nine $q\bar{q}$ flavour combinations (Table 4.1). In the framework of SU(3) there is no reason why the octet representations should be accompanied by singlets. In the quark model this is natural and the observation of nonets is a clear indicator of an underlying $\bar{q}$ structure.

There is only one question to be settled: What are the particular combinations of u$\bar{u}$, d$\bar{d}$ and s$\bar{s}$ that correspond to the three neutral states π^0, η^0, η'^0 or ρ^0, ω^0, ϕ^0? First of all consider the simpler case of SU(2) where the two neutrals will be u$\bar{u}$ and d$\bar{d}$. The $I = 0$ combination is u$\bar{u}$ + d$\bar{d}$ and $I = 1$ is $-$u$\bar{u}$ + d$\bar{d}$. The reason for these phases can be traced to equation (2.32) where we see that the doublet ($\bar{d}$, $-\bar{u}$) transforms like (u, d). The $I = 1$ state with $I_z = 0$ in (u, d) $\otimes$ (u, d) was given in Table 3.3 as

$$|I = 1, I_z = 0\rangle = \frac{1}{\sqrt{2}}(ud + du) \qquad (4.1)$$

Hence in (u, d) $\otimes$ ($\bar{d}$, $-\bar{u}$) we will have

$$|I = 1, I_z = 0\rangle = \frac{1}{\sqrt{2}}(-u\bar{u} + d\bar{d}) \qquad (4.2)$$

The orthogonal combination with $I = 0$ is therefore just the trace of
$\phi^+\phi = \frac{1}{\sqrt{2}}(\bar{u}u + \bar{d}d)$.

Proceeding to SU(3), the s and $\bar{s}$ have $I = 0$ in the SU(2) subgroup. Hence s$\bar{s}$ can occur in company with the isoscalar u$\bar{u}$ + d$\bar{d}$ but not with

TABLE 4.1
Meson nonet

	Charge	Strangeness	Examples	
u$\bar{d}$	+1	0	π^+	ρ^+
d$\bar{u}$	−1	0	π^-	ρ^-
u$\bar{u}$ ⎫			⎧ π^0	ρ^0
d$\bar{d}$ ⎬	0	0	⎨ η^0	ω^0
s$\bar{s}$ ⎭			⎩ η'^0	ϕ^0
u$\bar{s}$	+1 ⎫		⎧ K^+	K^{*+}
d$\bar{s}$	0 ⎭	+1	⎨ K^0	K^{*0}
$\bar{u}$s	−1 ⎫		⎧ K^-	K^{*-}
$\bar{d}$s	0 ⎭	−1	⎩ $\bar{K}^0$	$\bar{K}^{*0}$

the isovector d$\bar{d}$ − u$\bar{u}$. The SU(3) singlet is the obvious generalisation of the SU(2) singlet ($I = 0$) and hence, in notation (**SU**(3), **SU**(2)),

$$|\mathbf{1}, \mathbf{1}\rangle = \frac{1}{\sqrt{3}}(u\bar{u} + d\bar{d} + s\bar{s}) \tag{4.3}$$

Since the isovector contains no s$\bar{s}$ then

$$|\mathbf{8}, \mathbf{3}\rangle \equiv \frac{1}{\sqrt{2}}(d\bar{d} - u\bar{u}) \tag{4.4}$$

The third possible combination of u$\bar{u}$, d$\bar{d}$, s$\bar{s}$ orthogonal to these is[1]

$$|\mathbf{8}, \mathbf{1}\rangle \equiv \frac{1}{\sqrt{6}}(u\bar{u} + d\bar{d} - 2s\bar{s}) \tag{4.5}$$

---

[1] The **8** regular representation of SU(3) transforms as $\bar{\phi}\lambda_i\phi$ ($\bar{\phi}\lambda_3\phi$ in equation 4.4, $\bar{\phi}\lambda_8\phi$ in equation 4.5). The singlet is $\bar{\phi}\mathbf{1}\phi$ (equation 4.3). See equation (3.28).

For pseudoscalar and vector realisations of the above, a frequent notation is

$$|8, 3\rangle \equiv \pi^0, \rho^0$$
$$|8, 1\rangle \equiv \eta_8, \omega_8 \tag{4.6}$$
$$|1, 1\rangle \equiv \eta_1, \omega_1$$

The physical η and $X(\eta')$ or ω and ϕ are mixtures of $\eta_{1,8}$ and $\omega_{1,8}$ respectively. The η and η' are almost pure octet and singlet respectively. The physical ω and ϕ appear to be "ideal mixtures":

$$\phi = \frac{1}{\sqrt{3}}\omega_1 - \sqrt{\frac{2}{3}}\omega_8 \equiv s\bar{s}$$

$$\omega = \sqrt{\frac{2}{3}}\omega_1 + \sqrt{\frac{1}{3}}\omega_8 \equiv \frac{1}{\sqrt{2}}(u\bar{u} + d\bar{d}) \tag{4.7}$$

This question is discussed in more detail in sections 4.3.2 and 17.6.

4.1.1   G-PARITY

We should take care to make the labelling of quark and antiquark explicit in the SU(3) states. The $u\bar{d}$ state for instance can be either

$$\phi_S \equiv |u(1)\bar{d}(2) + \bar{d}(1)u(2)\rangle \frac{1}{\sqrt{2}} \tag{4.8}$$

or

$$\phi_A \equiv |u(1)\bar{d}(2) - \bar{d}(1)u(2)\rangle \frac{1}{\sqrt{2}} \tag{4.9}$$

the subscripts denoting the symmetry property under interchange of the labels 1 and 2. We shall in future take the labels to be understood in the ordering, hence

$$\phi_{S,A} \equiv |u\bar{d} \pm \bar{d}u\rangle \frac{1}{\sqrt{2}} \tag{4.10}$$

These two states are distinguished by their G-parity ($G \equiv C\, e^{i\pi T_2}$). With $T_2 \equiv \frac{1}{2}\tau_2$ then from equation (2.26) we have that $G = C i \tau_2 \equiv C \begin{pmatrix} 0 & 1 \\ -1 & 0 \end{pmatrix}$ where C is the charge conjugation operation. Hence

$$u \overset{G}{\to} \bar{d} \overset{G}{\to} -u; \qquad d \overset{G}{\to} -\bar{u} \overset{G}{\to} -d \qquad (4.11)$$

(see also section 2.2.1.3). Consequently

$$G\phi_S \to \frac{1}{\sqrt{2}}(-\bar{d}u - u\bar{d}) \equiv -\phi_S$$

$$(4.12)$$

$$G\phi_A \to \frac{1}{\sqrt{2}}(-\bar{d}u + u\bar{d}) \equiv +\phi_A$$

and so

$$\phi_S \to G = -1(\pi^+)$$

$$(4.13)$$

$$\phi_A \to G = +1(\rho^+)$$

The neutral partners are consequently

$$\phi_S: \quad \tfrac{1}{2}\{(d\bar{d} - u\bar{u}) + (\bar{d}d - \bar{u}u)\} = \pi^0$$

$$(4.14)$$

$$\phi_A: \quad \tfrac{1}{2}\{(d\bar{d} - u\bar{u}) - (\bar{d}d - \bar{u}u)\} = \rho^0$$

which are charge conjugation eigenstates

$$C\phi_S^0 = +\phi_S^0, \qquad C\phi_A^0 = -\phi_A^0 \qquad (4.15)$$

4.1.2 COMBINING WITH SPIN

How do mesons built from quarks and antiquarks obtain a spin or intrinsic angular momentum?

Suppose, first, that quarks and antiquarks were spinless. A quark and antiquark in a relative S-wave state of orbital angular momentum would form a meson with a spin zero and positive parity; hence a scalar meson. If the q and $\bar{q}$ were in a relative P-wave then a vector meson would be formed. The D-wave system forms $J = 2$ mesons and so on. In so far as one would expect the S-wave state to have the lowest energy then one would predict that a scalar meson nonet lies lowest in the spectrum. The

P-wave vectors would be expected to be the next heaviest states, the D-wave $J = 2$ mesons the next and so on.

Empirically the meson spectrum does not look at all like this. Vector mesons have comparable or even lighter masses than the scalar mesons. The lowest mass mesons are pseudoscalars ($J^P = 0^-$) and such states would not naturally arise if scalar quarks and antiquarks were building up the meson spectrum. The meson spectrum arises rather naturally if quarks have spin $\frac{1}{2}$. This will be seen in detail in Chapter 5. We will anticipate this and study the wavefunctions for q and $\bar{q}$ states formed from spin $\frac{1}{2}$ quarks.

The phenomenon of the lowest mass nonets being pseudoscalar and vector is consistent with spin $\frac{1}{2}$ quark and antiquark coupling to spin 0 and 1 respectively. The spin wavefunctions are immediately obtained from Table 3.3 if we replace u, d by $\uparrow$, $\downarrow$ (which denote the up and down S_z of the quarks' spins).

We will label the spin triplet wavefunction by χ_S and singlet by χ_A (the subscripts indicating their symmetry properties). Combining with the SU(3) wavefunctions we have the following possibilities for the symmetry properties of $\phi\chi$ under interchange of the labels 1 and 2:

$$\text{Symmetric:} \qquad \phi_S\chi_S, \quad \phi_A\chi_A \qquad (4.16)$$

$$\text{Antisymmetric:} \qquad \phi_S\chi_A, \quad \phi_A\chi_S \qquad (4.17)$$

Clearly the 0^-, 1^- system corresponds to the totally antisymmetric combinations since for the neutral charge conjugation eigenstates we have

$$\phi_S\chi_A = |C = +, S = 0\rangle \sim \pi^0$$

$$\phi_A\chi_S = |C = -, S = 1\rangle \sim \rho^0 \qquad (4.18)$$

Hence the meson states are

$$\phi_i^S \left| \frac{1}{\sqrt{2}}(\uparrow\downarrow - \downarrow\uparrow) \right\rangle$$

$$\phi_i^A |\uparrow\uparrow\rangle, \qquad \phi_i^A \left| \frac{1}{\sqrt{2}}(\uparrow\downarrow + \downarrow\uparrow) \right\rangle, \qquad \phi_i^A |\downarrow\downarrow\rangle \qquad (4.19)$$

where $i = 0$ (singlet), $i = 1 \ldots 8$ (octet) states are listed in Table 4.2. As an exercise verify that $G = C(-1)^I$ is satisfied for these states.

These properties have arisen as a consequence of the choice of antisymmetric $\phi_S\chi_A$ which corresponds to anticommutation of the creation or annihilation operators for the quarks (antiquarks).

TABLE 4.2
$q\bar{q}$ representations with explicit G-parity

$\dfrac{1}{\sqrt{2}}(u\bar{s} \pm \bar{s}u)$	$K^+ (K^{+*})$
$\dfrac{1}{\sqrt{2}}(d\bar{s} \pm \bar{s}d)$	$K^0 (K^{0*})$
$-\dfrac{1}{\sqrt{2}}(s\bar{u} \pm \bar{u}s)$	$K^- (K^{-*})$
$-\dfrac{1}{\sqrt{2}}(s\bar{d} \pm \bar{d}s)$	$\bar{K}^0 (\bar{K}^{0*})$
$\dfrac{1}{\sqrt{2}}(u\bar{d} \pm \bar{d}u)$	$\pi^+ (\rho^+)$
$-\dfrac{1}{\sqrt{2}}(d\bar{u} \pm \bar{u}d)$	$\pi^- (\rho^-)$
$\dfrac{1}{2}[(d\bar{d} - u\bar{u}) \pm (\bar{d}d - \bar{u}u)]$	$\pi^0 (\rho^0)$
$\dfrac{1}{2\sqrt{3}}[(u\bar{u} + d\bar{d} - 2s\bar{s}) \pm (\bar{u}u + \bar{d}d - 2\bar{s}s)]$	$\eta_8^0 (\omega_8^0)$
$\dfrac{1}{\sqrt{6}}[(u\bar{u} + d\bar{d} + s\bar{s}) \pm (\bar{u}u + \bar{d}d + \bar{s}s)]$	$\eta_1^0 (\omega_1^0)$

Note: These states have $C\pi^\pm = -\pi^\mp$, $C\rho^\pm = \rho^\mp$ and the strange states are defined analogously. $\phi_{S,A}$ have neutral charge conjugation eigenstates with $C = +1, -1$ respectively. Note that $\tau^- u = d$, $\tau^- \bar{d} = -\bar{u}$. The 0^- are ϕ_S and 1^- are ϕ_A.

4.2 Baryons

Taking the SU(3) fundamental representation (u, d, s) and combining it with the SU(2) ($\uparrow\downarrow$) one can form a six-dimensional fundamental representation of SU(6), $u\uparrow$, $d\uparrow$, $s\uparrow$, $u\downarrow$, $d\downarrow$, $s\downarrow$ (Gursey and Radicati, 1964). Physically in the quark model the intrinsic SU(3) degrees of freedom will be multiplied by the SU(2) spin of the quarks. For the mesons we have already anticipated this and seen the 0^- and 1^- nonets emerging and forming $\mathbf{1} \oplus \mathbf{35}$ of the $\mathbf{6} \otimes \bar{\mathbf{6}}$ of SU(6). The baryons are rather more complicated algebraically. To save a lot of work I shall just quote the following rules for combining states of different permutation symmetry and you can verify it by writing out the states explicitly. Denoting symmetric, mixed and antisymmetric states by S, M, A respectively, the symmetry properties that arise are shown in the matrix.

	S	M	A
S	S	M	A
M	M	S, M, A	M

(4.21)

Recalling that in SU(3) we found 10_S, 8_M and 1_A (equation 3.11 and Table 3.7), while in SU(2) (Table 3.4) 4_S and 2_M emerged, then the above rules imply, for instance, that the **10** with spin $\frac{3}{2}$ (**4** in SU(2)) will be totally symmetric; the **10** with spin $\frac{1}{2}$ (**2** in SU(2)) will be totally mixed and so forth.

To classify under SU(6) we collect together those states which are symmetric, then those which are mixed and finally those which are antisymmetric. These are listed below together with their (SU(3), SU(2)) subgroup dimensionalities. The total number of such states is given on the right (e.g. $(\mathbf{10} \otimes \mathbf{4}) \oplus (\mathbf{8} \otimes \mathbf{2}) = \mathbf{56}$).

$$\text{S:} \quad (\mathbf{10, 4}) \qquad + (\mathbf{8, 2}) \qquad\qquad = \mathbf{56} \qquad (4.22)$$

$$\text{M:} \quad (\mathbf{10, 2}) + (\mathbf{8, 4}) + (\mathbf{8, 2}) + (\mathbf{1, 2}) \qquad = \mathbf{70} \qquad (4.23)$$

$$\text{A:} \qquad\qquad\qquad (\mathbf{8, 2}) \qquad + (\mathbf{1, 4}) = \mathbf{20} \qquad (4.24)$$

We can immediately verify these results by using the Young diagram techniques. Combining three fundamental representations of SU(6) yields

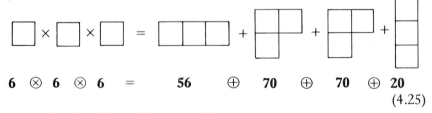

$$\mathbf{6} \otimes \mathbf{6} \otimes \mathbf{6} = \mathbf{56} \oplus \mathbf{70} \oplus \mathbf{70} \oplus \mathbf{20}$$

(4.25)

hence the $\mathbf{56}_S$, $\mathbf{70}_{M,S}$, $\mathbf{70}_{M,A}$, and $\mathbf{20}_A$ representations are seen.

We can explicitly write out these wavefunctions in the quark model exhibiting their SU(3) and SU(2) content. For example consider the Δ^+(uud) with $S_z = \frac{1}{2}(\uparrow\uparrow\downarrow)$. The SU(6) wavefunction being totally symmetric is

$$\frac{1}{3}\left\{\begin{array}{l} u\uparrow u\uparrow d\downarrow + u\uparrow d\uparrow u\downarrow + d\uparrow u\uparrow u\downarrow \\ + u\uparrow u\downarrow d\uparrow + u\uparrow d\downarrow u\uparrow + d\uparrow u\downarrow u\uparrow \\ + u\downarrow u\uparrow d\uparrow + u\downarrow d\uparrow u\uparrow + d\downarrow u\uparrow u\uparrow \end{array}\right\} \tag{4.26}$$

The $SU(3) \otimes SU(2)$ decomposition is

$$\frac{1}{\sqrt{3}}(uud + udu + duu)\frac{1}{\sqrt{3}}(\uparrow\uparrow\downarrow + \uparrow\downarrow\uparrow + \downarrow\uparrow\uparrow) \tag{4.27}$$

where the quarks are ordered $1, 2, 3$ and all possible combinations are taken to yield the wavefunction in equation (4.26).

With the full set of $SU(3)$ wavefunctions $\phi_{S,M_S,M_A,A}$ in Table 3.7 and $SU(2)\chi_{S,M_S,M_A}$ (Table 3.4) then the totality of combinations is listed in Table 4.3. Those states labelled S are totally symmetric (**56** of $SU(6)$).

An example of a 56plet of (broken) $SU(6)$ is the decuplet with $J^P = \frac{3}{2}^+$ containing the $\Delta(1236)$ and the octet with $J^P = \frac{1}{2}^+$ containing the nucleon. If the **56** is the lowest lying representation in the spectrum of baryons, then the correlation of $J = \frac{1}{2}$ with the octet and not decuplet emerges naturally as does the absence of the singlet state and consequent absence of nonet structure for the nucleon and its family.

TABLE 4.3

S:	$\phi_S\chi_S \equiv (\mathbf{10, 4})$	
	$\frac{1}{\sqrt{2}}(\phi_{M,S}\chi_{M,S} + \phi_{M,A}\chi_{M,A}) \equiv (\mathbf{8, 2})$	

M_S:	$\phi_S\chi_{M,S} \equiv (\mathbf{10, 2})$	M_A:	$\phi_S\chi_{M,A} \equiv (\mathbf{10, 2})$
	$\phi_{M,S}\chi_S \equiv (\mathbf{8, 4})$		$\phi_{M,A}\chi_S \equiv (\mathbf{8, 4})$
$\frac{1}{\sqrt{2}}(-\phi_{M,S}\chi_{M,S} + \phi_{M,A}\chi_{M,A}) \equiv (\mathbf{8, 2})$		$\frac{1}{\sqrt{2}}(\phi_{M,S}\chi_{M,A} + \phi_{M,A}\chi_{M,S}) \equiv (\mathbf{8, 2})$	
$\phi_A\chi_{M,A} \equiv (\mathbf{1, 2})$		$\phi_A\chi_{M,S} \equiv (\mathbf{1, 2})$	

A:	$\phi_A\chi_S \equiv (\mathbf{1, 4})$	
	$\frac{1}{\sqrt{2}}(\phi_{M,S}\chi_{M,A} - \phi_{M,A}\chi_{M,S}) \equiv (\mathbf{8, 2})$	

Note: $SU(3)$ states ϕ and $SU(2)$ states χ are combined and $SU(6)$ states are formed and classified by their symmetry behaviours, S being symmetric and having dimension **56**. Two mixed **70** dimensional representations and an antisymmetric **20** dimensional also arise. The explicit expressions for the $SU(3)$ and $SU(2)$ wavefunctions are given in Table 3.7 and Table 3.4 respectively.

The negative parity states around 1500–1700 MeV belong to a 70plet. At least for the strangeness zero members, the (SU(3), SU(2)) sub-structure of the **70** is seen with the following J^P states:

$$(\mathbf{10}, \mathbf{2})_{\frac{1}{2}}^{-}, \tfrac{3}{2}^{-} : (\mathbf{8}, \mathbf{2})_{\frac{1}{2}}^{-}, \tfrac{3}{2}^{-} : (\mathbf{8}, \mathbf{4})_{\frac{1}{2}}^{-}, \tfrac{3}{2}^{-}, \tfrac{5}{2}^{-} \tag{4.28}$$

while the $(\mathbf{1}, \mathbf{2})_{\frac{1}{2}}^{-}, \tfrac{3}{2}^{-}$ are seen, an example being the $\Lambda(1405)$ mentioned earlier as the lowest lying baryon singlet.

The SU(6) group structure does not answer the question why the **56** is the lowest mass multiplet, nor why the positive and negative parity multiplets alternate in mass as one moves up the scale of masses. These questions will be discussed in section 5.1 where a detailed description of the **70** states will also be given. First we will illustrate how to use the SU(6) wavefunctions (Table 4.3) in calculations.

4.3 Examples of simple calculations

4.3.1 THE NUCLEON'S CHARGES AND MAGNETIC MOMENTS

The nucleons are in a 56plet and the wavefunction for the octet states is given in Table 4.3, namely

$$\frac{1}{\sqrt{2}}(\phi_{M,S}\chi_{M,S} + \phi_{M,A}\chi_{M,A}) \tag{4.29}$$

where ϕ, χ refer to unitary and spin wavefunctions respectively, their explicit forms being given in Tables 3.7 and 3.4.

The nucleon charges are

$$\begin{pmatrix} \text{proton} \\ \text{neutron} \end{pmatrix} = \begin{pmatrix} 1 \\ 0 \end{pmatrix} = \sum_{i=1}^{3} \frac{1}{\sqrt{2}} \langle \phi_{M,S}\chi_{M,S} + \phi_{M,A}\chi_{M,A} | e_i | \phi_{M,S}\chi_{M,S}$$

$$+ \phi_{M,A}\chi_{M,A} \rangle \frac{1}{\sqrt{2}} \tag{4.30}$$

where ϕ will be that for the proton or the neutron respectively and e_i is the charge operator for the i-th quark.

It is traditional and simplifying at this point to exploit the fact that the total wavefunction is symmetric and hence to make the replacement

$$\sum_{i=1}^{3} \langle \dots | e_i | \dots \rangle \rightarrow 3 \langle \dots | e_{(3)} | \dots \rangle \tag{4.31}$$

since our wavefunctions have always been written with quark number 3 as special (i.e. they are S or A in quarks $(1, 2)$ but mixed in $(1, 3)$ or $(2, 3)$). We therefore have

$$\binom{e_P}{e_N} = \tfrac{3}{2}\{\langle \phi_{M,S}|e_{(3)}|\phi_{M,S}\rangle + \langle \phi_{M,A}|e_{(3)}|\phi_{M,A}\rangle\} \tag{4.32}$$

where we have exploited the fact that the spin states are orthonormal, i.e.

$$\langle \chi_{M,S}|\chi_{M,S}\rangle = \langle \chi_{M,A}|\chi_{M,A}\rangle = 1; \qquad \langle \chi_{M,S}|\chi_{M,A}\rangle = 0 \tag{4.33}$$

All that is left is to calculate the expectation values of the quark charge operator between the unitary wavefunctions. Writing

$$e_{(3)} \equiv Q(\tfrac{2}{3} \text{ or } -\tfrac{1}{3}) \tag{4.34}$$

with Q a scaled charge then

$$\langle \phi_{M,S}^P|e_{(3)}|\phi_{M,S}^P\rangle \equiv \tfrac{1}{6}\langle udu + duu - 2uud|e_{(3)}|udu + duu - 2uud\rangle$$
$$= \tfrac{1}{6}(\tfrac{2}{3} + \tfrac{2}{3} + 4(-\tfrac{1}{3}))Q = 0 \tag{4.35}$$

$$\langle \phi_{M,A}^P|e_{(3)}|\phi_{M,A}^P\rangle \equiv \tfrac{1}{2}\langle udu - duu|e_{(3)}|udu - duu\rangle$$
$$= \tfrac{1}{2}(\tfrac{2}{3} + \tfrac{2}{3})Q = \tfrac{2}{3}Q \tag{4.36}$$

Consequently

$$e_{\text{proton}} = \tfrac{3}{2}(0 + \tfrac{2}{3}Q) \equiv Q \tag{4.37}$$

and so the scaled charge is the same as that of the proton and will from now on always be set to unity. Similarly for the neutron we find

$$\langle \phi_{M,S}^N|e_{(3)}|\phi_{M,S}^N\rangle = \tfrac{1}{3}Q = \tfrac{1}{3} \tag{4.38}$$

$$\langle \phi_{M,A}^N|e_{(3)}|\phi_{M,A}^N\rangle = -\tfrac{1}{3}Q = -\tfrac{1}{3} \tag{4.39}$$

and hence

$$e_{\text{neutron}} = 0 \tag{4.40}$$

The vanishing of the matrix element $\langle \phi_{M,S}^P|e_{(3)}|\phi_{M,S}^P\rangle$ gives rise to some interesting selection rules (Moorhouse selection rules) in connection with the electromagnetic interactions of protons (Chapter 7) (Moorhouse, 1966).

As an exercise verify that the spin expectation values of the Pauli matrices are as follows:

$$\langle \chi^{\uparrow}_{M,S} | \sigma^{(3)}_+ | \chi^{\downarrow}_{M,S} \rangle = \langle \chi^{\uparrow}_{M,S} | \sigma^{(3)}_z | \chi^{\uparrow}_{M,S} \rangle = -\tfrac{1}{3}$$

$$\langle \chi^{\uparrow}_{M,A} | \sigma^{(3)}_z | \chi^{\uparrow}_{M,A} \rangle = \langle \chi^{\uparrow}_{M,A} | \sigma^{(3)}_+ | \chi^{\downarrow}_{M,A} \rangle = 1 \qquad (4.41)$$

$$\langle \chi^{\uparrow}_{M,A} | \sigma^{(3)}_z | \chi^{\uparrow}_{M,S} \rangle = 0$$

Then by considering the magnetic moment operator $\sum^{3}_{i=1} \mu e_i \boldsymbol{\sigma}_{iz}$ evaluated between proton wavefunctions or between neutron wavefunctions verify that $\mu \equiv \mu_P$ and

$$\frac{\mu_P}{\mu_N} = -\frac{3}{2} \qquad (4.42)$$

Note that this result is in remarkable agreement with the data which give

$$\frac{\mu_P}{\mu_N} = -\frac{2 \cdot 79}{1 \cdot 91} \qquad (4.43)$$

Baryon magnetic moments are described in more detail in section 7.1.

4.3.2 RADIATIVE VECTOR TO PSEUDOSCALAR MESON TRANSITIONS

The vector nonet wavefunctions are $\phi_A \chi_S$ and the pseudoscalars are $\phi_S \chi_A$ (equation 4.18). The fact that $\langle \chi_S | \chi_A \rangle \equiv 0$ demands that a spin operator be present in order to yield the spin transition. The $V \rightarrow P\gamma$ transition will therefore involve $\sum^{2}_{i=1} e_i \boldsymbol{\sigma}_i (\mathbf{S} \equiv \tfrac{1}{2}\boldsymbol{\sigma})$ and hence will be proportional to the quark magnetic moment (see also section 7.1, Chapter 12; Kokkedee, 1969, Becchi and Morpurgo, 1965b, Dalitz, 1965).

If $\boldsymbol{\varepsilon}$ is the polarisation vector of the emitted photon then the matrix element for $V \rightarrow P\gamma$ may be written

$$M \equiv \langle \phi_A \chi_S | \sum^{2}_{i=1} e_i \mu_i \boldsymbol{\sigma} \cdot \boldsymbol{\varepsilon} | \phi_S \chi_A \rangle \qquad (4.44)$$

where ϕ, χ are respectively the flavour and quark spin wavefunctions and μ_i the quark scale magnetic moment. We will take $\mu_u = \mu_d \equiv \mu$ but anticipate future developments on symmetry breaking and set $\mu_s \simeq \tfrac{3}{5}\mu_{u,d}$ when considering $K^* \rightarrow K\gamma$.

We will calculate explicitly the matrix elements for the V and γ to have $J_z = +1$, where the z-axis is the colinear axis defined by the $V \to P\gamma$ decay and the photon moves along the positive z-axis. Then

$$\boldsymbol{\varepsilon} = -\frac{1}{\sqrt{2}}(1, i, 0)$$

hence

$$\boldsymbol{\sigma} \cdot \boldsymbol{\varepsilon} \equiv -\frac{1}{\sqrt{2}}(\sigma_x + i\sigma_y) \equiv \sqrt{2}\sigma_+$$

We can now write out equation (4.44) explicitly for the case of $\omega \to \pi^0\gamma$ and find

$$M = -\sqrt{2}\mu\left\langle\left\{\frac{(d\bar{d}+u\bar{u})-(\bar{d}d+\bar{u}u)}{2}\right\}\uparrow\uparrow\middle| e_1\sigma_1^+ \right.$$
$$\left. + e_2\sigma_2^+\middle|\left\{\frac{(d\bar{d}-u\bar{u})+(\bar{d}d-\bar{u}u)}{2}\right\}\left(\frac{\uparrow\downarrow-\downarrow\uparrow}{\sqrt{2}}\right)\right\rangle \quad (4.45)$$

where we have assumed that ω is the "ideally mixed" combination of ω_1 and ω_8 that contains no s$\bar{s}$ (see equation 4.7 and section 17.6). Factorise this into

$$M = -\sqrt{2}\mu$$

$$\left\{ \left\langle\frac{(d\bar{d}+u\bar{u})-(\bar{d}d+\bar{u}u)}{2}\middle| e_1\middle|\frac{(d\bar{d}-u\bar{u})+(\bar{d}d-\bar{u}u)}{2}\right\rangle\left\langle\uparrow\uparrow\middle|\sigma_1^+\middle|\frac{\uparrow\downarrow-\downarrow\uparrow}{\sqrt{2}}\right\rangle\right. $$
$$\left. + (1 \leftrightarrow 2) \vphantom{\frac{1}{2}}\right\}$$

$$(4.46)$$

Then verify that

$$\langle\phi_A^0|e_1|\phi_S^0\rangle \equiv -\langle\phi_A^0|e_2|\phi_S^0\rangle = -\tfrac{1}{2} \quad (4.47)$$

$$\langle\chi_S(S_z = 1)|\sigma_1^+|\chi_A(S_z = 0)\rangle = -\langle\chi_S(S_z = 1)|\sigma_2^+|\chi_A(S_z = 0)\rangle = -\frac{1}{\sqrt{2}}$$

$$(4.48)$$

and hence

$$\langle e_1\sigma_1^+ + e_2\sigma_2^+\rangle \equiv 2\langle e_1\sigma_1^+\rangle \quad (4.49)$$

when the G-parity wavefunctions are used. Therefore

$$M(\omega_{J,z=1} \to \pi^0\gamma) \equiv -\mu \quad (4.50)$$

Notice that alternatively one need not write out the fully symmetrised G-parity wavefunctions so long as $e_1\sigma_1^+$ and $e_2\sigma_2^+$ are separately evaluated and equation (4.49) not used. This is the procedure that is most frequently met in the literature and so we illustrate its correspondence with the above treatment.

Write now for the SU(3) portion

$$\omega^0 \equiv \frac{1}{\sqrt{2}}(u\bar{u}+d\bar{d}); \qquad \pi^0 \equiv \frac{1}{\sqrt{2}}(d\bar{d}-u\bar{u}) \qquad (4.51)$$

so that ρ^0 and π^0 would be distinguished only by their spin wavefunctions. Equation (4.45) is therefore replaced by

$$M = -\sqrt{2}\mu\left\langle \frac{(u\bar{u}+d\bar{d})\uparrow\uparrow}{\sqrt{2}}\middle| e_1\sigma_1^+ + e_2\sigma_2^+ \middle| \frac{(d\bar{d}-u\bar{u})}{\sqrt{2}} \frac{(\uparrow\downarrow-\downarrow\uparrow)}{\sqrt{2}} \right\rangle$$

$$(4.52)$$

and

$$\langle e_1\sigma_1^+\rangle = \langle e_2\sigma_2^+\rangle = \frac{1}{2\sqrt{2}} \qquad (4.53)$$

hence again

$$M(\omega_{J_z=1} \to \pi^0\gamma) \equiv -\mu \qquad (4.50)$$

The analogous calculation for $\omega(J_z=-1)$ can be performed and, of course, this matrix element is μ also. The $\omega(J_z=0)$ cannot decay into a real photon (which is necessarily $J_z=\pm1$). Hence on summing over final states and averaging over the initial we have

$$\overline{\sum_{\text{spins}} |M|^2} = \tfrac{2}{3}\mu^2$$

which is equation (9) of Becchi–Morpurgo (1965b). To get from here to an absolute calculation of the rate requires discussion of whether nonrelativistic or relativistic phase space should be used (Becchi and Morpurgo, 1965b; Morpurgo, 1977). The observed rate is consistent with $\mu \equiv \mu_{\text{proton}}$, which agrees with the result found from the calculation of the proton magnetic moment. Such a relation between *baryon* magnetic moments and *meson* magnetic transitions is a distinct *quark model* result and outside any known symmetry scheme.

Before continuing we make a comment on the convention that we adopted at equation (4.51) *et seq.* As this convention is much easier on labour we shall adopt it hereon. Occasionally one encounters a calculation where the relative phases of two resonances in different multiplets are important (e.g. π and B in the $S = 0$, $L = 0, 1$ respectively, Tables 5.1 and 5.2) and then the G-parity wavefunctions must be used with no short cuts. Also do not conclude that one can always assume equation (4.53) for $\langle \hat{0}_1 \rangle = \langle \hat{0}_2 \rangle$ and hence hope to make another short cut; this is not true in general since, for example, the charge of π^0 is zero due to $\langle e_1 \rangle = -\langle e_2 \rangle$.

We can now compute the matrix elements for $V \to P\gamma$ throughout the nonet and compare with $\omega \to \pi\gamma$. First consider $\phi \to \pi\gamma$. With the ideal mixing assumed above for ω, then $\phi \equiv s\bar{s}$. Hence

$$M(\phi \to \pi\gamma) \propto \langle s\bar{s}|e_i|-u\bar{u} + d\bar{d}\rangle \equiv 0 \qquad (4.54)$$

The $\phi \to \pi\gamma$ is indeed very much suppressed relative to $\omega \to \pi\gamma$:

$$\frac{\Gamma(\phi \to \pi\gamma)}{\Gamma(\omega \to \pi\gamma)} \approx 0\cdot 6 \text{ per cent} \qquad (4.55)$$

and this is consistent with the ideal mixing and consequent equation (4.54).

For $\rho^0 \to \pi^0\gamma$ the only change relative to $\omega \to \pi^0\gamma$ is in the isospin state. Hence

$$\frac{M(\rho^0 \to \pi\gamma)}{M(\omega \to \pi\gamma)} \equiv \frac{(-u\bar{u} + d\bar{d}|e_1|-u\bar{u} + d\bar{d})}{(u\bar{u} + d\bar{d}|e_1|-u\bar{u} + d\bar{d})} = \frac{1}{3} \qquad (4.56)$$

Consequently one predicts

$$\Gamma(\rho^0 \to \pi^0\gamma) = \tfrac{1}{9}\Gamma(\omega \to \pi^0\gamma) \qquad (4.57)$$

since the phase spaces are identical ($m_\omega \simeq m_\rho$). The situation empirically is rather confused at present, there being some indication that $\rho \to \pi\gamma$ decays are too small. The theoretical significance has been studied by O'Donnel (1976), Edwards and Kamal (1976), and Randa and Donnachie (1977).

The decays $K^* \to K\gamma$ are interesting in connection with symmetry breaking. In particular we illustrate the neutral and charged ratio

$K^0(d\bar{s})$, $K^+(u\bar{s})$. If μ_i is the scale parameter for the magnetic moments of quarks then

$$\frac{M(K^{*0} \rightarrow K^0\gamma)}{M(K^{*+} \rightarrow K^+\gamma)} = \frac{\langle d\bar{s}\uparrow\uparrow|e_1\mu_1\sigma_1^+ + e_2\mu_2\sigma_2^+|d\bar{s}(\uparrow\downarrow-\downarrow\uparrow)\rangle}{\langle u\bar{s}\uparrow\uparrow|e_1\mu_1\sigma_1^+ + e_2\mu_2\sigma_2^+|u\bar{s}(\uparrow\downarrow-\downarrow\uparrow)\rangle} \qquad (4.58)$$

$$\equiv \frac{(\mu_s + \mu_d)}{(\mu_s - 2\mu_u)} \qquad (4.59)$$

Then if SU(3) were exact

$$\mu_s \equiv \mu_{u,d} \rightarrow \frac{\Gamma(K^{*0} \rightarrow K^0\gamma)}{\Gamma(K^{*+} \rightarrow K^+\gamma)} = 4 \qquad (4.60)$$

As an extreme example of SU(3) breaking consider $\mu_s \rightarrow 0$. Then

$$\frac{\Gamma(K^{*0} \rightarrow K^0\gamma)}{\Gamma(K^{*+} \rightarrow K^+\gamma)} = \frac{1}{4} \qquad (4.61)$$

(only the d and u quarks respectively will contribute). Empirically[1] $\mu_s \simeq \frac{3}{5}\mu_d$ and hence

$$\frac{\Gamma(K^{*0} \rightarrow K^0\gamma)}{\Gamma(K^{*+} \rightarrow K^+\gamma)} \simeq 1\cdot 3 \qquad (4.62)$$

The widths in keV are 75 ± 35 for $K^{*0} \rightarrow K^0\gamma$ and less than 80 for $K^{*+} \rightarrow K^+\gamma$.

This symmetry breaking is most interesting when one extends to SU(4) and charm. The charmed quark mass scale is so much larger than the strange and u, d mass scales that one expects $\mu_c \ll \mu_{s,u,d}$ (Chapter 17). Then one would predict

$$\frac{\Gamma(D^{*0}(c\bar{u}) \rightarrow D^0\gamma)}{\Gamma(D^{*+}(c\bar{d}) \rightarrow D^+\gamma)} = 4 \qquad (4.63)$$

since the charmed quark plays no role.

Within the SU(3) arena one can calculate $V \rightarrow P\gamma$ where P is η or η'. This requires knowledge of the singlet octet mixing in the $\eta - \eta'$ system. If we define

$$\eta = (\cos\theta)\eta_8 + (\sin\theta)\eta_1$$
$$X(\eta') = -(\sin\theta)\eta_8 + (\cos\theta)\eta_1 \qquad (4.64)$$

---

[1] See section 7.1.

then, taking $\mu_{u,d,s} = 1$, we can use the wavefunctions of Table 4.2 and the methods outlined above to obtain the matrix elements as follows:

$$\omega \to \pi\gamma = 1 \qquad \omega \to \eta\gamma = \frac{1}{3\sqrt{3}}(\cos\theta + \sqrt{2}\sin\theta)$$

$$\rho \to \pi\gamma = \tfrac{1}{3} \qquad \rho \to \eta\gamma = \frac{1}{\sqrt{3}}(\cos\theta + \sqrt{2}\sin\theta)$$

$$\phi \to \pi\gamma = 0 \qquad \phi \to \eta\gamma = \frac{2}{3\sqrt{3}}(-\sqrt{2}\cos\theta + \sin\theta)$$

$$\text{(4.65)}$$

$$X \to \omega\gamma = \frac{1}{3\sqrt{3}}(\sqrt{2}\cos\theta - \sin\theta)$$

$$X \to \rho\gamma = \frac{1}{\sqrt{3}}(\sqrt{2}\cos\theta - \sin\theta)$$

$$\phi \to X\gamma = \frac{2}{3\sqrt{3}}(\cos\theta + \sqrt{2}\sin\theta)$$

As an exercise verify these and see how they would be affected if $\mu_s \neq \mu_{u,d}$.

A particular case of interest is to choose

$$V_\gamma = \sqrt{\tfrac{3}{2}}(\tfrac{2}{3}u\bar{u} - \tfrac{1}{3}d\bar{d} - \tfrac{1}{3}s\bar{s}) \qquad \text{(4.66)}$$

which transforms like a photon. Then $M \to \gamma\gamma$ can be computed from $V_\gamma \to M\gamma$ ($M = \pi^0, \eta, X$). The matrix element reads (with the same approach as above)

$$M = -\sqrt{2}\mu\left\langle \sqrt{\frac{3}{2}}\left(\frac{2}{3}u\bar{u} - \frac{1}{3}d\bar{d} - \frac{1}{3}s\bar{s}\right)\uparrow\uparrow\left|e_1\sigma_1^+ + e_2\sigma_2^+\right|\left(\frac{d\bar{d}-u\bar{u}}{\sqrt{2}}\right)\left(\frac{\uparrow\downarrow-\downarrow\uparrow}{\sqrt{2}}\right)\right\rangle$$

$$= -\sqrt{2}\mu\left(\frac{\sqrt{3}}{2\sqrt{2}}\right)\left(\frac{1}{9}-\frac{4}{9}\right)(-1)\times 2 \qquad \text{(4.67)}$$

$$\underset{\langle e_1\rangle}{\uparrow} \quad \underset{\langle \sigma_1^+\rangle}{\uparrow} \quad \underset{\langle e_1\sigma_1^+\rangle = \langle e_2\sigma_2^+\rangle}{\uparrow}$$

and hence

$$M_{\pi^0 \to \gamma\gamma} \propto -\frac{\mu}{\sqrt{3}} \qquad \text{(4.68)}$$

Analogously for η_8 we find

$$M_{\eta,8 \to \gamma\gamma} = -\frac{1}{\sqrt{3}} M_{\pi^0 \to \gamma\gamma} \tag{4.69}$$

$$M_{\eta,1 \to \gamma\gamma} = -2\sqrt{\frac{2}{3}} M_{\pi^0 \to \gamma\gamma} \tag{4.70}$$

These are most easily checked by comparing the expectation value of e_1 between V_γ and the normalised $\eta_{1,8}$ and π states. For example

$$
\begin{aligned}
\frac{\langle V_\gamma | e_1 | \eta_8 \rangle}{\langle V_\gamma | e_1 | \pi^0 \rangle} &= \frac{(2u\bar{u} - d\bar{d} - s\bar{s}|e_1|u\bar{u} + d\bar{d} - 2s\bar{s})1/\sqrt{6}}{(2u\bar{u} - d\bar{d} - s\bar{s}|e_1| - u\bar{u} + d\bar{d})1/\sqrt{2}} \\
&= \frac{(4 + 1 - 2)1/\sqrt{6}}{(-4 + 1)1/\sqrt{2}}
\end{aligned} \tag{4.71}
$$

and hence equation (4.69) follows.

Then for an arbitrary mixture of η_1 and η_8 given by equation (4.64) the matrix element for $\eta \to \gamma\gamma$ relative to $\pi^0 \to \gamma\gamma$ will be

$$\frac{M(\eta \to \gamma\gamma)}{M(\pi^0 \to \gamma\gamma)} = -\frac{1}{\sqrt{3}}(\cos\theta + 2\sqrt{2}\sin\theta) \tag{4.72}$$

The phase space is proportional to the pseudoscalar mass cubed. Hence, with $(m_\eta/m_\pi)^3 \simeq 60$,

$$\frac{\Gamma(\eta \to \gamma\gamma)}{\Gamma(\pi^0 \to \gamma\gamma)} = 20(\cos\theta + 2\sqrt{2}\sin\theta)^2 \tag{4.73}$$

Empirically the ratio is about 30 which suggests $\eta \simeq \eta_8$. Similar conclusions arise by comparing $\eta' \to \gamma\gamma$ and $\eta' \to \rho\gamma$ where vector meson dominance is used. A good fit to all the radiative decays is found if $\theta \simeq 15°$.

4.4 Mass splittings in the hadron supermultiplets

Within the $SU(2)$ isospin multiplets there are small mass splittings, e.g. neutron and proton, $\pi^\pm$ and π^0. These are of the order of a per cent or so and in a quark model may be expected to arise from electromagnetic effects. For example, photon exchange between the constituent quarks will give a contribution to the overall energy (mass) proportional to the charges of the quarks involved. The contribution in the π^+ will therefore differ from that in the π^0 leading to a mass difference between these two states of order $\alpha \equiv e^2/4\pi \simeq 1$ per cent. These electromagnetic mass splittings are discussed in more detail in section 17.5.

Within SU(3) multiplets there are rather large mass splittings, e.g. the vector mesons ρ, K^*, ϕ are split by some 300 MeV in a mean mass of around 800 MeV while the $\Omega^- - \Delta$ baryon decuplet exibits a similar effect with a splitting of 450 MeV in a mean mass of around 1450 MeV. Qualitatively the pattern of SU(3) breaking in the mass spectrum seems to be that within a multiplet each time a strange quark replaces a nonstrange quark, about 150 MeV is added to the mass of the system. Examples of this are

$$\Delta(1235) - \Sigma^*(1385) - \Xi^*(1530) - \Omega^-(1670) \qquad (4.74)$$

and

$$\phi_{s\bar{s}}(1020) - K^*_{s\bar{u}}(890) - \rho, \omega(770) \qquad (4.75)$$

Hence in some sense (discussed in section 15.2.1) the strange quark has an effective mass some 150 MeV more than the u or d quarks.

There appears to be a spin–spin force between pairs of quarks. Such a spin–spin force emerges naturally in a theory where quarks interact by exchanging vector gluons (such a theory will be introduced in section 15.2 and detailed discussion of the mass splitting phenomenology in Chapter 17). This leads to shifts in the energy levels or masses which are proportional to the expectation value $\langle \mathbf{S}_i . \mathbf{S}_j \rangle$. This expectation value is dependent both on the spin state of the quark pair and the total spin of the whole system if more than two quarks are present.

In the mesons the $\mathbf{S} . \mathbf{S}$ force separates states with the same strangeness but with total quark spin $S = 0$ and $S = 1$ (e.g. 0^- and 1^-, π and ρ, K and K^*) and in the baryons $S = \frac{1}{2}$ and $S = \frac{3}{2}$ (e.g. N and Δ).

This $\mathbf{S} . \mathbf{S}$ force also leads to a splitting of the Σ and Λ masses in the octet. In these states the pair of nonstrange quarks are in $I = 1$ and $I = 0$ respectively and hence, by the overall symmetry of the 56plet wavefunction, are in $S = 1$ and $S = 0$ respectively. Since the strange quark has a larger "mass" than the nonstrange, then one might anticipate that the $\mathbf{S} . \mathbf{S}$ coupling involving strange quarks differs from that involving nonstrange by analogy with the mass dependence of the magnetic interaction familiar in QED. In such a case, the differing total spin of the u, d pair in the Σ^0 and Λ^0 gives rise to the mass splitting through the **S . S** force.

The above discussion shows that we are able to qualitatively understand the pattern of mass splittings in the baryon 56plet and the pseudoscalar and vector mesons. A quantitative discussion of the spin–spin splittings will be deferred until Chapter 17 after the charmed spectroscopy has been introduced.

Of particular interest are the mass splittings within the vector mesons and the psuedoscalars. Together these form **1 + 35** dimensional SU(6) supermultiplets but the pattern of the mass splittings is quite different in the two SU(3) nonets of 0^- and 1^- mesons. This has significant implications for field theories of strong interactions which are introduced in sections 15.2 and 17.6.

4.4.1   VECTOR MESON MASSES AND THE ZWEIG RULE

The $K^*(892)$ has a mass that is almost exactly midway between $\omega(783)$ and $\phi(1020)$. The vector mesons are all $q\bar{q}$ in S-wave coupled to spin of one and the masses of the system will, to first approximation, be the same if $m_u = m_d = m_s$.

Strange baryons are more massive than their nonstrange counterparts which suggests that $m_s > m_{u,d}$. In turn the separation in masses of ω, K^*, ϕ suggest that they contain no, one, two strange quarks respectively:

$$\omega\left(\frac{u\bar{u} + d\bar{d}}{\sqrt{2}}\right), \qquad K^*(d\bar{s}), \qquad \phi(s\bar{s}) \qquad (4.76)$$

The expectation value of the Hamiltonian $\mathcal{H}$ between $q\bar{q}$ vector meason wavefunctions will receive contributions from the explicit quark content and also a contribution M_1 common to the $^3S_1$ nature of the system (e.g. spin–spin splittings etc., Chapter 17). Hence the mass of $K^*(d\bar{s})$ can be written

$$\langle d\bar{s}|\mathcal{H}|d\bar{s}\rangle = M_1 + d + s \qquad (4.77)$$

(where the d, s labels on the right-hand side refer to the flavour contributions to the mass).

The equal separation of ω, K^* and ϕ masses emerges if the ω and ϕ have quark content as in equation (4.76) since then

$$\langle s\bar{s}|\mathcal{H}|s\bar{s}\rangle = M_1 + 2s \tag{4.78}$$

$$\left\langle \frac{u\bar{u}+d\bar{d}}{\sqrt{2}}\left|\mathcal{H}\right|\frac{u\bar{u}+d\bar{d}}{\sqrt{2}}\right\rangle = M_1 + 2u \tag{4.79}$$

(where we ignore difference in u and d masses) and hence

$$m_{K^*} = \frac{m_\phi + m_\omega}{2} \tag{4.80}$$

which is well satisfied and

$$\phi - K^* = K^* - \omega \equiv u - s \approx 120 \text{ MeV} \tag{4.81}$$

to be compared with the baryons in equation (4.74).

Since the ρ has $I = 1$ it cannot contain other than $u\bar{u}$ and $d\bar{d}$ and so will have the same mass as ω $[(u\bar{u}+d\bar{d})/\sqrt{2}]$. In fact these states are nearly degenerate but the ω does appear to be about 10 MeV heavier than the ρ, suggesting that there may be some small admixture of $s\bar{s}$ in its wavefunction.

The states $\phi(s\bar{s})$ and ω $[(u\bar{u}+d\bar{d})/\sqrt{2}]$ are known as "ideally mixed", that is they are in a mixture of **8** and **1** of SU(3) since

$$s\bar{s} \equiv \frac{1}{\sqrt{3}}\left(\frac{u\bar{u}+d\bar{d}+s\bar{s}}{\sqrt{3}}\right) - \sqrt{\frac{2}{3}}\left(\frac{u\bar{u}+d\bar{d}-2s\bar{s}}{\sqrt{6}}\right) \tag{4.82}$$

and hence

$$\phi(s\bar{s}) = \frac{1}{\sqrt{3}}\omega_1 - \sqrt{\frac{2}{3}}\omega_8 \tag{4.83}$$

$$\omega\left(\frac{u\bar{u}+d\bar{d}}{\sqrt{2}}\right) = \sqrt{\frac{2}{3}}\omega_1 + \frac{1}{\sqrt{3}}\omega_8 \tag{4.84}$$

Apart from the mass systematics there is another argument that supports (near) ideal mixing. This is connected to the fact that $\phi \to 3\pi$ is suppressed relative to $\phi \to K\bar{K}$ even though phase space favours the former.

If the ϕ is dominantly $s\bar{s}$ then the quark diagrammatic representation of the decays $\phi \to K\bar{K}$ or $\pi\rho$ will be as in Fig. 4.1. Now invent a rule (the OZI or Zweig rule, Okubo, 1963; fig. 12 in Zweig, 1964b; Iizuka, 1966) that demands that disconnected diagrams like Fig. 4.1(b) are very much suppressed relative to connected ones like Fig. 4.1(b). An analogy can be made with bar magnets (Harari, 1974). A bar magnet has north and

(a)   (b)

FIG. 4.1. Quark diagrams for decay vertices. (a) Disconnected or "forbidden". (b) Connected or "allowed".

south poles. Cut it into two (Fig. 4.2) and two new poles are created— north and south poles do not annihilate and then two new ones pop up from nowhere. If a meson was like a bar magnet, e.g. a string with charge (quark–antiquark) at the end points, then on cutting the string two new ends are created (Fig. 4.2) just like the $\phi \to K\bar{K}$ diagram.

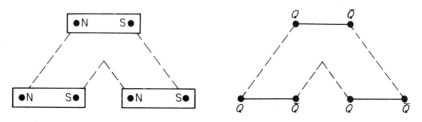

FIG. 4.2. Bar magnet and string analogies for allowed meson decay diagrams.

One can imagine a world where the OZI rule is exact (the disconnected diagram absolutely forbidden) the ϕ being pure s$\bar{\text{s}}$, the ω and ρ being degenerate and $\phi \to K\bar{K}$ always. The real world differs slightly from this ideal case, the ω and ρ are not exactly degenerate and $\phi \to 3\pi$ some-times.[1] This is not surprising since we know that $\phi \to K\bar{K}$ and $K\bar{K}$ can go into $\rho\pi$. Each of these separate processes is described by a Zweig allowed diagram (Fig. 4.1(b)). Hence if hadronic interactions are described by the flow of quark lines then quark diagrams like Fig. 4.3 must exist because unitarity requires $\phi \to K\bar{K} \to \rho\pi$.

[1] $\Gamma(\phi \to 3\pi) \simeq 0\cdot07\Gamma(\omega \to 3\pi)$ is a measure of the validity of the OZI rule (phase space favours ϕ decay) and also its violation ($0\cdot07 \neq$ zero).

If one imagines the $q\bar{q}$ connected by a string then diagrams like Fig. 4.1(b) are planar whereas those with crossed lines (Fig. 4.3) will be twisted. Empirically it seems that the twisted diagrams are suppressed relative to the planar. This has been investigated in the "dual uni-tarisation models" (for an introduction see Chan and Tsou, 1977), which contain a small expansion parameter so that a perturbative approach can be made for hadrodynamics. This small parameter is related to the twists in diagrams and it appears that at high energies the nonplanar twisted diagrams are negligible relative to planar. Hence the OZI rule would be exact in the limit $m_\phi \to \infty$. Consequently it is argued that the rule is better for $\psi(c\bar{c})$ than $\phi(s\bar{s})$ due to $m_\psi > m_\phi$.

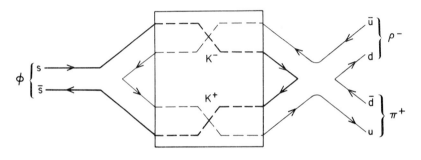

FIG. 4.3 Quark diagram for $\phi \to K\bar{K} \to \rho\pi$.

Intuitively it may perhaps be reasonable that twisted diagrams are suppressed and we illustrate this by returning to the bar magnet or string example. After the bar magnet has fractured into two new magnets (Fig. 4.2), then if both of these magnets rotate the original north and south poles can annihilate to leave us with a single, new magnet (Fig. 4.4). This sequence of events should be less probable than the straightforward single fracture.

If quarks are bound by interacting with gluons then one necessarily expects amplitudes to exist whereby

$$q_1\bar{q}_1 \to \text{gluons} \to q_2\bar{q}_2 \qquad (4.85)$$

These will allow $s\bar{s} \to$ gluons $\to u\bar{u}$ and $\phi \to \rho\pi$ can occur (Fig. 4.5). If the quark–gluon coupling is small enough then $\phi \to \rho\pi$ will be suppressed.

However $\phi \to K\bar{K}$ also presumably occurs by gluon production of the nonstrange pair. Why is this not also suppressed? A possible answer to this question emerges in quantum chromodynamics where coloured quarks interact with coloured vector gluons. This theory is introduced

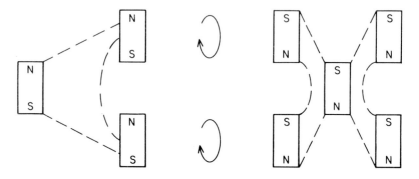

FIG. 4.4. Bar magnet analogy for $\phi \to K\bar{K} \to \rho\pi$.

in section 15.2 and its relation with the OZI rule is discussed when the newly discovered charmonium spectroscopy ψ, $\chi(c\bar{c})$ is described. This is a spectroscopy of massive states (3 to 4 GeV) containing a fourth flavour of quark (charmed quark) and their decays into hadrons violate

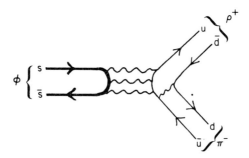

FIG. 4.5. $\phi(s\bar{s}) \to$ gluons $\to \rho\pi$ ($u\bar{u}$, $d\bar{d}$).

the OZI rule. As such they provide an interesting laboratory for studying the OZI rule and learning about hadrodynamics at quark level. This is dealt with in sections 16.1 and 16.2.

4.4.2 PSEUDOSCALAR MESONS

If we ignore the small mass splitting between charged and neutral pions and kaons then there are four masses that characterise the psuedoscalar nonet. These are

$$\pi(140), \qquad K(496), \qquad \eta(550), \qquad \eta'(960) \qquad (4.86)$$

The π has $I = 1$ and therefore contains no strange quarks if made from $q\bar{q}$. Clearly the η and η' must contain more than simply $u\bar{u}$ and $d\bar{d}$ since they are both much heavier than the π. Consequently the ideal mixing that we encountered in the vector mesons is far from being realised in the pseudoscalars; indeed the radiative decays of η and η' suggested that $\eta \simeq \eta_8$ and $\eta' \simeq \eta_1$ (section 4.3.2). The masses also qualitatively fit in with this assignment; however there appear to be some complications in this nonet which will become apparent.

First of all we will derive the Gell-Mann–Okubo formula for the octet states by a procedure analogous to that employed in the case of vector mesons (Gell-Mann, 1961; Okubo, 1962). As in equation (4.77) we have for the pseudoscalar K^+:

$$(u\bar{s}|\mathcal{H}|u\bar{s}) = M_0 + u + s \qquad (4.87)$$

(where M_0 is the $^1S_0$ analogue of M_1 for $^3S_1$). For the π^+

$$(u\bar{d}|\mathcal{H}|u\bar{d}) = M_0 + u + d \equiv M_0 + 2u \qquad (4.88)$$

while the η_8 yields

$$\left\langle \frac{1}{\sqrt{6}}(u\bar{u} + d\bar{d} - 2s\bar{s}) \left| \mathcal{H} \right| \frac{1}{\sqrt{6}}(u\bar{u} + d\bar{d} - 2s\bar{s}) \right\rangle = M_0 + \frac{4u + 8s}{6}$$

$$(4.89)$$

These yield the Gell-Mann–Okubo mass formula

$$4K - \pi = 3\eta_8 \qquad (4.90)$$

Inserting the known K and π masses yields

$$m(\eta_8) = 613 \text{ MeV} \qquad (4.91)$$

to be compared with $\eta(550)$ and $\eta'(960)$.

Notice that there is a nonvanishing matrix element of the mass operator between the SU(3) singlet and the isoscalar member of the octet, namely

$$\left\langle \frac{1}{\sqrt{3}}(u\bar{u}+d\bar{d}+s\bar{s}) \middle| \mathcal{H} \middle| (u\bar{u}+d\bar{d}-2s\bar{s})\frac{1}{\sqrt{6}} \right\rangle = \frac{2\sqrt{2}}{3}(u-s) \quad (4.92)$$

Hence in general **1** and **8** isoscalars are not eigenstates of the mass matrix; instead the physical states will be **1, 8** mixtures. If we define the mixing angle θ by

$$\eta = \cos\theta\, \eta_8 + \sin\theta\, \eta_1$$
$$\eta' = -\sin\theta\, \eta_8 + \cos\theta\, \eta_1 \quad (4.93)$$

and postulate that η_8 obeys the Gell-Mann–Okubo mass formula, then we can calculate the magnitude of θ. Since

$$\eta_8 = \eta\, \cos\theta - \eta'\, \sin\theta \quad (4.94)$$

(where η, η' are eigenstates of the mass matrix) then

$$4K - \pi = 3(\eta\, \cos^2\theta + \eta'\, \sin^2\theta) \equiv 3\eta_8 \quad (4.95)$$

The $\eta_8(613)$ is satisfyingly between the physical mass eigenstates $\eta(550)$ and $\eta'(960)$ and hence a solution can be found. This is

$$\tan^2\theta = \frac{4K - \pi - 3\eta}{3\eta' - 4K + \pi} \approx 0.2 \quad (4.96)$$

where we inserted the masses in the final step. Hence η is dominantly octet but does have some singlet admixture. This agrees with the conclusions drawn from radiative transitions section 4.3.2.

All of the foregoing discussion involved only η_8 and nothing was said about the mass expectation value of η_1. In SU(3) this reduced matrix element is quite independent of the octet. In the quark model the singlet is specified and if only three quark flavours exist

$$\eta_1 \equiv \frac{1}{\sqrt{3}}(u\bar{u}+d\bar{d}+s\bar{s}) = \eta'\, \cos\theta + \eta\, \sin\theta \quad (4.97)$$

Hence

$$\langle \eta_1 | \mathcal{H} | \eta_1 \rangle = \frac{1}{\sqrt{3}}(4u + 2s) + M_0$$

$$\equiv \langle \eta' | \mathcal{H} | \eta' \rangle \cos^2\theta + \langle \eta | \mathcal{H} | \eta \rangle \sin^2\theta \quad (4.98)$$

and

$$\langle \eta_8 | \mathcal{H} | \eta_8 \rangle = M_0 + \frac{2u + 4s}{3}$$

$$\equiv \langle \eta' | \mathcal{H} | \eta' \rangle \sin^2 \theta + \langle \eta | \mathcal{H} | \eta \rangle \cos^2 \theta \qquad (4.99)$$

Consequently

$$\eta_1 + \eta_8 \equiv \eta + \eta' = 2(u + s + M_0) \equiv 2K \qquad (4.100)$$

The left-hand side is 40 per cent larger than the right which suggests that the $\eta - \eta'$ system receives some contribution (SU(3) singlet?) not present in the K(SU(3) octet). Notice that for the vector nonet we would have obtained as analogue of equation (4.100):

$$\omega_1 + \omega_8 \equiv \omega + \phi = 2(u + s + M_1) \equiv 2K^* \qquad (4.101)$$

which is well satisfied. Indeed it was from this relation and the equal mass separation of ωK^* and ϕ that we concluded that the vector mesons were well described by ideally mixed combinations of just $u\bar{u}$, $d\bar{d}$ and $s\bar{s}$.

Since the vector mesons are so nearly ideally mixed, then a fourth flavour of quark c will form a state $^3S_1(c\bar{c})$ that by orthogonality will be uncoupled from $u\bar{u}$, $d\bar{d}$ and $s\bar{s}$ in $^3S_1$. The $^1S_0$ states are not ideally mixed and, moreover, appear problematic with regard to their masses. Hence it is possible that $^1S_0(c\bar{c})$ may be mixed in to the physical η and η' (Gaillard *et al.*, 1975; Harari, 1976a). This question is discussed further in section 17.6 to which the interested reader may prefer to continue immediately.

5 $SU(6) \times 0(3)$: Quarks with Orbital Excitation

A collection of quarks couple their spins to a total S. If there are three flavours of quark (u, d, s) the resulting system belongs to some $SU(6)$ representation. Now place the quarks in a potential, e.g. a harmonic oscillator. In this well they will have orbital angular momentum L. Angular momentum conservation emerges from rotational invariance in three-dimensional space, the resulting group structure being $0(3)$. The full symmetry group structure of the quarks in a potential is then $SU(6) \otimes 0(3)$. Finally coupling $L \oplus S = J$ generates the total angular momentum of the system which is identified with that of the hadron thus formed.

5.1 Baryons

For the three quark baryon states we impose a rule which seems to work (this will be discussed in Chapter 8), namely that only $(SU(6) \otimes 0(3))_{\text{symmetric}}$ representations exist. This is the so-called "symmetric" quark model, the "symmetric" arising due to the restriction that is imposed on the $SU(6) \otimes 0(3)$ states. The 56plet lying lowest and the alternating signs of parity as one proceeds upwards in mass now emerge naturally as consequences of this $(SU(6) \otimes 0(3))_{\text{sym}}$ structure of the spectrum.

5.1.1 THE GROUND STATE

Suppose that all three quarks sit in (1s) states of a harmonic oscillator potential. The 0(3) state may be represented by

$$(1s)(1s)(1s) \equiv (1s)^3 \rightarrow L^P = 0^+ \text{ (angular momentum, parity)} \quad (5.1)$$

and is clearly symmetric under interchange of any pair since all three are in the same state (1s).

We have demanded that the $(SU(6) \otimes 0(3))$ state be totally symmetric. This means that the SU(6) and 0(3) pieces must be both S, both M or both A as all other combinations yield M or A symmetry properties for the $SU(6) \otimes 0(3)$ state (see for instance the multiplication table for symmetries (equation 4.21)). In the present example the 0(3) state $(1s)^3$ is symmetric, S, and so the SU(6) state must also be S. The symmetric SU(6) representation is a 56plet (equation 4.22) and so it is indeed the 56plet that lies lowest in the spectrum.

The 56plet contains Δ with $S = \frac{3}{2}$ and N with $S = \frac{1}{2}$. Since the ground state has $L = 0$, and hence by convention positive parity, then we find that the lowest mass states in the spectrum must be

$$\mathbf{10}, J^P = \tfrac{3}{2}^+; \qquad \mathbf{8}, J^P = \tfrac{1}{2}^+ \quad (5.2)$$

which is indeed satisfactory empirically. Of course, this has arisen as a direct consequence of our restriction to $(SU(6) \otimes 0(3))_{sym}$ states. The theoretical significance of this restriction will be discussed in section 8.2. First we will investigate the further consequences of it, in particular, what are we now forced to predict must be the first excited state?

5.1.2 THE FIRST EXCITED STATE

We excite one quark from (1s) to (1p) and the 0(3) state becomes

$$(1s)(1s)(1p) = (1s)^2(1p) \rightarrow L^P = 1^- \quad (5.3)$$

Since one state now differs from the others (namely 1p as against 1s) we can form both symmetric and mixed symmetry 0(3) states. For example, imagine that $\psi(\mathbf{r}_1, \mathbf{r}_2, \mathbf{r}_3)$ is the $(1s)^3$ state, completely symmetric in the quark coordinates $\mathbf{r}_1, \mathbf{r}_2, \mathbf{r}_3$ relative to some origin, e.g. the centre of mass. The $(1s)^2(1p)$ state can be described by wavefunctions $\mathbf{r}_1\psi$, $\mathbf{r}_2\psi$ or

$\mathbf{r}_3\psi$ depending upon which of the quarks is excited. The mixed symmetry combinations are then

$$\psi_{M,A} \equiv \frac{1}{\sqrt{2}}(\mathbf{r}_1 - \mathbf{r}_2)\psi \tag{5.4}$$

$$\psi_{M,S} \equiv \frac{1}{\sqrt{6}}(\mathbf{r}_1 + \mathbf{r}_2 - 2\mathbf{r}_3)\psi \tag{5.5}$$

while the symmetric is

$$\psi_S \equiv \frac{1}{\sqrt{3}}(\mathbf{r}_1 + \mathbf{r}_2 + \mathbf{r}_3)\psi \tag{5.6}$$

Having excited only one quark then no antisymmetric wavefunction can be formed.

If we choose the origin to be the centre of mass of the three-quark system then

$$\mathbf{r}_1 + \mathbf{r}_2 + \mathbf{r}_3 \equiv 0 \tag{5.7}$$

and so the state ψ_S vanishes, only mixed symmetry 0(3) states existing at the first excited level.

You may be wondering what happens if a different origin is chosen since the symmetric state would no longer vanish. This symmetric state would indeed exist but is not a genuine internal excitation of the three-quark system, rather it is a state where three quarks in an s-state (like nucleon) have their centre of mass rotating about the observer with one unit of angular momentum. Hence it is not a genuine excitation of the three-quark system but is known as a centre of mass excitation or "spurious state". Consequently only the mixed symmetry states are of physical interest in the present context of baryon spectroscopy.

In order that the overall wavefunction formed on combining the 0(3) and SU(6) states be totally symmetric, then the mixed symmetry 0(3) state requires that the SU(6) state also be of mixed symmetry. Hence the first excited level is predicted to be a 70plet in SU(6).

The **70** contains SU(3) **1**, **8**, and **10** with quark total spin coupled to $\frac{1}{2}$ and also an **8** when the quarks couple to spin $\frac{3}{2}$ (equation 4.23).

Combining $S = \frac{1}{2}$ or $\frac{3}{2}$ with $L = 1$ yields the following negative parity states

$$^2\mathbf{10} \quad (S = \tfrac{1}{2}) \oplus (L = 1) \to J^P = \tfrac{1}{2}^-, \tfrac{3}{2}^- \tag{5.8}$$

$$^2\mathbf{8} \quad (S = \tfrac{1}{2}) \oplus (L = 1) \to J^P = \tfrac{1}{2}^-, \tfrac{3}{2}^- \tag{5.9}$$

$$^4\mathbf{8} \quad (S = \tfrac{3}{2}) \oplus (L = 1) \to J^P = \tfrac{1}{2}^-, \tfrac{3}{2}^-, \tfrac{5}{2}^- \tag{5.10}$$

$$^2\mathbf{1} \quad (S = \tfrac{1}{2}) \oplus (L = 1) \to J^P = \tfrac{1}{2}^-, \tfrac{3}{2}^- \tag{5.11}$$

The singlet representations appear to be seen

$$\begin{aligned} \tfrac{1}{2}^- &: \quad S_{01}\Lambda(1405) \\ \tfrac{3}{2}^- &: \quad D_{03}\Lambda(1520) \end{aligned} \tag{5.12}$$

The nonstrange members of all the other representations have also been isolated and appear to be

$$^2\mathbf{10}: \quad S_{31}(1650), \ D_{33}(1670) \tag{5.13}$$

$$^2\mathbf{8}: \quad S_{11}(1535), \ D_{13}(1520) \tag{5.14}$$

$$^4\mathbf{8}: \quad S_{11}(1700), \ D_{13}(1700), \ D_{15}(1670) \tag{5.15}$$

Isolating strange baryons is a harder task experimentally but candidates for the available slots in the **70** are emerging, e.g. Λ states S_{01} (1670) and S_{03}(1690) and Σ states D_{15}(1765), D_{13}(1670, 1580), S_{11}(1620, 1750). The Σ masses suggest that some mixing may be taking place between $^2\mathbf{8}$ and $^4\mathbf{8}$ for instance.

5.1.3 HIGHER STATES

In a harmonic oscillator potential the $(1s)^2(1d)$ excitation is degenerate with both $(1s)^2(2s)$ and $(1s)(1p)^2$ for a three-quark system. In general any one of these will have a spurious component which will be indicated by its dependence on $\mathbf{R}$. However we can eliminate these by appropriate linear combinations. For example, one can construct two symmetric representations with $L = 0^+$:

$$\psi(1s)^2(2s) \propto \frac{\sqrt{2}}{3}\left(\frac{9}{2} - \alpha^2(r_1^2 + r_2^2 + r_3^2)\right)\psi_0 \tag{5.16}$$

$$\psi(1s)(1p)^2 \propto \frac{2}{3}\alpha^2(\mathbf{r}_1 . \mathbf{r}_2 + \mathbf{r}_2 . \mathbf{r}_3 + \mathbf{r}_3 . \mathbf{r}_1)\psi_0 \tag{5.17}$$

where α^2 has dimensions GeV^2. The combination that is independent of R is therefore

$$\psi(L = 0^+) = \sqrt{\tfrac{2}{3}}(1s)^2(2s) + \sqrt{\tfrac{1}{3}}(1s)(1p)^2 \tag{5.18}$$

To see this define the orthogonal basis

$$\mathbf{R} = \frac{1}{3}(\mathbf{r}_1 + \mathbf{r}_2 + \mathbf{r}_3)$$

$$\boldsymbol{\rho} \equiv \frac{1}{\sqrt{2}}(\mathbf{r}_1 - \mathbf{r}_2) \tag{5.19}$$

$$\boldsymbol{\lambda} \equiv \frac{1}{\sqrt{6}}(\mathbf{r}_1 + \mathbf{r}_2 - 2\mathbf{r}_3)$$

and hence

$$\mathbf{r}_1 \equiv \mathbf{R} + \frac{1}{\sqrt{2}}\boldsymbol{\rho} + \frac{1}{\sqrt{6}}\boldsymbol{\lambda}$$

$$\mathbf{r}_2 \equiv \mathbf{R} - \frac{1}{\sqrt{2}}\boldsymbol{\rho} + \frac{1}{\sqrt{6}}\boldsymbol{\lambda} \tag{5.20}$$

$$\mathbf{r}_3 \equiv \mathbf{R} \qquad\qquad - \frac{2}{\sqrt{6}}\boldsymbol{\lambda}$$

Consequently

$$\psi(1s)(1p)^2 \propto \tfrac{2}{3}\alpha^2(3R^2 - \tfrac{1}{2}(\rho^2 + \lambda^2))\psi_0 \tag{5.21}$$

$$\psi(1s)^2(2s) \propto \frac{\sqrt{2}}{3}\left(\frac{9}{2} - \alpha^2(3R^2 + \rho^2 + \lambda^2)\right)\psi_0 \tag{5.22}$$

and so

$$\psi(L = 0^+) \propto \left(\frac{9}{2} - \frac{3\alpha^2}{2}(\rho^2 + \lambda^2)\right)\psi_0 \tag{5.23}$$

The orthogonal combination corresponds to a state with the internal motion in the ground state and the centre of mass in the (2s) excitation.

Since $\psi(L = 0^+)$ is spatially symmetric it combines with the symmetric **56** of SU(6).

By consulting a text of harmonic oscillator wavefunctions construct the following combinations, express them in terms of $\mathbf{R}, \boldsymbol{\rho}, \boldsymbol{\lambda}$ as above,

verify that **R** dependence drops out and that the symmetry properties
are such that the SU(6) representations are as indicated.

$$\psi(\mathbf{56}, L = 0^+) = \sqrt{\tfrac{2}{3}}(1s)^2(2s) + \sqrt{\tfrac{1}{3}}(1s)(1p)^2 \qquad (5.24)$$

$$\psi(\mathbf{70}, L = 0^+) = \sqrt{\tfrac{1}{3}}(1s)^2(2s) + \sqrt{\tfrac{2}{3}}(1s)(1p)^2 \qquad (5.25)$$

$$\psi(\mathbf{56}, L = 2^+) = \sqrt{\tfrac{2}{3}}(1s)^2(1d) - \sqrt{\tfrac{1}{3}}(1s)(1p)^2 \qquad (5.26)$$

$$\psi(\mathbf{70}, L = 2^+) = \sqrt{\tfrac{1}{3}}(1s)^2(1d) - \sqrt{\tfrac{2}{3}}(1s)(1p)^2 \qquad (5.27)$$

$$\psi(\mathbf{20}, L = 1^+) = (1s)(1p)^2 \qquad (5.28)$$

The orthogonal combinations are the **56**, $L = 0^+$ spurious ones discussed already, a spurious **56**, $L = 2^+$ with internal ground state and
(1d) centre of mass, and spurious **70**, $L = 0^+$, 1^+, 2^+ with (1s)(1p)
internal and (1p) centre of mass (Karl and Obryk, 1968; Faiman and
Hendry, 1968).

There are many positive parity baryon resonance in the 2 GeV mass
region which are candidates for these representations. Candidates for
the strangeness zero states of the **56**, $L = 2^+$ seem to be seen, namely

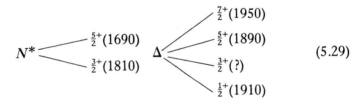

$$\qquad (5.29)$$

A detailed discussion of the assignments of resonances may be found in
the "Baryon Resonance" session report of International Conferences
and these should be referred to. The situation in this region is controversial at present.

One question that has to be settled is whether the full spectroscopy
(equations 5.24–5.28) is realised or whether the simpler pattern of **56**
with positive parity and **70** with negative is the true situation in Nature.
The latter emerges if one freezes one of the internal degrees of freedom
so that two of the quarks form a "diquark" (Lichtenberg, 1968; Capps,
1974, 1975). The centre of mass of the diquark is at $\tfrac{1}{2}(\mathbf{r}_1 + \mathbf{r}_2)$ and the
relative coordinate between this and the remaining quark is

$$\tfrac{1}{2}(\mathbf{r}_1 + \mathbf{r}_2) - \mathbf{r}_3 \equiv \boldsymbol{\lambda}\sqrt{\tfrac{3}{2}} \qquad (5.30)$$

Then if ψ_0 is the ground state which is symmetric, and hence
56, $L = 0^+$, then for N excitations of $\boldsymbol{\lambda}$ we have a symmetric state for N

even and mixed symmetry for N odd. Hence the sequence would be $(\mathbf{56}, 0^+)$, $(\mathbf{70}, 1^-)$, $(\mathbf{56}, 2^+)$ etc.

Possible candidates for $(\mathbf{70}, 3^-)$ are ${}^4\mathbf{8}(\frac{9}{2}^-(2140))$, ${}^2\mathbf{8}$ or ${}^4\mathbf{8}(\frac{7}{2}^-(2190, 2000); \frac{5}{2}^-(2100?\ldots))$ and ${}^2\mathbf{10}(\frac{7}{2}^-(2200)\frac{5}{2}^-(1960??))$. A$(\mathbf{56}, 4^+)$ probably exists containing ${}^2\mathbf{8}(N^*(2220)\frac{9}{2}^+)$ and ${}^4\mathbf{10}(\Delta(2420)\frac{11}{2}^+)$ (see Horgan, 1974; Cashmore *et al.*, 1975a; Dalitz, 1977).

5.1.4 RADIAL EXCITATIONS

In the harmonic oscillator potential a radially excited (2s) $\mathbf{56}, 0^+$ is expected degenerate with the $L = 2$ (1d) levels. There are candidates for $(\mathbf{56}, 0^+)^*$ and $(\mathbf{56}, 0^+)^{**}$, namely at $(\mathbf{56}, 0^+)$ are the $P_{11}(1470)$, $P_{33}(1690)$, while at $(\mathbf{56}, 0^+)^{**}$ are $P_{11}(1780)$, $P_{33}(2080)$. The mass separation of P_{33} and P_{11} is of the order of $\Delta(1236)$ and $N(940)$. However the masses overall appear rather low since, for example, the $(\mathbf{56}, 0^+)^*$ is expected to lie around 1700 degenerate with the $L = 2$(1d) levels rather than in the vicinity of the $L = 1$ states, (1p), around 1500 MeV.

5.2 Mesons

A fermion–antifermion system is an eigenstate of parity with

$$P = (-)^{L+1} \tag{5.31}$$

(recall that fermion and antifermion have opposite intrinsic parities—hence equation 5.31). The charge conjugation applied to neutral systems with orbital angular momentum L and total spin S is

$$C = (-)^{L+S} \tag{5.32}$$

Applied to the $q\bar{q}$ system we have a series of states with $CP = +$ for $S = 1$ and $CP = -$ for $S = 0$. For $S = 0$ the system's total angular momentum $J \equiv L$ and so

$$C = (-)^J = -P \tag{5.33}$$

hence $J^{PC} = 0^{-+}, 1^{+-}, 2^{-+}, \ldots$ for this sequence. One cannot form $CP = -$ states with

$$P = (-)^J = -C \tag{5.34}$$

These latter states with $J^{PC} = 0^{+-}, 1^{-+}, 2^{+-}, \ldots$ are known as "exotics" or sometimes "exotics of the second kind" to distinguish them from the $qq\bar{q}\bar{q}$ states in **27** etc. which are "exotics of the first kind" (i.e. states which are not in SU(3) **1** or **8**).

The resulting J^{PC} of the mesons are therefore as listed in Table 5.1. For each J^{PC} a nonet of mesons is predicted composed of uds and $\bar{u}\bar{d}\bar{s}$ flavoured quarks.

TABLE 5.1
J^{PC} for $q\bar{q}$ mesons

	$S = 0$	$S = 1$
$L = 0$	0^{-+}	1^{--}
$L = 1$	1^{+-}	0^{++}
		1^{++}
		2^{++}
$L = 2$	2^{-+}	1^{--}
		2^{--}
		3^{--}
$\vdots$		$\vdots$

The nonets at $L = 0$ are well established. The empirical situation for $L \geqslant 1$ is summarised in Table 5.2. Radially excited vectors ($\rho'(1600)$, $\omega'(1780)$, $K^{*\prime}(1650)$) and pseudo-scalars ($E(1420)$?, $K'(1400)$) may also have been seen.

TABLE 5.2
Possible classification of $q\bar{q}$ mesons

J^{PC}	$I = 1$	$I = 0$	$I = 0$	$I = \frac{1}{2}$	
0^{-+}	$\pi(140)$	$\eta(550)$	$\eta'(960)$	$K(495)$	
1^{--}	$\rho(770)$	$\omega(780)$	$\phi(1020)$	$K^*(890)$	
1^{+-}	$B(1235)$	—	—	$Q(1300)$	
0^{++}	$\delta(970)$	$\varepsilon(1200 \text{ or } 700)$	$S^*(993)$	$\kappa(1250)$	mix
1^{++}	$A_1(1100)$	$D(1285)$?	—	$Q(1400)$	
2^{++}	$A_2(1310)$	$f(1270)$	$f'(1514)$	$K^{**}(1420)$	
2^{-+}	$A_3(1640)$				
1^{--}					
2^{--}					
3^{--}	$g(1680)$	$\omega(1675)$		$K^{***}(1780)$	

5.3 Mass splittings in $L > 0$ supermultiplets

A spin–spin force acting between quark pairs in baryons will separate the $S = \frac{1}{2}$ and $S = \frac{3}{2}$ baryon masses. Similarly a spin–spin force between quark and antiquark separates $S = 0$ and 1 meson masses. A quantitative discussion requires acquaintance with the details and role of the colour degree of freedom which is described in Chapters 8 and 15. Hence we defer discussion of these spin–spin splittings until Chapter 17 and at present proceed with a discussion of the patterns of splittings in the $L = 1$ and $L = 2$ supermultiplets arising from spin–orbit forces between the quarks.

5.3.1 $L = 1$ MESONS

For $q\bar{q}$ and qqq systems with $L > 0$ there is the possibility of a spin–orbit force between the quarks,

$$\sum_{i,j} \mathbf{L}_i \cdot \mathbf{S}_j \tag{5.37}$$

This will lead to separations in masses of states with the same L and S but differing J as follows. Since $\mathbf{J} = \mathbf{L} \oplus \mathbf{S}$ then

$$2\mathbf{L} \cdot \mathbf{S} \equiv \mathbf{J}^2 - \mathbf{L}^2 - \mathbf{S}^2 \tag{5.38}$$

and so

$$2\langle \mathbf{L} \cdot \mathbf{S} \rangle_{J,L,S} = J(J+1) - L(L+1) - S(S+1) \tag{5.39}$$

As a first example of the effect of this force we shall examine the $L = 1$ mesons. For a $q\bar{q}$ system the parity is $P = (-)^{L+1}$ and the charge conjugation is $C = (-1)^{L+S}$. Hence when $S = 0$ the $L = 1$ state has $J^{PC} = 1^{+-}$ while for $S = 1$ we have $J^{PC} = 0^{++}, 1^{++}, 2^{++}$. The magnitude of $\langle \mathbf{L} \cdot \mathbf{S} \rangle$ for each of these states is as follows

J^{PC}	0^{++}	1^{++}	1^{+-}	2^{++}
$\langle \mathbf{L} \cdot \mathbf{S} \rangle$	-2	-1	0	1

$$\tag{5.40}$$

Hence the characteristic pattern of $\mathbf{L} . \mathbf{S}$ forces here is that

$$\Delta m(2^{++} - 1^{++}) = 2\Delta m(1^{++} - 0^{++}) \tag{5.41}$$

(the $J^{PC} = 1^{+-}$ state can be shifted relative to the $S = 1$, 0^{++}, 1^{++}, 2^{++} states by the $\mathbf{S} . \mathbf{S}$ force). The empirical situation with respect to the 0^{++} and 1^{++} states is somewhat confused. It is possible that the $I = 1$ states exist as follows:

$$2^{++} A_2(1310); \quad 1^{++} A_1(1100); \quad 1^{+-} B(1235); \quad 0^{++} \delta(970) \tag{5.42}$$

which fit in rather well with $\mathbf{L} . \mathbf{S}$ force splittings and only a small effect from the $\mathbf{S} . \mathbf{S}$ force. This is quite different from the $L = 0$ case where it is the $\mathbf{S} . \mathbf{S}$ force that accounts for the $0^- - 1^-$ splitting. There is not necessarily a contradiction in this. For example, if the $\mathbf{S} . \mathbf{S}$ and $\mathbf{L} . \mathbf{S}$ forces are arising from vector gluon exchange between quark pairs, then the potential that is generated will be very similar in form to the familiar case of the hydrogen atom in QED. In this familiar example there are two places that spin–spin couplings arise. One is a contact interaction of form

$$\mathbf{S}_i . \mathbf{S}_j \delta(\mathbf{r}_i - \mathbf{r}_j) \tag{5.43}$$

and the other is the tensor force which is of form

$$\mathbf{S}_i . \mathbf{S}_j - \frac{3\mathbf{S}_i . \mathbf{r}\mathbf{S}_j . \mathbf{r}}{|\mathbf{r}|^2} \tag{5.44}$$

In S-waves $\langle \mathbf{S}_i . \mathbf{S}_j \rangle = 3\langle \mathbf{S}_i . \mathbf{r}\mathbf{S}_j . \mathbf{r}/r^2 \rangle$ and so the latter contribution is absent, the hyperfine splitting arising entirely from the contact interaction. In fact the contact interaction can *only* contribute in the S-wave case due to the presence of the delta function $\delta^3(\mathbf{r})$, since for small distance separation in a state of orbital angular momentum L the wavefunction is proportional to $|\mathbf{r}|^L$, hence vanishing for all but S-waves. The conclusion from the spectroscopy therefore appears to be that, at least for $L = 1$ mesons, the tensor interaction is small or absent (but see also the $c\bar{c}$ spectroscopy section 16.3).

5.3.2 L = 1 BARYONS

The $L = 1$ baryon situation is rather more complicated than the meson case since SU(3) $\mathbf{1}, \mathbf{8}$ and also $\mathbf{10}$ are present. The $^4\mathbf{8}$ states $\frac{1}{2}^-, \frac{3}{2}^-$ are more massive than their $^2\mathbf{8}$ counterparts due to the $\mathbf{S}.\mathbf{S}$ force which pushes up the $S = \frac{3}{2}$ states relative to the $S = \frac{1}{2}$ just as in the $L = 0$ cases of N and Δ. This splitting of $S = \frac{1}{2}$ and $S = \frac{3}{2}$ is shown in the first column of Fig. 5.1.

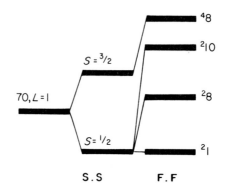

FIG. 5.1. $\mathbf{70}$ $L = 1$ baryon supermultiplet. The $\mathbf{S}.\mathbf{S}$ force splits $S = \frac{1}{2}$ and $S = \frac{3}{2}$ states. An SU(3)-dependent $\mathbf{F}_i.\mathbf{F}_j$ force splits $\mathbf{1}, \mathbf{8}$ and $\mathbf{10}$ multiplets.

The separation of the $\mathbf{1}, \mathbf{8}$ and $\mathbf{10}$ states can arise from a force that is dependent on the unitary spin. If there is such a force between pairs of quarks i, j having form $F_i.F_j$ (where $F_{i,j}$ are SU(2) generators as described in section 2.2.2.1) then the $F_i.F_j$ can be calculated by noting that

$$2\langle F_{(i)}.F_{(j)}\rangle \equiv F_{(i+j)}^2 - F_{(i)}^2 - F_{(j)}^2 \qquad (5.45)$$

The eigenvalues of the F^2 are known as the Casimir operators of the group and depend upon the representation. They have been evaluated in section 2.2.2.3) (equation 2.57 et seq.) and for the cases of current interest are

$$\mathbf{1}: \quad F^2 = 0, \qquad \mathbf{8}: \quad F^2 = 3, \qquad \mathbf{10}: \quad F^2 = 6 \qquad (5.46)$$

while for the quarks

$$\mathbf{3}: \quad F^2 = \frac{4}{3} \qquad (5.47)$$

Hence in $\mathbf{1}, \mathbf{8}, \mathbf{10}$ the expectation value $2\langle F_{(i)}.F_{(j)}\rangle$ takes on values

$$\mathbf{1}: \quad 0 - \frac{8}{3} \equiv -\frac{8}{3}, \qquad \mathbf{8}: \quad 3 - \frac{8}{3} \equiv \frac{1}{3}, \qquad \mathbf{10}: \quad 6 - \frac{8}{3} \equiv \frac{10}{3} \qquad (5.48)$$

and so there will be equal splitting between **1, 8, 10**. The mean masses of these multiplets are in reasonable agreement with this, i.e.

$$\mathbf{1} \simeq 1570 \text{ MeV}, \qquad \mathbf{8} \simeq 1680 \text{ MeV}, \qquad \mathbf{10} \simeq 1800 \text{ MeV} \quad (5.49)$$

where we have taken the N and Δ states in **8** and **10** and added 150 MeV for the extra unit of strangeness in comparing with the singlet Λ.

This unitary spin-dependent interaction is at present rather *ad hoc*. A more natural way to split the **1, 8** and **10** states comes from the contact interaction of equation (5.43).

Notice that only one quark is excited when forming the $L = 1$ baryons (quark number three using Table 3.7 and equation 5.5); hence quarks numbers one and two remain in S-wave. As a result a contact spin-dependent interaction between this pair can contribute to the 70plet mass splittings. If this pair have spin zero ($\chi_{M,A}$ states in Table 3.7), then $\langle \mathbf{S}_1 . \mathbf{S}_2 \rangle = -\frac{3}{4}$. For spin one ($\chi_{M,S}$ or χ_S) then $\langle \mathbf{S}_1 . \mathbf{S}_2 \rangle = +\frac{1}{4}$. Consequently

$$\langle \mathbf{S}_1 . \mathbf{S}_2 \rangle \, \alpha \, +\tfrac{1}{4} \text{ for } (\mathbf{10, 2}) \text{ and } (\mathbf{8, 4})$$

$$\alpha \, -\tfrac{1}{4} \text{ for } (\mathbf{8, 2})$$

$$\alpha \, -\tfrac{3}{4} \text{ for } (\mathbf{1, 2})$$

using the SU(6) representations of Table 4.3. The approximately equal splitting of **1, 8** and **10** has emerged as a result of the way the 70plet symmetry correlates these unitary spin representations with the quarks' spin states.

The splitting in masses between the states of different J in each of the supermultiplets shows that more than just $\mathbf{L} . \mathbf{S}$ forces are present. For example, in the ${}^4\mathbf{8}$

$J =$	$\frac{5}{2}$	$\frac{3}{2}$	$\frac{1}{2}$
$\langle \mathbf{L} . \mathbf{S} \rangle =$	3	-2	-5

$$(5.50)$$

and so the masses should be ordered $\frac{1}{2}^-, \frac{3}{2}^-, \frac{5}{2}^-$ with the former pair split by $\frac{3}{5}$ of the latter. However, the empirical masses appear to be jumbled. The $\frac{1}{2}^-$ and $\frac{3}{2}^-$ are comparable and heavier than the $\frac{5}{2}^-$. Similarly in the ${}^2\mathbf{8}$ the $\frac{1}{2}^-$ is heavier than the $\frac{3}{2}^-$. In the ${}^2\mathbf{1}$, however, the $\frac{3}{2}^-$ is heavier than the $\frac{1}{2}^-$. Isgur and Karl (1978) have shown that the tensor in equation

(5.44) plays an important role. In particular it mixes the $^2\mathbf{8}$ and $^4\mathbf{8}$ states with the same J.

5.3.3 L=2 EXCITATIONS

At $L = 2$ in the baryon spectrum one expects that a set of positive parity baryons will exist. The even L enables the spatial wavefunction to be symmetric without being a spurious state (contrast $L = 1$) and so a spin–unitary spin 56plet can exist here (equations 5.26 and 5.29).

In the 56plet, the $S = \frac{1}{2}(N)$ couple with $L = 2$ to yield states with $J^P = \frac{3}{2}^+, \frac{5}{2}^+$ while the $S = \frac{3}{2}(\Delta)$ couple to $J^P = \frac{1}{2}^+, \frac{3}{2}^+, \frac{5}{2}^+, \frac{7}{2}^+$. Just as in the $L = 0$ case the $S = \frac{3}{2}(\Delta)$ states are more massive than their $S = \frac{1}{2}(N)$ counterparts; so again we find this pattern repeated here, the members of the supermultiplet with zero stangeness having been quoted at equation (5.29). As in the $L = 1$, $S = \frac{3}{2}$ states, we again see that the pattern of masses of the J^P states with $S = \frac{3}{2}$ do not appear to follow an $\mathbf{L}.\mathbf{S}$ pattern. An $\mathbf{L}.\mathbf{S}$ splitting would require

$$\Delta m\left(\frac{7}{2}^+ - \frac{5}{2}^+\right):\Delta m\left(\frac{5}{2}^+ - \frac{3}{2}^+\right):\Delta m\left(\frac{3}{2}^+ - \frac{1}{2}^+\right) = (7\times9-5\times7):(5\times7-3\times5):$$
$$(3\times5-1\times3)$$
$$= 7:5:3$$

whereas the observed masses do not have a monotonic behaviour with spin.

More detailed discussion of mass splitting systematics in the quark model can be found in Dalitz (1977), de Rujula et al. (1975), and Isgur and Karl (1978).

6 Quark-Current Interactions: Symmetries and Dynamics of Decay Vertices

6.1 U(6)×U(6) and SU(6)

We have seen how hadron states *at rest* appear to be quark–antiquark ($q\bar{q}$) or three-quark (qqq) composites with orbital excitation and we discussed the structure of the resulting classification symmetry. We will now develop these ideas more formally in order to discuss the symmetry properties of transitions such as $N^* \to N\gamma$, $N^* \to N\pi$ and so forth.

A U(6) algebra is generated by the 35 generators

$$\frac{\lambda_\alpha}{2}, \quad \frac{\sigma_j}{2}, \quad \frac{\lambda_\alpha \sigma_j}{2} \quad \begin{cases} \alpha = 1\ldots 8 \\ j = 1, 2, 3 \end{cases} \tag{6.1}$$

where λ_α are the SU(3) matrices (section 2.2) and σ_j the Pauli spin matrices. A quark belongs to the fundamental **6** representation of $U(6)_q$ while an antiquark belongs to **6** of $U(6)_{\bar{q}}$. In turn a quark is a singlet under $U(6)_{\bar{q}}$ and an antiquark is **1** under $U(6)_q$. If we consider the group $U(6)_q \times U(6)_{\bar{q}}$ generated by

$$\sum_q \frac{\lambda_\alpha}{2}, \quad \frac{\sigma_j}{2}, \quad \frac{\lambda_\alpha \sigma_j}{2}$$

$$\sum_{\bar{q}} \frac{\lambda'_\alpha}{2}, \quad \frac{\sigma'_j}{2}, \quad \frac{(\lambda_\alpha \sigma_j)'}{2} \tag{6.2}$$

then

$$q \in (\mathbf{6}, \mathbf{1})$$
$$\bar{q} \in (\mathbf{1}, \mathbf{6})$$

(6.3)

A baryon made of qqq can therefore belong to $(\mathbf{56}, \mathbf{1})$, $(\mathbf{70}, \mathbf{1})$, $(\mathbf{20}, \mathbf{1})$ representations of $U(6)_q \times U(6)_{\bar{q}}$. If we imagine that the quarks in the hadron have relative orbital angular momentum in the hadron rest frame then one has the symmetry group $U(6)_q \times U(6)_{\bar{q}} \times 0(3)$ as a classification scheme for hadron spectroscopy. The lowest lying baryons are therefore $(\mathbf{56}, \mathbf{1})L = 0$ and $(\mathbf{70}, \mathbf{1})L = 1$ representations of this symmetry. The $q\bar{q}$ mesons will belong to $(\mathbf{6}, \bar{\mathbf{6}})$ representations of $U(6)_q \times U(6)_{\bar{q}}$ with the lowest-lying states being in the $(\mathbf{6}, \bar{\mathbf{6}})$, $L = 0$ representation of $U(6) \times U(6) \times 0(3)$.

Whereas the $U(6) \times U(6)$ algebra is generated by

$$\frac{\lambda_\alpha}{2}, \quad \frac{\lambda_\alpha \sigma_j}{2}, \quad \frac{\sigma_j}{2}, \quad \frac{\lambda'_\alpha}{2}, \quad \frac{\lambda'_\alpha \sigma'_j}{2}, \quad \frac{\sigma'_j}{2}$$

(6.4)

the *subgroup* $SU(6)$ is generated by

$$\Lambda_\alpha \equiv \sum_Q \frac{\lambda_\alpha}{2} + \sum_{\bar{Q}} \frac{\lambda'_\alpha}{2}$$

$$S_j = \sum_Q \frac{\sigma_j}{2} + \sum_{\bar{Q}} \frac{\sigma'_j}{2}$$

$$(\Lambda S)_{\alpha j} \equiv \sum_Q \frac{\lambda_\alpha \sigma_j}{2} + \sum_{\bar{Q}} \frac{\lambda'_\alpha \sigma'_j}{2}$$

(6.5)

To appreciate the relation of $SU(6)$ to $U(6) \times U(6)$ we show the classifications of various objects under these groups.

	$U(6) \times U(6)$	$SU(6)$
Quark:	$(\mathbf{6}, \mathbf{1})$	$\mathbf{6}$
Antiquark:	$(\mathbf{1}, \mathbf{6})$	$\bar{\mathbf{6}}$
Baryons:	$(\mathbf{56}, \mathbf{1})$ etc.	$\mathbf{56}$ etc.
Mesons:	$(\mathbf{6}, \bar{\mathbf{6}})$	$\mathbf{1} + \mathbf{35}$

(6.6)

The fact that $SU(6)_S$ is a subgroup of $U(6) \times U(6)$ is best illustrated by the classification of the 36 meson states. Under $SU(6)$ the $\mathbf{1}$ ($\simeq \eta'$) and

35 (η...) are completely different representations and their interactions with other hadrons are not related by the Clebsch–Gordan
coefficients of SU(6). However the η, η' ... are all in the same **(6, 6)** of
U(6) × U(6) and hence their properties are related by Clebsch–Gordans
of that group (compare the discussion of masses in section 17.6).

The observed hadronic spectroscopy is indeed suggestive of a broken
$SU(6)_S \times 0(3)$ symmetry describing the states at rest and this emerges
naturally in the quark model. By allowing spin–spin and spin–orbit
forces between the constituent quarks the breaking of this symmetry
can be well described. This is discussed in section 5.3 and Chapter 17.

Are decay processes also described by an $SU(6) \otimes 0(3)$ symmetry? It
is easy to see that they are not.

$SU(6)_S$ contains the SU(2) subgroup generated by σ_j and one can
think of this as the "intrinsic spin" of the hadrons or the quark system's
total spin (before coupling with the orbital angular momentum in the
system). If $SU(2)_S$ is a symmetry of a decay amplitude then the initial
state and final N particle system must be in the same representation.
Consequently the "intrinsic spin" of the hadrons is conserved. This
forbids some well-known decays:

$$\underset{(S=3/2)}{\Delta} \longrightarrow \underset{(S=\frac{1}{2})}{N} \otimes \underset{(S=0)}{\pi} \tag{6.7}$$

and

$$\underset{(S=1)}{\rho} \longrightarrow \underset{(S=0)}{\pi} \otimes \underset{(S=0)}{\pi} \tag{6.8}$$

Therefore although $SU(6)_S$ appears to be a good description of particles
at rest it is not a symmetry of decay vertices.

6.2 SU(6)$_W$ and states in motion with $p_z \neq 0$

Empirically $SU(2)_S$ is not a symmetry of decay processes even though it
appears to be approximately a good symmetry for classifying states at
rest. Physically this yields the clue as to the source of the failure: decays
involve particles in motion. How do the states, in particular the quarks,
transform under Lorentz boosts?

Boosts are generators of the Lorentz group and quarks belong to the
spin $\frac{1}{2}$ representation of the space rotational subgroup. Recalling the

transformation properties of spinors under spatial rotations (equations 2.6 and 2.7)

$$q' = e^{\frac{1}{2}i\theta n.\sigma} q \tag{6.9}$$

then, since $\frac{1}{2}\alpha_i$ are the Lorentz boost generators acting on the spin $\frac{1}{2}$ representatiion, we have (Bjorken and Drell, 1964, p. 22; Gasiorowicz, 1967, p. 38)

$$q \rightarrow q' = e^{\frac{1}{2}\omega n.\alpha} q \tag{6.10}$$

for the spinor transformation under a Lorentz boost from rest to a frame with velocity $|v|/c = \tanh \omega$.

First we will show the explicit effect of this boost. After boosting by ω_i ($i = x, y, z$)

$$q'(p_i) = \cosh \frac{\omega}{2}\left(1 + \alpha_i\left(\tanh \frac{\omega}{2}\right)_i\right) q(0) \tag{6.11}$$

In terms of E, m, p

$$\cosh \omega \equiv E/m ; \qquad \sinh \omega \equiv |\mathbf{p}|/m \tag{6.12}$$

and hence $\cosh^2 \omega - \sinh^2 \omega = 1$. In turn these yield

$$\cosh \frac{\omega}{2} = \sqrt{\left(\frac{E+m}{2m}\right)} ; \qquad \left(\sinh \frac{\omega}{2}\right)_i = \frac{p_i}{E+m}\sqrt{\left(\frac{E+m}{2m}\right)} \tag{6.13}$$

(hence $2 \cosh \omega/2 \sinh \omega/2 = p/m \equiv \sinh \omega$).

At rest a Dirac four-component spinor may be formally written

$q \equiv \begin{pmatrix} \chi \\ 0 \end{pmatrix}$ where χ is

$$\chi_u \equiv \begin{pmatrix} 1 \\ 0 \end{pmatrix}, \qquad \chi_d \equiv \begin{pmatrix} 0 \\ 1 \end{pmatrix} \tag{6.14}$$

for the spin up and down configurations respectively. Note that χ is a Pauli two-component spinor. The rest spinor is normalised so that $q^+q = 1$. The combination $\bar{q}q \equiv q^+\beta q$ equals unity when the spinor q represents a particle at rest.

With the representation

$$\alpha_i \equiv \begin{pmatrix} 0 & \sigma_i \\ \sigma_i & 0 \end{pmatrix}, \qquad \beta \equiv \begin{pmatrix} 1 & 0 \\ 0 & -1 \end{pmatrix} \qquad (6.15)$$

where σ_i are the i = 1, 2, 3 Pauli 2×2 matrices, then the boost equation (6.11) becomes

$$q'(p_i) = \begin{pmatrix} \cosh\dfrac{\omega}{2} & \sigma_i\left(\sinh\dfrac{\omega}{2}\right)_i \\ \sigma_i\left(\sinh\dfrac{\omega}{2}\right)_i & \cosh\dfrac{\omega}{2} \end{pmatrix} q(0) \qquad (6.16)$$

Written out explicitly this yields

$$q(p_i) = \begin{pmatrix} \cosh\dfrac{\omega}{2} \\ \sigma_i\left(\sinh\dfrac{\omega}{2}\right)_i \end{pmatrix} \equiv \sqrt{\left(\dfrac{E+m}{2m}\right)} \begin{pmatrix} 1 \\ \dfrac{\boldsymbol{\sigma}\cdot\mathbf{p}}{E+m} \end{pmatrix} \qquad (6.17)$$

The normalisations are

$$\left\{ \begin{matrix} q^+q \\ \bar{q}q \end{matrix} \right\} = \frac{E+m}{2m}\left(1 \pm \frac{p^2}{(E+m)^2}\right) \Rightarrow \left\{ \begin{matrix} E/m \\ 1 \end{matrix} \right\} \qquad (6.18)$$

The $\bar{q}q$ is therefore referred to as the covariant norm. The spinor ψ normalised so that $\psi^+\psi = 1$ is therefore $\psi \equiv \sqrt{(m/E)}q$. Hence under boosts

$$\psi(p) = (\cosh\omega)^{-\frac{1}{2}} \begin{pmatrix} \cosh\dfrac{\omega}{2} \\ \sigma_i\left(\sinh\dfrac{\omega}{2}\right)_i \end{pmatrix} \frac{1}{\sqrt{2}}\begin{pmatrix} 1 \\ \sigma_i \end{pmatrix} \qquad (6.19)$$

The quark with spin up or down along the z-axis forms a two-dimensional fundamental representation of SU(2). Under z boosts the quark preserves its up–down character but the boost operator α_3 does not commute with $\sigma_{x,y}$ which are two of the generators of the group SU(2)$_S$. An operator that does not commute with some or all generators of a group will in general cause transitions between different representations of that group. Hence a boost along the z (or any) direction will lead to transitions between representations of SU(2)$_S$. Consequently

$SU(2)_S$ is not a suitable symmetry group for moving systems since the classification of a state will differ in different frames.

It was then noticed that an $SU(2)$ group ($SU(2)_W$) is generated by $(1, \beta\sigma_x, \beta\sigma_y, \sigma_z)$ all of which commute with α_3. Consequently representations of this group are well defined under z boosts (Lipkin and Meshkov, 1965, 1966; Barnes *et al.*, 1965).

It may be helpful to gain a feeling for the way that the $\beta\sigma_x$, $\beta\sigma_y$ which distinguish $SU(2)_W$ from $SU(2)_S$ enter the picture physically.

The generators of a group G give rise to transformations of the form

$$\psi' = e^{i\theta G}\psi \tag{6.20}$$

where θ is a measure of the size of the transformation. For $SU(2)_W$ the rotations are generated by

$$R_z(\theta) = e^{i\theta\sigma_z}, \qquad R_{x,y}(\theta) = e^{i\theta\beta\sigma_{x,y}} \tag{6.21}$$

The particular case $\theta = \pi$ is easy to illustrate. We will show that a spin $\frac{1}{2}$ system moving along the z-axis with $\mathbf{J} = \mathbf{L} + \boldsymbol{\sigma}/2$ has its momentum p_z left invariant under the above $SU(2)_W$ rotations. Clearly R_z leaves p_z invariant. A rotation about x for $\theta = \pi$ sends $p_z \to -p_z$ and when supplemented by a parity reflection p_z is again recovered.

Hence p_z is left invariant under the combined action of parity and rotation about x or y. Now the x, y rotation is generated by $\exp(i\pi J_x) \equiv \exp[i\pi(\sigma_x/2)]$ and combined with parity we have ($P_{\text{int}} \equiv$ the intrinsic parity of the system)

$$P_{\text{int}} \exp[i\pi(\sigma_x/2)] \equiv P_{\text{int}}\left(\cos\frac{\pi}{2} + i\sigma_x \sin\frac{\pi}{2}\right) \tag{6.22}$$

since $\sigma_x^2 \equiv 1$. In turn this quantity is simply $P_{\text{int}} i\sigma_x$. Then since $P_{\text{int}}^2 \equiv 1$, we obtain

$$P_{\text{int}} i\sigma_x \equiv \exp\left(i\pi P_{\text{int}} \frac{\sigma_x}{2}\right) \tag{6.23}$$

For a single quark the intrinsic parity is $+1(-1)$ for the upper (lower) components. The intrinsic parity operator is therefore the same as

$$\beta \equiv \begin{pmatrix} 1 & 0 \\ 0 & -1 \end{pmatrix}$$

It is in this way that the β has entered in with the $\sigma_{x,y}$ and leads to the "W spin" generators

$$W_{x,y} = \frac{\beta \sigma_{x,y}}{2}, \qquad W_z = \frac{\sigma_z}{2} \qquad (6.24)$$

In combination with the SU(3) degrees of freedom we have the SU(6)$_W$ group generated by

$$W^i = \sum_{q,\bar{q}} \left(\frac{\lambda_i}{2} + \frac{\lambda_i'}{2} \right) \qquad i = 1 \ldots 8$$

$$W^i_{x,y} = \sum_{q,\bar{q}} \left(\frac{\lambda_i}{2} + \frac{\lambda_i'}{2} \right) \beta \sigma_{x,y} \qquad i = 0 \ldots 8 \qquad (6.25)$$

$$W^i_z = \sum_{q,\bar{q}} \left(\frac{\lambda_i}{2} + \frac{\lambda_i'}{2} \right) \sigma_z \qquad i = 0 \ldots 8$$

in contrast to SU(6)$_S$ generators in equation (6.5).

By changing from SU(6)$_S$ to SU(6)$_W$ we have found a group that classifies particles either at rest or in motion along the z-axis. Consequently this group is in principle a candidate for describing collinear or decay processes. To see if it is phenomenologically successful we must first see how the hadron states transform under SU(6)$_W$.

Both SU(6)$_S$ and SU(6)$_W$ are subgroups of the rest symmetry U(6) × U(6) and so $\mathbf{S}$ and $\mathbf{W}$ are well defined in the rest frame. Since

$$\beta q(\bar{q}) = +q(-\bar{q}) \qquad (6.26)$$

then we have that

$$\text{quarks:} \qquad \mathbf{W}_q = \mathbf{S}_q$$
$$\text{antiquarks:} \quad W_z = S_z; \quad W_{x,y} = -S_{x,y} \qquad (6.27)$$

Since $\mathbf{W} \equiv \mathbf{S}$ for quarks, then baryon states made of qqq will transform in the same way for SU(6)$_W$ as they do for SU(6)$_S$; for example N and Δ are in **56** of both SU(6)$_S$ and of SU(6)$_W$. For mesons, which contain $\bar{q}$, the SU(6)$_W$ classification differs from that of SU(6)$_S$.

A $q\bar{q}$ state with $S_z = 1$ or $W_z = 1$ will be

$$|S = 1, S_z = +1\rangle = |W = 1, W_z = 1\rangle \qquad (6.28)$$

The states $|S = 1, S_z = 0\rangle$ and $|W = 1, W_z = 0\rangle$ are obtained by acting on the $|11\rangle$ states with the spin lowering operators $\sum_{q,\bar{q}} (S_x - iS_y)$ and

$\sum_{q,\bar{q}} (W_x - iW_y)$ respectively. Since

$$|S = 1, S_z = 1\rangle = |W = 1, W_z = 1\rangle = q\uparrow\bar{q}\uparrow \qquad (6.29)$$

then

$$|S = 1, S_z = 0\rangle = q\uparrow\bar{q}\downarrow + q\downarrow\bar{q}\uparrow \qquad (6.30)$$

whereas

$$|W = 1, W_z = 0\rangle = -q\uparrow\bar{q}\downarrow + q\downarrow\bar{q}\uparrow \qquad (6.31)$$

because $(W_x - iW_y)\bar{q} = -(S_x - iS_y)\bar{q}$. Consequently we find that

$$|W = 1, W_z = 1\rangle = |S = 1, S_z = 1\rangle$$
$$|W = 1, W_z = 0\rangle = -|S = 0, S_z = 0\rangle$$
$$|W = 0, W_z = 0\rangle = -|S = 1, S_z = 0\rangle \qquad (6.32)$$
$$|W = 1, W_z = -1\rangle = -|S = 1, S_z = -1\rangle$$

Hence the $q\bar{q}$ system has separated into a triplet and singlet under each SU(2) but the $S_z = 0$, $W_z = 0$ states have flipped:

$$S = 1 \rightleftarrows W = 0, \qquad S = 0 \rightleftarrows W = 1 \qquad (6.33)$$

The $L = 0$ meson states of $U(6) \times U(6)$ are therefore classified in $\mathbf{1} + \mathbf{35}$ of $SU(6)_S$ and $SU(6)_W$ as follows.

$$SU(6)_S \begin{cases} \mathbf{1}: \ S = 0, (\eta_1)\mathbf{1} \\ \mathbf{35}: \begin{cases} S = 0, (\pi\eta_8)\mathbf{8} \\ S = 1, S_z = \pm 1, (\rho\omega_8)\mathbf{8} + (\omega_1)\mathbf{1} \\ S = 1, S_z = 0, (\rho\omega_8)\mathbf{8} + (\omega_1)\mathbf{1} \end{cases} \end{cases}$$

$$SU(6)_W \begin{cases} \mathbf{1}: \ W = 0, (\omega_1)\mathbf{1} \\ \mathbf{35}: \begin{cases} W = 0, (\rho\omega_8)\mathbf{8} \\ W = 1, W_z = \pm 1, (\rho\omega_8)\mathbf{8} + (\omega_1)\mathbf{1} \\ W = 1, W_z = 0, (\pi\eta_8)\mathbf{8} + (\eta_1)\mathbf{1} \end{cases} \end{cases}$$

The decays $\Delta \to N\pi$ and $\rho \to \pi\pi$ which were forbidden in $SU(6)_S$ are allowed by $SU(6)_W$. The π is now $|W = 1, W_z = 0\rangle$ and so

$$\underset{W=\frac{3}{2}}{\Delta} \to \underset{(W=\frac{1}{2})}{N} \otimes \underset{(W=1)}{\pi}$$

$$\qquad (6.34)$$

$$\underset{(W=0, W_z=0)}{\rho} \to \underset{(W=1)}{\pi} \otimes \underset{(W=1)}{\pi}$$

Consequently $SU(6)_W$ avoids the obvious difficulties that were found in $SU(6)_S$. Notice that under boosts L_z is conserved but L is not. Hence the symmetry is referred to as $SU(6)_W \otimes 0(2)_{L,z}$ (the L_z subscript is really superfluous since W_z conservation implies L_z conservation through $J_z = L_z + S_z = W_z + L_z$).

6.3 A model with $SU(6)_W \times 0(2)_{L,z}$ vertex symmetry: duality diagrams with spin

It is instructive to present a model of decays which does indeed have $SU(6)_W \times 0(2)_{L,z}$ as a vertex symmetry. This symmetry emerges naturally if meson emission from baryons takes place by creation of a $q\bar{q}$ pair from the vacuum (Fig. 6.1) and the quarks have *no momentum transverse to the collinear z-axis* defined by the B→MB decay.

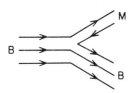

FIG. 6.1. B→BM by $q\bar{q}$ production from the vacuum.

This model is triply instructive.

i. It connects naturally with dual models where "duality diagrams" are constructed from the elementary vertices of the topology in Fig. 6.1 (Harari, 1969; Rosner, 1969). Hadronic 2-2 processes involving the exchange of internal quantum numbers are empirically allowed if they can be constructed from such vertices without twisting (e.g. Fig. 6.2(a)). Processes which cannot be so constructed (e.g. $\pi^+ p \rightarrow K^+ \Sigma^+$ as against $K^- p \rightarrow \pi^- \Sigma^+$) are empirically much suppressed (Fig. 6.2(b)) (technically, it is the imaginary part of the $t = 0$ amplitude which is suppressed).

ii. The model naturally illustrates why $SU(6)_W$ as against $SU(6)_S$ emerges as the vertex symmetry.

iii. It highlights the crucial role of quark momenta transverse to the z-axis. In particular we shall see that if these are *not* negligible then the

$SU(6)_W \times 0(2)_{L,z}$ is broken as a vertex symmetry even though $SU(6)_W$ can be successful as a classification symmetry.

In the picture where baryons are qqq and mesons are $q\bar{q}$ composites, decay vertices $B_1 \rightarrow MB_2$ (B = baryon, M = meson) appear to involve the

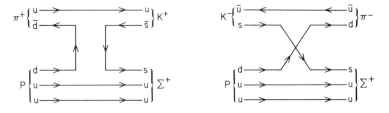

FIG. 6.2. Quark (duality) diagrams for $\pi^+p \rightarrow K^+\Sigma^+$ and $K^-p \rightarrow \pi^-\Sigma^+$.

creation of a $q\bar{q}$ pair representable by Fig. 6.2. These pictorial representations of $SU(3)$ invariant vertices can be constructed subject to the following rules.

1. Each q or $\bar{q}$ has a directed line $\rightarrow$ for q and $\leftarrow$ for $\bar{q}$.
2. A baryon is three lines pointing to the right, $\rightrightarrows$, a meson is q to the right and $\bar{q}$ to the left, $\rightleftarrows$.
3. The Zweig rule (sections 4.4 and 16.1) demands that the q and $\bar{q}$ of a single meson cannot be disconnected, e.g. Fig. 4.1(a) is not allowed.

These rules are the basis of quark graphs for general n-point functions which seem to be a faithful diagrammatic representation of the constraints that duality imposes. They are known as duality diagrams.

Carlitz and Kislinger (1970) suggested how one might incorporate spin into this picture. Consider $B \rightarrow MB$, Fig. 6.2, in a frame where the BMB momenta are collinear along the z-axis. The spin orientations of quarks that flow through are supposed to remain unchanged. The spin correlation of the $q\bar{q}$ pair is to be determined. If the pair is created from the vacuum and hence has $C = P = +$, then $S = 1$ and $L = 1$, i.e. the $q\bar{q}$ are in a ${}^3P_0$ configuration (Micu, 1969).

If one now makes the additional assumption that the momenta of the quarks transverse to the z-axis can be neglected then $L_z = S_z = 0$ is constrained upon the ${}^3P_0$, i.e. this becomes a ${}^3P_0(L_z = 0, S_z = 0)$ model. The fact that the $q\bar{q}$ are created with $S = 1$, $S_z = 0$ means that they have

$W = 0$, $W_z = 0$. Hence Wspin is conserved and $SU(2)_W$ is the spin symmetry of the vertex. Combining with the $SU(3)$ degree of quark freedom implies that the vertex symmetry of this particular model is $SU(6)_W \times 0(2)_{L,z}$ (Carlitz and Kislinger, 1970). Notice in passing that $SU(6)_S$ is not a good vertex symmetry because it would require the $q\bar{q}$ to be created with $S = 0$, $S_z = 0$ which requires L even by charge conjugation and hence violates parity.

Notice that the $SU(6)_W \times 0(2)_{L,z}$ vertex symmetry has emerged as a consequence of the $p_T = 0$ assumption. The neglect of p_T in large q^2 transitions may be justifiable as the Q value of $\gamma N \to N^*$ is large and hence p_T is small compared to Q. In the present example of B $\to$ BM (where M could be a real photon $q\bar{q} \to V \to \gamma$) the Q value is in general small and so it is not at all clear that it is justifiable to neglect p_T in resonance decay vertices.

If we relax this constraint then the $^3P_0$ $(q\bar{q})$ can have $L_z = \pm 1$, $S_z = \mp 1$ present in addition to the $L_z = 0$, $S_z = 0$ so far allowed. This generates the $^3P_0$ model of Colglazier and Rosner (1971). The $L_z = \pm 1$, $S_z = \mp 1$ component breaks the $SU(6)_W \times 0(2)_{L,z}$ vertex symmetry because L_z is no longer conserved (such a model is discussed in detail in section 6.7).

Hence to the extent that $\langle |p_T| \rangle \neq 0$, i.e. the quarks produced from the vacuum are not lined up along the collinear axis, then $SU(6)_W \times 0(2)_{L,z}$ will *not* be a vertex symmetry even though $SU(6)_W$ will remain a classification symmetry for the initial state with $p_T = 0$. Notice in this connection that even in the rest frame of the initial hadron the $SU(6)_S$ will only be a good classification symmetry if the p_T of the quarks is negligible. Since the confining of the quarks to the proton will give them internal momenta as a result of the uncertainty principle, then the nucleon will not be purely $L = 0$, **56** of the Pauli $SU(2)$ ($SU(6)$), but will be in a mixture of configurations due to the Dirac nature of the quarks. This configuration mixing has been investigated algebraically by Bucella *et al.* (1970) and by Melosh (1964). It is particularly explicit in the quark model discussion of Le Youanc *et al.* (1974b) described in section 6.6. These effects will lead to deviations from naive nonrelativistic computations of static properties of the nucleon ($g_A/g_V \neq \frac{5}{3}$) and will also show up as breaking $SU(6)_W \times 0(2)_{L,z}$ vertex symmetry in decay processes (see Chapter 7, in particular section 7.3). It is therefore not a surprise to learn from the following examples that $SU(6)_W \times 0(2)_{L,z}$ is empirically not a good decay symmetry in general.

6.3.1 SU(6)$_W$ AND DECAYS

Empirically the decay B $\rightarrow \omega\pi$ occurs dominantly in the mode where the ω has helicity ± 1, the helicity zero production being much suppressed (Ascoli *et al.*, 1968). This phenomenon is in contradiction to the predictions of SU(6)$_W \times 0(2)_{L_z}$ and is easily demonstrated. The ω and π have $L = 0$ and hence the $\pi\omega$ system has $J_z \equiv W_z$. Hence $\pi\omega(\lambda = \pm 1)$ has $W_z = \pm 1$ and conservation of W_z implies that B has $W_z = \pm 1$. However the B is a $q\bar{q}$ state with $S = 0$, hence $S_z = 0$, or in W spin basis $W = 1$, $W_z = 0$. Therefore the B does not exist with $W_z = 1$ and so the $\pi\omega(\lambda = \pm 1)$ mode is forbidden contrary to data.

Similarly, in photoproduction, L_z conservation implies that only $L_z = 0$ final state baryons can be excited and hence the D$_{13}$(1520) and F$_{15}$(1690) which have $S \equiv W = \frac{1}{2}$ can only be photoproduced in helicity $\frac{1}{2}$. Empirically this amplitude is almost zero if a proton target is used (Walker, 1969). What is happening in this case is that the matrix element involves an important contribution from quark orbital flip (electric or convection current—equation 7.31) which violates SU(6)$_W$ (see also section 7.3).

6.4 SU(6)$_W$ of currents

In weak and electromagnetic interactions the time components $V_0^\alpha(\mathbf{x}, t)$, $A_0^\alpha(\mathbf{x}, t)$ of the vector and axial currents measured there can be integrated over all space to define 16 vector and axial-vector charges (Gell-Mann, 1962; Adler and Dashen, 1968):

$$\left. \begin{aligned} Q^\alpha(t) &= \int d^3x V_0^\alpha(\mathbf{x}, t) \\ Q_5^\alpha(t) &= \int d^3x A_0^\alpha(\mathbf{x}, t) \end{aligned} \right\} = 1 \dots 8 \qquad (6.35)$$

At equal times these charges satisfy the commutation relations

$$\begin{aligned} [Q^\alpha(t), Q^\beta(t)] &= if^{\alpha\beta\gamma}Q^\gamma(t) \\ [Q^\alpha(t), Q_5^\beta(t)] &= if^{\alpha\beta\gamma}Q_5^\gamma(t) \\ [Q_5^\alpha(t), Q_5^\beta(t)] &= if^{\alpha\beta\gamma}Q^\gamma(t) \end{aligned} \qquad (6.36)$$

Hence the right- and left-handed charges $Q^\alpha \pm Q_5^\alpha$ each form an SU(3) group (Chapter 2) and these commute with each other. Consequently the above algebra is a chiral SU(3) × SU(3); explicitly if $Q_\pm = \frac{1}{2}(Q^\alpha \pm Q_5^\alpha)$ then

$$[Q_+^\alpha, Q_+^\beta] = if^{\alpha\beta\gamma}Q_+^\gamma$$
$$[Q_-^\alpha, Q_-^\beta] = if^{\alpha\beta\gamma}Q_-^\gamma \qquad (6.37)$$
$$[Q_+^\alpha, Q_-^\beta] = 0$$

From these commutation relations sum rules can be derived (Adler, 1965a, b; Weisberger, 1965) by inserting a complete set of hadronic intermediate states into the commutator.

Now attempt to enlarge the SU(3) × SU(3) algebra. In a current quark model with canonical equal time commutation relations among the quark fields a U(12) algebra emerges. This algebra has 144 generators which are the integrals over the local densities $\bar{q}(x)\Gamma(\lambda^\alpha/2)q(x)$ ($\alpha = 0 \ldots 8$). Here Γ is any of the 16 Dirac covariants.

To exploit the commutation relations and derive sum rules it is in practice of use to sandwich them between states with infinite momentum (Fubini and Furlan, 1965; Adler and Dashen, 1968). When $p_z \to \infty$ many of the U(12) operators' matrix elements vanish between finite mass states. The surviving operators are known as "good operators".

It is straightforward to verify that only 35 independent good operators survive in the $p_z \to \infty$ limit. We give a few illustrations.

Under boosts (equation 6.11) the spinors transform as

$$q(p) = \cosh\frac{\omega}{2}\left(1 + \alpha_i \tanh_i \frac{\omega}{2}\right) q(0) \qquad (6.38)$$

hence

$$q(p_i) = \begin{pmatrix} \cosh\dfrac{\omega}{2} & \chi \\[2mm] \sigma_i \sinh_i \dfrac{\omega}{2} & \chi \end{pmatrix} \qquad (6.39)$$

where $\chi \equiv q(0)$ is a two-component spinor. In the particular limit of $p_z \to \infty$

$$q(p_z \to \infty) = \cosh\frac{\omega}{2}\begin{pmatrix} \chi \\ \sigma_z\chi \end{pmatrix} \qquad (6.40)$$

We can now compute the properties of the various $\bar{q}\Gamma(\lambda/2)q$ densities. Dropping the explicit λ SU(3) dependence formally we have, for example (α_z is given in equation 6.15)

$$q^+\alpha_z q \equiv \cosh^2\frac{\omega}{2}\chi^+(1, \sigma_z)\begin{pmatrix} 0 & \sigma_z \\ \sigma_z & 0 \end{pmatrix}\begin{pmatrix} 1 \\ \sigma_z \end{pmatrix}\chi \qquad (6.41)$$

$$= 2\cosh^2\frac{\omega}{2}\chi^+ 1\chi$$

$$\equiv q^+ 1q$$

Notice that $q^+\beta\sigma_x q$ and $q^+\sigma_x q$ have quite different behaviours

$$q^+\begin{pmatrix}\beta \\ 1\end{pmatrix}\sigma_x q \equiv \cosh^2\frac{\omega}{2}\chi^+(1, \sigma_z)\begin{pmatrix} \sigma_x & 0 \\ 0 & \mp\sigma_x \end{pmatrix}\begin{pmatrix} 1 \\ \sigma_z \end{pmatrix}\chi \qquad (6.42)$$

and so $q^+\sigma_x q \xrightarrow{p_z\to\infty} 0$ whereas $q^+\beta\sigma_x q$ remains finite. The analogous result holds for $q^+\sigma_y q$ and $q^+\beta\sigma_y q$ in the $p_z \to \infty$ limit.

If you carry this exercise through you will find that the good operators form an $SU(6)_W$ algebra due to their $q^+\lambda_\alpha(1, \beta\sigma_x, \beta\sigma_y, \sigma_z)q$ structure (Dashen and Gell-Mann, 1965). Explicitly

$$F^\alpha(t) = \int d^3x V_0^\alpha(t) = \int d^3x q^+(x)\frac{\lambda^\alpha}{2}q(x)$$

$$\simeq \int d^3x q^+(x)\alpha_z\frac{\lambda^\alpha}{2}q(x)$$

$$= \int d^3x V_3^\alpha(x) \qquad (\alpha = 1\ldots 8) \qquad (6.43)$$

$$F_x^\alpha(t) = \int d^3x V_{23}^\alpha(t) = \int d^3x q^+(x)\beta\sigma_x\frac{\lambda^\alpha}{2}q(x)$$

$$\simeq \int d^3x q^+(x)\beta\alpha_y\frac{\lambda^\alpha}{2}q(x) = \int d^3x V_{20}^\alpha(x) \qquad (6.44)$$

$$F_y^\alpha(t) = \int d^3x V_{31}^\alpha(t) = \int d^3x q^+(x)\beta\sigma_y\frac{\lambda^\alpha}{2}q(x)$$

$$\alpha = (0\ldots 8)$$

$$\simeq \int d^3x q^+(x)\beta\alpha_x\frac{\lambda^\alpha}{2}q(x) = \int d^3x V_{01}^\alpha(x) \qquad (6.45)$$

$$F_z^\alpha(t) = \int d^3x A_3^\alpha(t) = \int d^3x q^+(x)\sigma_z\frac{\lambda^\alpha}{2}q(x)$$

$$\simeq \int d^3x q^+(x)\gamma_5\frac{\lambda^\alpha}{2}q(x) = \int d^3x A_0^\alpha(x) \qquad (6.46)$$

These $\hat{F}^{\alpha}$ quantities are thereby recognised as being the generators of an $SU(6)_W$ algebra which is called $SU(6)_{W,\text{currents}}$ since it arose when we investigated the transformation properties of the quark currents.

We have already seen that single particle states in motion along the z-axis can be classified into $SU(6)_W \times 0(2)_{L,z}$ supermultiplets. This $SU(6)_W$ is referred to as $SU(6)_{W,\text{classification}}$ and is generated by $\hat{W}^{\alpha}$ which close on an $SU(6)_W$ algebra.

6.4.1 ARE $SU(6)_{W,\text{cla}}$ AND $SU(6)_{W,\text{cur}}$ RELATED?

The generators $\hat{F}^{\alpha}$ of $SU(6)_{W,\text{cur}}$ are the integrals of local measurable currents. Are they related to the $\hat{W}^{\alpha}$ generators of $SU(6)_{W,\text{cla}}$? Dashen and Gell-Mann (1966a) suggested that they might be related by a unitary transformation V:

$$\hat{W}^{\alpha} = V\hat{F}^{\alpha}V^{-1} \tag{6.47}$$

Can $V \equiv 1$ so that $\hat{W}^{\alpha} \equiv \hat{F}^{\alpha}$? The answer is no. For example the axial charge is a generator of $SU(6)_{W,\text{cur}}$ and if it was also a generator of $SU(6)_{W,\text{cla}}$ then it would not give rise to transitions between different multiplet representations of the group—a generator can by definition only cause transitions within a multiplet (Chapter 2). Then by PCAC one would deduce that no resonance can decay into πN and $\pi\Delta$. Other contradictions can be found. One would predict $g_A/g_V = 5/3$, the nucleon anomalous magnetic moment would vanish (Dashen and Gell-Mann, 1965b) and the form factors of the ρ-meson would all vanish (Bell and Hey, 1974). Hence $V \neq 1$ and the $SU(6)_{W,\text{cla}}$ and $SU(6)_{W,\text{cur}}$ have different generators.

The hadrons that form multiplets of $SU(6)_{W,\text{cla}}$ are built from quarks which are often referred to as classification or constituent quarks. The generators of $SU(6)_{W,\text{cur}}$ arise from quark currents. The quarks here are often referred to as current quarks. Hadrons can be classified in multiplets of $SU(6)_{W,\text{cur}}$, i.e. in a current quark basis, and their wavefunctions denoted $|\phi_c\rangle$. In the constituent quark basis they will be in multiplets and their wavefunctions denoted $|\psi_q\rangle$. These representations are related by the hypothesised unitary transformation V:

$$|\phi_c\rangle = V|\psi_q\rangle \tag{6.48}$$

Hence for a nucleon $|\psi_q\rangle \equiv |56, L_z = 0\rangle$. After V acts on this we will have a mixture of configurations

$$|\text{Nucleon}\rangle = \alpha|56, L_z = 0\rangle + \beta|56, L_z = +1\rangle + \gamma|70, L_z = 0\rangle + \cdots \tag{6.49}$$

As a result the $\pi N \to N^*$ transition will occur.

If we have an explicit form for V then the $\alpha, \beta, \gamma \ldots$ can be calculated and the current matrix elements between hadron states would emerge by calculating

$$\langle \phi_c|\hat{0}_c|\phi_c\rangle \tag{6.50}$$

Alternatively, one can transform the operator

$$\langle \phi_c|\hat{0}_c|\phi_c\rangle = \langle \psi_q|V^{-1}\hat{0}_c V|\psi_q\rangle = \langle \psi_q|\hat{0}_q|\psi_q\rangle$$
$$\hat{0}_q \equiv V^{-1}\hat{0}_c V \tag{6.51}$$

From the resulting transformation properties of $\hat{0}_q$ plus the known $SU(6)_{W,\text{cla}}$ behaviour of ψ_q, the matrix element can be calculated.

6.4.2 A PARTICULAR UNITARY TRANSFORMATION BETWEEN $SU(6)_{W,\text{cur}}$ AND $SU(6)_{W,\text{cla}}$

Buccella *et al.* (1970) proposed a phenomenological candidate for V. Writing

$$V_B = \prod_{i=1}^{3} V_B^{(i)} \tag{6.52}$$

then

$$V_B^{(i)} = \exp(-i\theta Z) \tag{6.53}$$

where

$$Z \equiv (\boldsymbol{\sigma} \times \mathbf{L})_3 = i(\sigma_+ L_- - \sigma_- L_+) \tag{6.54}$$

with $\sigma_\pm$ the spin raising or lowering operators in quark spin space while $L_\pm$ raise or lower the z component of angular momentum of the quarks. The angle θ is arbitrary.

Within the framework of the free quark model Melosh (1974) attempted to provide some theoretical justification for this. The structure of V is similar to that of a transverse Lorentz boost—see also section 7.3 in this respect. A detailed discussion of Melosh's work and the relation between his transformation V and similar transformations by Foldy, Wouthuysen and others is given in Bell (1974) which the reader interested in this subject should study.

In a classification quark basis the nucleon is a member of the representation $(\mathbf{8}, \mathbf{2})$ $\mathbf{56}(L=0)$ of $(SU(3), SU(2)) \in SU(6)(0(3))$. For a nucleon with spin projection $+\frac{1}{2}$ along the z-axis then its wavefunction may be written

$$\psi^{(0)} = \frac{1}{\sqrt{2}} [|8\rangle_\alpha |\tfrac{1}{2}\uparrow\rangle_\alpha + |8\rangle_\beta |\tfrac{1}{2}\uparrow\rangle_\beta]_0 \qquad (6.54)$$

where the subscript after the bracket denotes the value of L_z and for ease of notation we use α, β in place of $M_{S,A}$ (Table 3.7). Now act on this classification space state with the operator V_B. The resulting nucleon wavefunction in the current basis is found to be

$$|\phi_c\rangle = \cos\theta|\psi\rangle + \sin\theta|\delta\psi\rangle \qquad (6.55)$$

$$\equiv \frac{1}{\sqrt{2}} \cos\theta [|8\rangle_\alpha |\tfrac{1}{2}\uparrow\rangle_\alpha + |8\rangle_\beta |\tfrac{1}{2}\uparrow\rangle_\beta]_0$$

$$+ \frac{1}{\sqrt{2}} \sin\theta \left\{ \left[|8\rangle_\beta |\tfrac{1}{2}\downarrow\rangle_\beta - \tfrac{1}{2}|8\rangle_\alpha |\tfrac{1}{2}\downarrow\rangle_\alpha + \frac{\sqrt{2}}{3}|8\rangle_\alpha |\tfrac{3}{2}\downarrow\rangle_s \right]_1 \right.$$

$$\left. + [\sqrt{\tfrac{2}{3}}|8\rangle_\alpha |\tfrac{3}{2}\uparrow\rangle_s]_{-1} \right\} \qquad (6.56)$$

which is identical to the chiral $SU(3) \otimes SU(3)$ wavefunction of the nucleon derived by Buccella *et al.*

With this current space wavefunction for the nucleon one can take matrix elements of current operators. For example, the expectation value of $I_3\sigma_z$ between proton states (which is relevant to g_A/g_V, equation 6.88) yields

$$\frac{g_A}{g_V} = \frac{5}{3}(\cos^2\theta - \sin^2\theta) \equiv \frac{5}{3}(1 - 2\sin^2\theta) \qquad (6.57)$$

In the absence of configuration mixing induced by V, $\cos\theta = 1$ and hence $g_A/g_V = \frac{5}{3}$. Empirically $g_A/g_V \simeq \frac{5}{4}$ and so $\sin^2\theta \simeq \frac{1}{8}$.

The above has shown the calculation of $\langle\phi_c|\hat{O}_c|\psi_c\rangle$. From equation (6.51) we can equivalently compute $\langle\psi_q|\hat{O}_q|\psi_q\rangle$ by deriving

$$\hat{O}_q \equiv V^{-1}\hat{O}_c V \qquad (6.58)$$

and using $\psi_q \equiv \psi(0)$ of equation (6.54). As before we will illustrate the calculation of g_A/g_V for which

$$\hat{0}_c = \sum_{j=1}^{3} (\sigma_z I_3)^{(j)}$$

Then we have that

$$\hat{0}_q = \sum_{j=1}^{3} [V^{-1}(\sigma_z I_3) V]_j$$

$$\equiv \sum_{j=1}^{3} I_3^{(j)} [\cos\theta + i(\boldsymbol{\sigma} \times \mathbf{L})_3 \sin\theta] \sigma_z [\cos\theta - i(\boldsymbol{\sigma} \times \mathbf{L})_3 \sin\theta]$$

$$\equiv \sum_{j=1}^{3} I_3^{(j)} \sigma_z [\cos\theta - i(\boldsymbol{\sigma} \times \mathbf{L})_3 \sin\theta][\cos\theta - i(\boldsymbol{\sigma} \times \mathbf{L})_3 \sin\theta]$$

$$\equiv \sum_{j=1}^{3} I_3^{(j)} \sigma_z [\cos 2\theta - i \sin 2\theta (\boldsymbol{\sigma} \times \mathbf{L})_3]$$

$$\equiv \sum_{j=1}^{3} I_3^{(j)} (\sigma_z \cos 2\theta + \sin 2\theta \, \boldsymbol{\sigma}_T . \mathbf{L}_T) \tag{6.59}$$

Between nucleon classification states which have $L = 0$ the latter term involving $L_{\pm}$ cannot contribute. From the first term we again find

$$\frac{g_A}{g_V} = \frac{5}{3} \cos 2\theta \tag{6.60}$$

in agreement with equation (6.57).

6.5   The SU(3) × SU(3) subalgebra of SU(6)$_W$

The generators of SU(6)$_{W,\mathrm{cur}}$ in equations (6.43) to (6.46) include $\int d^3x V_0^\alpha(x, t)$ and $\int d^3x A_0^\alpha(x, t)$ which are respectively $Q^\alpha(t)$ and $Q_5^\alpha(t)$. The right-handed charges $Q^\alpha + Q_5^\alpha$ and left-handed $Q^\alpha - Q_5^\alpha$ generate two commuting SU(3) algebras and so the chiral SU(3) × SU(3) is contained within SU(6)$_{W,\mathrm{cur}}$. It is conventional to label states or operators by their transformation properties under this SU(3) × SU(3) subalgebra (Gilman and Harari, 1968; Harari, 1968).

If a state is in representation A of the $Q^\alpha + Q_5^\alpha$ SU(3) and B of $Q^\alpha - Q_5^\alpha$ then we label it

$$\{(A, B)_{S,z}, L_z\} \tag{6.61}$$

where S_z is the eigenvalue of the axial-vector charge Q_5^0 (which is identically equal to the z component at infinite momentum, equation 6.43) amd L_z is then defined as $J_z - S_z$. The direct product $A \otimes B$ then gives the ordinary SU(3), Q^α, content of the state.

As examples we can illustrate the transformation properties of Q^α and Q_5^α themselves. Since

$$Q^\alpha \equiv \frac{(Q^\alpha + Q_5^\alpha)}{2} + \frac{(Q^\alpha - Q_5^\alpha)}{2} \tag{6.62}$$

then it transforms as

$$\{(\mathbf{8}, \mathbf{1})_0, 0\} + \{(\mathbf{1}, \mathbf{8})_0, 0\} \tag{6.63}$$

Similarly Q_5^α transforms as

$$\{(\mathbf{8}, \mathbf{1})_0, 0\} - \{(\mathbf{1}, \mathbf{8})_0, 0\} \tag{6.64}$$

Of particular relevance are the representations of the quarks and antiquarks in SU(3)×SU(3). We define the current quarks by their transformation properties as follows. When $p_z \to \infty$ a quark with spin up has positive helicity. The convention that $V - A$ is left handed for leptons implies the positive helicity quark is a triplet of $Q^\alpha + Q_5^\alpha$ and so

$$q^\uparrow \equiv \{(\mathbf{3}, \mathbf{1})_{1/2}, 0\}; \qquad q_\downarrow \equiv \{(\mathbf{1}, \mathbf{3})_{-1/2}, 0\} \tag{6.65}$$

while

$$\bar{q}^\uparrow \equiv \{(\mathbf{1}, \bar{\mathbf{3}})_{1/2}, 0\}; \qquad \bar{q}_\downarrow \equiv \{(\bar{\mathbf{3}}, \mathbf{1})_{-1/2}, 0\} \tag{6.66}$$

These also illustrate the behaviours under the charge conjugation and natural parity operations (Jacob and Wick, 1959; Harari, 1968; Gilman and Harari, 1968):

$$[(A, B)_{S,z}, L_z] \xrightarrow{\mathcal{N}} [(B, A)_{-S,z}, -L_z] \eta_{AB}^N \tag{6.67}$$

$$[(A, B)_{S,z}, L_z] \xrightarrow{\mathscr{C}} [(\bar{B}, \bar{A})_{S,z_2} L_z] \eta_{AB}^C$$

where the phase factors depend upon the orbital momentum of the state. If we are combining q and $\bar{q}$ the η factors also depend upon the W spin of the resulting state.

Now combine q and $\bar{q}$ with $L_z = 0$. Then for the $L = 0$ level

$$|J^{PC} = 1^{--}, J_z = +1\rangle \equiv q^\uparrow \bar{q}^\uparrow = |\{(3, \bar{3})_1, 0\}\rangle$$
$$|J^{PC} = 1^{--}, J_z = -1\rangle \equiv q_\downarrow \bar{q}_\downarrow = |\{\bar{3}, 3)_{-1}, 0\}\rangle$$

(6.68)

since $(3, 1) \otimes (1, \bar{3}) = (3, \bar{3})$ etc. Under the "normal" SU(3), Q^α, we find $3 \otimes \bar{3} = 1 \oplus 8$.

For the $J_z = 0$ we are combining $(3, 1)_{1/2} \times (\bar{3}, 1)_{-1/2}$ which yields $(8, 1)_0 \oplus (1, 1)_0$. For the octet states

$$|8J^{PC} = 1^{--}, J_z = 0\rangle = \frac{1}{\sqrt{2}}|\{(8, 1)_0, 0\} + \{(1, 8)_0, 0\}\rangle$$

$$|8J^{PC} = 0^{-+}, J_z = 0\rangle = \frac{1}{\sqrt{2}}|\{(8, 1)_0, 0\} - \{(1, 8)_0, 0\}\rangle$$

(6.69)

Notice that the $J^{PC} = 1^{--}$ transforms like Q^α and the $J^{PC} = 0^{-+}$ like Q_5^α. By acting on equation (6.69) with $\mathscr{C}$ we see that the convention is $|\{(8, 1)_0, 0\}\rangle = -|\{(1, 8)_0, 0\}\rangle$ for $L = 0$, hence $\eta_{AB}^C(L = 0) = -$.

For these $L = 0$ states we can give a translation dictionary between the classification of SU(3) × SU(3) and SU(6)$_W$.

$$|J^{PC} = 0^{-+}, J_z = 0\rangle \leftrightarrow |W = 1, W_z = 0; J_z = 0\rangle \qquad (6.70)$$

$$|J^{PC} = 1^{--}, J_z = 0\rangle \leftrightarrow |W = 0, W_z = 0; J_z = 0\rangle \qquad (6.71)$$

$$|J^{PC} = 1^{--}, J_z = \pm 1\rangle \leftrightarrow |W = 1, W_z = \pm 1; J_z = \pm 1\rangle \qquad (6.72)$$

where the SU(3) ⊗ SU(3) representations of the states are given in equations (6.68) and (6.69).

Now consider three quarks at infinite momentum with $J_z = \frac{1}{2}$. Suppose that the nucleon acted under the algebra of currents as if it were two current quarks with $S_z = \frac{1}{2}$ each and coupled symmetrically with a third current quark with $S_z = -\frac{1}{2}$. We would then have a state

$$\{(3, 1)_{1/2} \otimes (3, 1)_{1/2} \otimes (1, 3)_{-1/2}\}_{\text{sym}} \qquad (6.73)$$

(where we have temporarily suppressed the $L_z = 0$). Now $3 \otimes 3 = 6 \oplus \bar{3}$, and of these the symmetrical combination is 6. Hence

$$|S = \frac{1}{2} \text{ or } \frac{3}{2}, S_z = \frac{3}{2}\rangle_{qqq} \equiv \{(6, 3)_{1/2}, 0\} \qquad (6.74)$$

Under Q^α we have $\mathbf{6} \otimes \mathbf{3} = \mathbf{8} \oplus \mathbf{10}$ and hence the nucleon octet and Δ decuplet emerge. If all three quarks had carried $S_z = +\frac{1}{2}$ then we would have

$$|S = \tfrac{3}{2}, S_z = \tfrac{3}{2}\rangle_{qqq} = \{(\mathbf{3}, \mathbf{1}) \otimes (\mathbf{3}, \mathbf{1}) \otimes (\mathbf{3}, \mathbf{1})\}_{sym} = \{(\mathbf{10}, \mathbf{1})_{3/2}, 0\} \quad (6.75)$$

Since Q_5^α is a generator of $SU(3) \times SU(3)$ it cannot cause transitions from one representation to another. Hence if the physical nucleon were $\{(\mathbf{6}, \mathbf{3})_{1/2}, 0\}$ at infinite momentum then Q_5^α would only connect it to itself or the Δ (since the Δ also has $\{(\mathbf{6}, \mathbf{3})_{1/2}, 0\}$ representation). This would predict $g_A/g_V = \frac{5}{3}$ and, via PCAC, forbid $\pi N \to N^*$.

For $\pi N \to N^*$ to occur both N and N^* must have common representations in $SU(3) \otimes SU(3)$ of currents. As a result phenomenological mixing schemes were suggested:

$$|N\rangle = \alpha\{(\mathbf{6}, \mathbf{3})_{1/2}, 0\} + \beta\{(\bar{\mathbf{3}}, \mathbf{3})_{1/2}, 0\} + \gamma\{(\mathbf{3}, \bar{\mathbf{3}})_{-1/2}, 1\} + \cdots \quad (6.76)$$

(Harari, 1968; Gilman and Harari, 1968; Buccella et al., 1970).

In order to bring some systematics to this phenomenology a unitary transformation V was sought which transforms an irreducible representation of the algebra of currents into the physical state, e.g.

$$|N\rangle = V|\{(\mathbf{6}, \mathbf{3})_{1/2}, 0\}\rangle \quad (6.77)$$

A candidate for V was suggested by Buccella et al. and the effect of this on the wavefunction (equation 6.54) has been exhibited in equation (6.56). Within the framework of a free-quark model a transformation V has been derived by Melosh and discussed by Eichten et al. (1973). For our purposes it is sufficient to suppose that V is a single quark operator, and hence depends only on the coordinates of a single quark and does not create connected $q\bar{q}$ pairs. The V developed by the authors above indeed has this property.

Then if we wish to evaluate the matrix element of a current, say Q_5,

$$\langle \text{Hadron}|Q_5|\text{Hadron}\rangle$$

$$= \langle \text{Irreducible representation of currents}|V^{-1}Q_5 V|\text{I.R. currents}\rangle \quad (6.78)$$

we need only calculate the transformation properties of $V^{-1}Q_5 V$. These will be the transformation properties of the most general linear combination of single quark operators that can be written down consistent with Lorentz invariance and SU(3).

Consider then $V^{-1}Q_5V$. We know that Q_5 transforms as $\{(\mathbf{8}, \mathbf{1})_0 - (\mathbf{1}, \mathbf{8})_0, 0\}$ which has $J_z = 0$. What other combinations of $q\bar{q}$ exist with $J_z = 0$? The candidate $\{(\mathbf{8}, \mathbf{1})_0 + (\mathbf{1}, \mathbf{8})_0, 0\}$ is Q^α and so is no good, and we are left with $\{(\mathbf{3}, \bar{\mathbf{3}})_1, -1\}$ and $\{(\bar{\mathbf{3}}, \mathbf{3})_{-1}, 1\}$. To have the axial transformation property it is the combination $\{(\mathbf{3}, \bar{\mathbf{3}})_1, -1\} - \{(\bar{\mathbf{3}}, \mathbf{3})_{-1}, 1\}$ that is relevant. (Act on Q_5^α with $\mathcal{N}$ and compare with $\mathcal{N}$ acting on this combination of $(\mathbf{3}, \bar{\mathbf{3}})$ and $(\bar{\mathbf{3}}, \mathbf{3})$.) Hence

$$V^{-1}Q_5^\alpha V = \{(\mathbf{8}, \mathbf{1})_0 - (\mathbf{1}, \mathbf{8})_0, 0\} \quad \text{and} \quad \{[(\mathbf{3}, \bar{\mathbf{3}})_1 - 1] - [(\bar{\mathbf{3}}, \mathbf{3})_{-1}, 1]\}$$

(6.79)

which behave as components of $\mathbf{35}$'s of the full $\mathrm{SU}(6)_W$ of currents (Gilman et al., 1973, 1974; Hey et al., 1973). These terms are often referred to as W_z and $(W_+L_- - W_-L_+)$ respectively in the $\mathrm{SU}(6)_W$ language (compare equation 6.59).

For real photoabsorbtion or emission the dipole operator

$$D_\pm \equiv \int \mathrm{d}^3x \, \frac{(x \pm iy)}{\sqrt{2}} \, V_0(\mathbf{x}, t)$$

(6.80)

is relevant to the transitions. This transforms as (compare equation 6.69)

$$D_\pm^\alpha = \{(\mathbf{8}, \mathbf{1})_0 + (\mathbf{1}, \mathbf{8})_0, \pm 1\}$$

(6.81)

By analogous arguments to the above one can see that the most general form of single quark operator with $J_z = 1$ will consist of four terms:

$$\{(\mathbf{8}, \mathbf{1})_0 + (\mathbf{1}, \mathbf{8})_0, 1\}, \quad \{(\mathbf{3}, \bar{\mathbf{3}})_1, 0\}, \quad \{(\mathbf{8}, \mathbf{1})_0 - (\mathbf{1}, \mathbf{8})_0, 1\},$$

$$\{(\bar{\mathbf{3}}, \mathbf{3})_{-1}, 2\}$$

(6.82)

These are sometimes written as L_+, W_+, W_zL_+, $W_-L_+L_+$ respectively.

For the $J_z = 0$ Q_5^α natural parity $\mathcal{N}$ constrained the four possibilities to just two because $J_z = 0$ is an eigenstate. The $J_z = +1$ or -1 is not an eigenstate of natural parity $\mathcal{N}$ and so all four possibilities can be present (and appear to be needed phenomenologically: section 7.3; Hey and Weyers, 1974; Close, 1974b; Cashmore et al., 1975).

Matrix elements for π emission or $\gamma N \to N^*$ can now be calculated. The irreducible representations for N and N^* are known, and the $V^{-1}Q_5^\alpha V$ contains two pieces with arbitrary weight. The $\pi N \to N^*$

matrix elements will therefore contain two reduced matrix elements weighted by appropriate Clebsch–Gordan coefficients. Similarly $\gamma N \rightarrow N^*$ will involve four reduced matrix elements. These will be exhibited in Table 7.1 in section 7.2.2.

6.6 SU(3) × SU(3) configuration mixing in the harmonic oscillator quark model

The SU(6) hadron wavefunctions are traditionally written in terms of Pauli two-component spinors for the constituent quarks. This receives motivation from the nonrelativistic quark model where the quarks are approximately at rest in the hadron rest frame, i.e. the hadron has negligible internal momenta, and hence the four-component Dirac quark spinor approximates to a two-component Pauli Spinor.

Is the momentum of a quark in a hadron really negligible? In models one tends to predict that it can be sizeable. One illustration of this is given in the MIT bag (Chapter 18) where a free quark is confined to a sphere of radius R. Similarly, in a harmonic oscillator quark model $(\langle p_i^2 \rangle)^{1/2} \sim R^{-1}$ and for a quark with effective mass of around 300 MeV one finds

$$(\langle p_i^2 \rangle)^{1/2} \simeq m_q$$

This has the consequence that the "small" components of the Dirac four-spinor are of comparable size to the large. These sizeable "small" components are the origin of many of the corrections to naive SU(6) nonrelativistic quark model results. This is illustrated in the bag model (Chapter 18).

Le Youanc *et al.* (1974b) hypothesised that in the SU(6) wavefunctions the Pauli spinors should be replaced by Dirac spinors, the latter taking account of the internal quark motions in the hadron. Hence the Pauli spinors χ_i for the i-th quark are replaced with the Dirac spinor

$$u_i(s) = \begin{pmatrix} \chi_i(s) \\ \mu_q \boldsymbol{\sigma}_i \cdot p_i \chi_i(s) \end{pmatrix} \tag{6.83}$$

where μ_q is approximately the normal quark magnetic moment $\mu_q \simeq \frac{1}{2} m_q$.

This hypothesis is implicit in many quark model calculations. For example, the electromagnetic current interaction of the current (Dirac) quark is $\bar{u}\gamma_\mu u A^\mu$. This can be written out explicitly in 2×2 matrix form for $\mu = 1, 2, 3$ using the spinors in equation (6.17) and $\gamma_0 \equiv \beta$, $\gamma_i \equiv \beta\alpha_i$ (equation 6.15). Hence

$$\sqrt{\left[\frac{(E'+m)(E+m)}{4EE'}\right]}\left(\chi^+, \frac{\boldsymbol{\sigma}_i \cdot \mathbf{p}_i'}{E'+m}\chi^+\right)\begin{pmatrix} 0 & \boldsymbol{\sigma} \cdot \mathbf{A} \\ \boldsymbol{\sigma} \cdot \mathbf{A} & 0 \end{pmatrix}\begin{pmatrix} \chi \\ \dfrac{\boldsymbol{\sigma} \cdot \mathbf{p}_i}{E+m}\chi \end{pmatrix} \tag{6.84}$$

where $\mathbf{p}_i' \equiv \mathbf{p}_i + \mathbf{k}$ with $\mathbf{k}$ the three momentum of the electromagnetic field. Expanding out we have the form of the interaction for two-component spinors

$$\sqrt{\left[\frac{(E'+m)(E+m)}{4EE'}\right]}\chi^+\left(\boldsymbol{\sigma} \cdot \mathbf{A}\frac{\boldsymbol{\sigma} \cdot \mathbf{p}_i}{E+m} + \frac{\boldsymbol{\sigma} \cdot \mathbf{p}_i'}{E'+m}\boldsymbol{\sigma} \cdot \mathbf{A}\right)\chi \tag{6.85}$$

Now use the identity of Pauli matrices

$$\boldsymbol{\sigma} \cdot \mathbf{a}\,\boldsymbol{\sigma} \cdot \mathbf{b} = \mathbf{a} \cdot \mathbf{b} + i\boldsymbol{\sigma} \cdot \mathbf{a} \times \mathbf{b} \tag{6.86}$$

and if $E \simeq E' \simeq m$ we find the familiar structure

$$\frac{1}{2m_q}\chi^+[(\mathbf{p}+\mathbf{p}') \cdot \mathbf{A} + i\boldsymbol{\sigma} \cdot \mathbf{k} \times \mathbf{A}]\chi \tag{6.87}$$

of the nonrelativistic electromagnetic interaction. The spin dependent term conserves L_z and is the only term that would be allowed in $SU(6)_W$. The first term has $L_z = \pm 1$ if a transversely polarised photon is absorbed and consequently breaks $SU(6)_W \times 0(2)_{L,z}$. It has arisen as a result of the non-negligible "small" components in the quark spinor. This model is discussed again with reference to photoproduction in Chapter 7.

Having made this hypothesis one can similarly study its effect in the axial matrix elements, e.g. g_A/g_V. For a nucleon at rest g_A/g_V is given by the coefficient of $\bar{\psi}\gamma_3\gamma_5\psi$ (expand out $\bar{\psi}\gamma_\mu\gamma_5\psi$ in two-component spinors and verify that the $\mu = 3$ component is finite when $\mathbf{p}' = \mathbf{p} \to 0$). At quark level $\mathbf{p}' = \mathbf{p} \neq 0$ and expanding out the quark current we have

$$\bar{\psi}\gamma_3\gamma_5\psi = \frac{E+m}{2E}\left(\chi^+, \frac{\boldsymbol{\sigma} \cdot \mathbf{p}}{E+m}\chi^+\right)\begin{pmatrix} \sigma_3 & 0 \\ 0 & \sigma_3 \end{pmatrix}\begin{pmatrix} \chi \\ \dfrac{\boldsymbol{\sigma} \cdot \mathbf{p}}{E+m}\chi \end{pmatrix}$$

$$= \left(\frac{E+m}{2E}\right)\chi^+\left(\sigma_3 + \frac{\boldsymbol{\sigma} \cdot \mathbf{p}}{(E+m)^2}(p_3 - i(\boldsymbol{\sigma} \times \mathbf{p})_3)\right)\chi \tag{6.88}$$

which becomes

$$\chi^+\left[\sigma_3\left(1-\frac{p_T^2}{E(E+m)}\right)+\frac{\boldsymbol{\sigma}_T\cdot\mathbf{p}_T p_z}{E(E+m)}\right]\chi \qquad (6.89)$$

after manipulating the $\boldsymbol{\sigma}$ matrices using equation (6.86). This has the same formal structure as equation (6.59) and for g_A/g_V we find

$$\frac{g_A}{g_V}=\frac{5}{3}\left(1-\frac{p_T^2}{E(E+m)}\right) \qquad (6.90)$$

The depletion of g_A/g_V relative to $\frac{5}{3}$ is clearly seen to be due to the "small" components in the quark spinors, in particular upon the transverse momenta. Since transverse momenta break the $SU(6)_W \times 0(2)_{L,z}$ then the deviation of g_A/g_V from the $SU(6)_W$ value is understood.

Formally there is an intimate relation between these results and those arising from the transformation V of section 6.4. The reason is the algebraic similarity between V (equations 6.53 and 6.54) and the Lorentz boost $L(p)$ (equations 6.10 and 6.11) that generates the Dirac four-component spinor of momentum p from the rest Pauli spinor of the nonrelativistic $SU(6)$ model. This relationship has been made more precise by Le Youanc et al. (1974b). They take the spinor (equation 6.16) and explicitly boost it to the *infinite momentum frame*.

In the $p_z \to \infty$ boost the spatial part of the wavefunction is Lorentz contracted following the model of Licht and Pagnamenta (1970) and the relative time dependence is ignored (compare Feynman et al., 1971, section 7.4)

$$\psi_{p_z=\infty}\left(\left\{\frac{p_{iz}}{\sqrt{1-\beta^2}},\mathbf{p}_{iT}\right\}\right)=\psi_{\text{rest}}(\{p_{iz},\mathbf{p}_{iT}\}) \qquad (6.91)$$

The spinors transform as follows. For the i-th quark

$$Lu_i(s,\mathbf{p})=\cosh\frac{\omega}{2}\begin{pmatrix}1 & \sigma_z\\ \sigma_z & 1\end{pmatrix}\begin{pmatrix}\chi_i\\ \mu_q\boldsymbol{\sigma}\cdot\mathbf{p}\chi_i\end{pmatrix}$$

$$\equiv\cosh\frac{\omega}{2}\begin{pmatrix}\psi_i\\ \sigma_{iz}\psi_i\end{pmatrix} \qquad (6.92)$$

where

$$\psi_i\equiv[1+\mu_q\sigma_{iz}(\boldsymbol{\sigma}_i\cdot\mathbf{p}_i)]\chi_i \qquad (6.93)$$

Now for ease of notation we temporarily drop the i subscript and concentrate on the column vector. This may be written

$$\begin{pmatrix} (1+\sigma_z\sigma_\perp \cdot p_\perp + p_z)\chi \\ (1+\sigma_\perp \cdot p_\perp\sigma_z + p_z)\sigma_z\chi \end{pmatrix} \qquad (6.94)$$

Then using $\sigma_\perp\sigma_z = -\sigma_z\sigma_\perp$ and writing

$$\Sigma \equiv \begin{pmatrix} \sigma & 0 \\ 0 & \sigma \end{pmatrix}$$

we have for the boosted spinor

$$Lu_i(s, \mathbf{p}) = \cosh \frac{\omega}{2}[1 + \mu_q p_{iz} + \mu_q\beta_i\Sigma_i^{(+)}p_i^{(-)} + \mu_q\beta_i\Sigma_i^{(-)}p_i^{(+)}]\begin{pmatrix} \chi_i \\ \sigma_{iz}\chi_i \end{pmatrix} \qquad (6.95)$$

where $\Sigma_i^\pm \equiv \frac{1}{2}(\Sigma_x \pm i\Sigma_y)$ and $p_i \equiv (p_x^\mp + ip_y)$.

First of all note what would be the result if the internal momenta were small. The infinite momentum spinor in equation (6.95) would be

$$\cosh\frac{\omega}{2}\begin{pmatrix} \chi_i \\ \sigma_{iz}\chi_i \end{pmatrix} \qquad (6.96)$$

The generators of $SU(3) \otimes SU(3)$ are $\frac{1}{2}(1 \pm \Sigma_{iz})$ and the Dirac spinor in equation (6.96) transforms identically to a two-component Pauli spinor χ_i. Hence the $SU(3) \times SU(3)$ representation of the nucleon that arises from the standard $SU(6)$ Pauli spinor classification is preserved in the boost to $p_z = \infty$ if the quarks' momenta (transverse to z) are negligible.

However, if the quark momenta are not small the nucleon will have a different $SU(3) \times SU(3)$ classification when $p_z \to \infty$ than at rest. This is obvious from the form of the boosted spinor in equation (6.95). We notice that the $\beta\Sigma$ are the W spin generators and that the W spin representation of the quark is being altered due to the presence of the $W^+ p^-$ and $W^- p^+$ operators in the boost in equation (6.95). Only if the quark momentum is along the direction of the z boost is its W spin representation preserved. The presence of transverse quark momenta lead to $SU(3) \times SU(3)$ representation mixing. One can legitimately regard this mixing as a consequence of quark confinement since this will give the quarks a momentum even when the parent hadron is at rest (compare Chapter 18). Equivalently this is the dynamical origin of the $SU(6)_W$ violating decays like $B \to \omega\pi$.

The chiral $SU(3) \times SU(3)$ wavefunction for the nucleon at infinite momentum can now be calculated essentially as in section 6.4 (equation 6.56). The difference is that we previously left $\cos \theta$ and $\sin \theta$ undetermined *a priori*. In the present approach the mixing is given in terms of $\langle p_T^2 \rangle$ and μ_q^2.

For example the calculation of g_A/g_V involves

$$Q_5^\alpha = \sum_i \sigma_{iz}(1 - \mu_q^2 p_i^2) \qquad (6.97)$$

(compare equations 6.57, 6.60 and 6.90) where we have dropped $p^\pm$ contributions which vanish between nucleon wavefunctions with $L_z = 0$. Hence

$$\frac{g_A}{g_V} = \frac{5}{3}(1 - 2\delta) \qquad (6.98)$$

where $\delta \simeq \frac{1}{6}$ for $R^2/\mu_q^2 \simeq 3$ which is determined from the harmonic oscillator parameters fitted to the ground state ($m_q \simeq 300$ MeV, $R^2 \simeq 5$ to 10 GeV^{-2}). Hence the magnitude of g_A/g_V is in good agreement with data. In particular note the structure of equation (6.90) for g_A/g_V and compare with the MIT bag and results of Bogobluibov (1968) discussed in section 18 (in particular equations 18.37 and 18.43).

6.7 Quark pair creation and SU(6)$_W$ breaking

In section 6.3 we saw how a $^3P_0$ model of quark pair creation from the vacuum gave rise to an $SU(6)_W \times 0(2)_{L,z}$ vertex symmetry if the momenta of the quarks transverse to the z direction were ignored. If the possibility of transverse momenta is admitted, in particular in the $^3P_0$ pair, then the $SU(6)_W$ is broken. The magnitude and systematics of its breaking will in general depend upon the assumed potential in which the constituents exist since this leads to the form of the wavefunctions that control the quarks' momentum distributions in the hadrons.

By assuming Gaussian wavefunctions for the quarks Le Youanc *et al.* (1973, 1974a, 1975a) have carried out a systematic programme of investigation in the quark pair creation model. For the details and explicit phenomenology the reader is referred to those papers. As introduction some of the more general features are outlined here.

The basic dynamics of the model is that the decay $A \rightarrow BC$ takes place as the consequence of a $q\bar{q}$ pair created with vacuum ($^3P_0$) quantum numbers. If ABC are mesons the essential diagram for the process is drawn in Fig. 6.3.

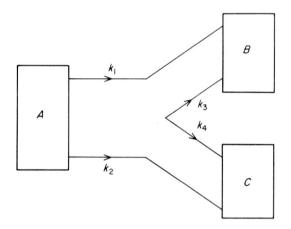

FIG. 6.3. $A \rightarrow BC$ via $^3P_0$ $q\bar{q}$ creation.

The $^3P_0$ pair is created with $(\mathbf{L}, L_z) = (1, m)$ and $(\mathbf{S}, S_z) = (1, -m)$ and $\mathbf{L}, \mathbf{S}$ are coupled to give $\mathbf{J} = 0$. To introduce the ideas, consider all ABC as S-wave $q\bar{q}$ states (e.g. $\rho(J_z = 0) \rightarrow \pi^+ \pi^-$). The spin-dependent contribution to the matrix element may be factored out and so

$$(M_B M_C | T | M_A) = \gamma \sum_m (1m, 1-m|00)(\chi_B \chi_C | \chi_A \chi_{-m}) I_{A,BC}^m \quad (6.99)$$

where γ is a number representing the amplitude for creating the $q\bar{q}$ pair from the vacuum, and is *a priori* unknown. The quanity $I_{A,BC}^m$ is the spatially dependent contribution to the matrix element. With the momenta as shown in Fig. 6.3 then

$$I^m(A; BC) \equiv I_{A,BC}^m(\mathbf{k}_B, \mathbf{k}_C; \mathbf{k}_A) = \int d^3\mathbf{k}_1 d^3\mathbf{k}_2 d^3\mathbf{k}_3 d^3\mathbf{k}_4 \delta(\mathbf{k}_1 + \mathbf{k}_2 - \mathbf{k}_A)$$

$$\times \delta(\mathbf{k}_2 + \mathbf{k}_4 - \mathbf{k}_C) \delta(\mathbf{k}_1 + \mathbf{k}_3 - \mathbf{k}_B) \delta(\mathbf{k}_3 + \mathbf{k}_4) Y_1^M(\mathbf{k}_3 - \mathbf{k}_4)$$

$$\times \psi_A(\mathbf{k}_1, \mathbf{k}_2) \psi_B(\mathbf{k}_1, \mathbf{k}_3) \psi_C(\mathbf{k}_2, \mathbf{k}_4) \quad (6.100)$$

The origin of these terms should be obvious. Three delta functions constrain the internal momenta to yield the mesons' momenta and the fourth shows that the $q\bar{q}$ are produced from the vacuum with no momentum. The $\psi_{A,B,C}$ are the spatial wavefunctions of the $q\bar{q}$ states A, B, C (we shall suppose them to be harmonic oscillator wavefunctions as in other calculations, e.g. Chapter 7). Finally the $Y_1^M(\mathbf{k}_3 - \mathbf{k}_4)$ is the spatial dependence of the $^3P_0$ pair.

To simplify the calculation, specialise to the centre of the mass frame (rest frame of A). We shall route the internal momenta as follows:

$$\mathbf{k}_1 = \tfrac{1}{2}(\mathbf{k} + \mathbf{k}_B) = -\mathbf{k}_2$$
$$\mathbf{k}_3 = \tfrac{1}{2}(-\mathbf{k} + \mathbf{k}_B) = -\mathbf{k}_4 \qquad (6.101)$$

consistent with the delta function constraints when $\mathbf{k}_A = 0$ and $\mathbf{k}_C = -\mathbf{k}_B$. The integral $I_m(A; BC)$ is transformed into

$$I_m(A; BC) = \tfrac{1}{8}\delta(\mathbf{k}_B + \mathbf{k}_C) \int d^3k\, Y_1^m(\mathbf{k}_B - \mathbf{k})\psi_A(\mathbf{k}_B + \mathbf{k})\psi_B(-\mathbf{k})\psi_C(\mathbf{k}) \qquad (6.102)$$

The harmonic oscillator wavefunctions for ground state mesons when suitably normalised may be written

$$\psi_A(\mathbf{k}_1, \mathbf{k}_2) = \left(\frac{R_A^2}{\pi}\right)^{3/4} \exp\left(\frac{(\mathbf{k}_1 - \mathbf{k}_2)^2 R_A^2}{8}\right) \qquad (6.103)$$

and so on, where R_A has dimensions of inverse momenta or length and so are a measure of the "size" of the state. Substituting into equation (6.102) yields

$$I_m(A; BC) = \frac{1}{8}\delta(\mathbf{k}_B + \mathbf{k}_C) \int d^3k\, Y_1^m(\mathbf{k}_B - \mathbf{k})\left(\frac{R_A^2 R_B^2 R_C^2}{\pi^3}\right)^{3/4}$$
$$\times \exp\left[-\frac{(\mathbf{k}_B + \mathbf{k})^2 R_A^2}{8}\right] \exp\left[-\frac{k^2 R_B^2}{8}\right] \exp\left[-\frac{k^2 R_C^2}{8}\right] \qquad (6.104)$$

The way to deal with this integral is a standard procedure. First eliminate the cross-terms in the exponent by making the substitution

$$\mathbf{k}' = \mathbf{k} + \frac{R_A^2 \mathbf{k}_B}{(R_A^2 + R_B^2 + R_C^2)}; \qquad d\mathbf{k}' \equiv d\mathbf{k} \qquad (6.105)$$

so that the exponential becomes

$$\exp\left[-\frac{\mathbf{k}'^2(R_A^2+R_B^2+R_C^2)}{8}\right]\exp\left[-\frac{\mathbf{k}_B^2 R_A^2(R_B^2+R_C^2)}{8(R_A^2+R_B^2+R_C^2)}\right] \tag{6.106}$$

The integral therefore becomes

$$I_m(A;BC)=\frac{1}{8}\delta(\mathbf{k}_B+\mathbf{k}_C)\exp\left[-\frac{\mathbf{k}_B^2 R_A^2(R_B^2+R_C^2)}{8(R_A^2+R_B^2+R_C^2)}\right]\frac{(R_A R_B R_C)^{3/2}}{\pi^{3/4}}\frac{1}{\pi^{3/2}}$$

$$\times\int d^3\mathbf{k}\, Y_1^m(\mathbf{k}_B-\mathbf{k})\exp\left[-\frac{\mathbf{k}'^2(R_A^2+R_B^2+R_C^2)}{8}\right] \tag{6.107}$$

All that remains is to manipulate

$$Y_1^m(\mathbf{k}_B-\mathbf{k})\equiv-\boldsymbol{\varepsilon}_m\cdot(\mathbf{k}_B-\mathbf{k})\sqrt{\left(\frac{3}{4\pi}\right)} \tag{6.108}$$

into a form dependent on $\mathbf{k}'$. Substitute equation (6.105) into this and we have finally

$$I_m(A;BC)=-\frac{1}{8}\delta(\mathbf{k}_B+\mathbf{k}_C)\exp\left[-\frac{\mathbf{k}_B^2(R_B^2+R_C^2)R_A^2}{8(R_A^2+R_B^2+R_C^2)}\right]$$

$$\times\left(\frac{R_A R_B R_C}{\pi}\right)^{3/2}\frac{1}{\pi^{3/4}}\sqrt{\frac{3}{4\pi}}\,\boldsymbol{\varepsilon}_m\cdot\mathbf{k}_B\left(1+\frac{R_A^2}{R_A^2+R_B^2+R_C^2}\right)$$

$$\times\int 4\pi k'^2\, dk'\exp\left(-\frac{k'^2(R_A^2+R_B^2+R_C^2)}{8}\right) \tag{6.109}$$

where we have eliminated the $\boldsymbol{\varepsilon}\cdot\mathbf{k}'$ piece (since upon integration it vanishes by parity).

The Gaussian integral is standard (equation 7.4.4 of Abramowitch and Stegun)

$$\int_0^\infty k'^2\, dk'\exp\left(-k'^2\frac{\Sigma R^2}{8}\right)=2\sqrt{\pi}\left(\frac{2}{\Sigma R^2}\right)^{3/2} \tag{6.110}$$

so that the integral finally becomes

$$I_m(A;BC)=\sqrt{\frac{3}{4\pi}}\frac{\boldsymbol{\varepsilon}_m\cdot\mathbf{k}_C}{\pi^{3/4}}\exp\left[-\frac{\mathbf{k}_C^2 R_A^2(R_B^2+R_C^2)}{8(R_A^2+R_B^2+R_C^2)}\right]$$

$$\times\left(\frac{2R_A R_B R_C}{R_A^2+R_B^2+R_C^2}\right)^{3/2}\left(\frac{2R_A^2+R_B^2+R_C^2}{R_A^2+R_B^2+R_C^2}\right)\delta(\mathbf{k}_B+\mathbf{k}_C) \tag{6.111}$$

In the cases where $R_B = R_C$ (e.g. $\rho \to \pi^+ \pi^-$, Le Youanc, 1973) define $x = R_A^2$, $r = R_B^2 = R_C^2$ and the integral may be written in the useful form

$$I_m = \sqrt{\left(\frac{3}{4\pi}\right)} \frac{\boldsymbol{\varepsilon}_m \cdot \mathbf{k}}{\pi^{3/4}} 2\sqrt{8} [R_A^{-3/4} F(x, r, k^2)]|_{x = R_A^2, r = R_{B,C}^2} \quad (6.112)$$

with

$$F(x, r, k^2) \equiv \left(\frac{xr}{x+2r}\right)^{3/2} \frac{x+r}{x+2r} \exp\left[\frac{-k^2 xr}{4(x+2r)}\right] \quad (6.113)$$

6.6.1 DECAYS OF RADIALLY EXCITED STATES

The above calculation would have applied to the decay $\psi \to D\bar{D}$ if the decay had been kinematically allowed. Barbieri *et al.* (1975) have used the model to study $\psi'' \to D\bar{D}$ where ψ'' is the second radial excitation of ψ. All that needs modification relative to the example above is that ψ_A is the relevant radial as against the ground state wavefunction. Instead of calculating the integral afresh it is more immediate to notice that the radial wavefunctions can be related to the ground state by differentiation.

For example

$$[\psi_{2s}(k)(\pi R^2)^{3/4}] \equiv \sqrt{\frac{3}{2}}\left(1 - \frac{4}{3} R^2 \frac{d}{dR^2}\right)[\psi_{1s}(k)(\pi R^2)^{3/4}] \quad (6.114)$$

with the result that

$$I'_m \equiv \sqrt{\frac{3}{2}}\left(1 - \frac{4}{3} x \frac{d}{dx}\right) I_m \quad (6.115)$$

where I_m is given as a function of x in equation (6.112). Similarly for the second radial excitation one has (Barbieri *et al.*, 1975)

$$I''_m \equiv \sqrt{\left(\frac{15}{8}\right)}\left(1 - \frac{8}{15} x \frac{d}{dx} + \frac{16}{15} x^2 \frac{d^2}{dx^2}\right) I_m \quad (6.116)$$

6.6.2 $L = 1$ MESON DECAYS AND $B \to \omega \pi$

Having developed the basic ideas of the model we can study now the $SU(6)_W$ breaking in decays of $L = 1$ mesons. To recapitulate: the problem of interest is the helicity structure of decays like $B \to \omega \pi$. The final pair each have $L_z = 0$ and L_z conservation $(SU(6)_W \times 0(2)_{L,z})$ would therefore require B to have $L_z = 0$ and hence $J_z = 0$. The B in $J_z = \pm 1$ helicity would therefore be forbidden to decay into $\omega\pi$ (contrary to data) (section 6.3.1).

In the pair creation model the $\omega\pi$ have $L_z = 0$ but the ${}^3P_0$ can have $m \equiv L_z = \pm 1$ or 0. The B with helicity $J_z = \pm 1$ therefore decays with an $m = \pm 1$ ${}^3P_0$ pair, while the $J_z = 0$ decays with an $m = 0$ pair. The restriction to the latter in the Carlitz–Kislinger model (section 6.3) generated the $SU(6)_W$ vertex symmetry of that model and forbade the $J_z = \pm 1$ decay of $B \to \omega\pi$. The $m = \pm 1$ pair break the $SU(6)_W$ vertex symmetry. We shall therefore be interested in the relative size of the breaking and conserving amplitudes, or equivalently

$$\frac{I_{m=\pm 1}(A; BC)}{I_{m=0}(A; BC)} \tag{6.117}$$

In contrast to the previous examples, the initial particle A is now an $L = 1$ state. If it has $m \equiv J_z = \pm 1$ or 0 then its wavefunction is (take $\pi \equiv$ particle labelled B in $A \to BC$)

$$\psi_A^m(\mathbf{k} + \mathbf{k}_\pi) = \left(\frac{R_A^2}{\pi}\right)^{3/4} i \sqrt{\frac{2}{3}} R_A Y_1^m(\mathbf{k} + \mathbf{k}_\pi) \exp\left[-\frac{R_A^2(\mathbf{k} + \mathbf{k}_\pi)^2}{8}\right] \tag{6.118}$$

and so we are interested in

$$\frac{I_1}{I_0} = \frac{\int d^3\mathbf{k}\, Y_1^1(\mathbf{k}_B - \mathbf{k}) Y_1^{-1}(\mathbf{k} + \mathbf{k}_\pi) \exp\{-[(\mathbf{k} + \mathbf{k}_\pi)^2 R_A^2 - \mathbf{k}^2(R_\pi^2 + R_\omega^2)]/8\}}{\int d^3\mathbf{k}\, Y_1^0(\mathbf{k}_B - \mathbf{k}) Y_1^0(\mathbf{k} + \mathbf{k}_\pi) \exp\{-[(\mathbf{k} + \mathbf{k}_\pi)^2 R_A^2 - \mathbf{k}^2(R_\pi^2 + R_\omega^2)]/8\}} \tag{6.119}$$

The same transformation to k' as the dummy variable is used as before (equation 6.105). The spherical harmonics are

$$Y_1^{\pm m}(\mathbf{k}_\pi \mp \mathbf{k}) = -\varepsilon_{\pm m} \sqrt{\frac{3}{4\pi}} \cdot (\mathbf{k}_\pi\{1 \pm [R_B^2/(R_\pi^2 + R_B^2 + R_\omega^2)]\} \mp \mathbf{k}') \tag{6.120}$$

If we choose to define the quantisation axis by $\hat{\mathbf{k}}_B$ then

$$Y_1^1(\mathbf{k}_\pi - \mathbf{k}) Y_1^{-1}(\mathbf{k} + \mathbf{k}_\pi) = \frac{3}{4\pi}\left\{\mathbf{k}_\pi^2\left(1 - \frac{R_B^4}{(R_\pi^2 + R_B^2 + R_\omega^2)}\right) - k_z'^2 \right.$$
$$\left. + \text{parity odd terms}\right\} \tag{6.121}$$

$$Y_1^0(\mathbf{k}_\pi - \mathbf{k}) \, Y_1^0(\mathbf{k} + \mathbf{k}_\pi) = \frac{3}{4\pi} \left\{ -\frac{(k_x'^2 + k_y'^2)}{2} + \text{parity odd terms} \right\} \quad (6.122)$$

Noting that $k_z'^2 = k_x'^2 = k_y'^2 \equiv \tfrac{1}{3}\mathbf{k}'^2$ in the integral then it is straightforward to integrate the exponentials and you should verify that

$$\left(\frac{I_0}{I_1} \right)_{B \to \omega\pi} = 1 - \frac{\mathbf{k}_\pi^2(R_\pi^2 + R_B^2 + R_\omega^2)}{4} \left\{ 1 - \frac{R_B^4}{(R_B^2 + R_\pi^2 + R_\omega^2)^2} \right\}$$
$$(6.123)$$

This result follows at once by exploiting the fact that

$$\int k^4 \, dk \, \exp(-ak^2) = \frac{3}{2a} \int k^2 \, dk \, \exp(-ak^2) \quad (6.124)$$

where $a \equiv \tfrac{1}{8}\Sigma R^2$ in our example.

If the B, ω and π are assumed to have the same size R, then

$$\left(\frac{I_0}{I_1} \right)_{B \to \omega\pi} = 1 - \frac{2}{3} R^2 \mathbf{k}_\pi^2 \quad (6.125)$$

The magnitude of R in a harmonic oscillator potential is related to the separation in mass between S and P, P and D etc. levels. With $R^2 \simeq 8 \, \text{GeV}^{-2}$ (Le Youanc *et al.*, 1973) then

$$\left(\frac{I_0}{I_1} \right)_{B \to \omega\pi} = 0.36 \quad (6.126)$$

which is consistent with the observed value (Ascoli *et al.*, 1968; Werbrouck *et al.*, 1970) of $0.47^{+0.20}_{-0.30}$ but somewhat smaller than more recent data (Ascoli *et al.*, 1970) which give 0.68 ± 0.12. In any event the model and data are agreed that the $SU(6)_W \, J_z = \pm 1$ forbidden amplitude in fact *dominates* over the allowed $J_z = 0$.

7 Electromagnetic Interactions and Radiative Transitions in Explicit Models

The question of how quarks contribute to radiative transitions between hadrons has been investigated for many years. The standard hypothesis is that a single quark in the hadron emits or absorbs a photon and the transition $B_1 \to B_2 \gamma$ is thereby triggered, a calculation of the amplitude involving a sum over all quarks and antiquarks in the hadrons concerned. The central point of controversy has been the nature of the quark's electromagnetic interaction.

Early nonrelativistic models imagined the quarks to have a large mass with the consequence that they had large anomalous moments (Becchi and Morpurgo, 1965a,b; Faiman and Hendry, 1969). Such a picture could be motivated intuitively by imagining each quark to be surrounded by a meson cloud due to the strong interactions that are responsible for its binding within the hadron. For interactions of the quark with long wavelength electromagnetic waves this internal structure is not probed, consequently one views the quark as having these charge and magnetic moment distributions attached rigidly to it.

Calculations of resonance photoproduction, on the other hand, seemed to work very well phenomenologically if the quark was light, of the order 300 MeV (Copley *et al.*, 1969a,b). With such a mass the nonrelativistic nature of the model becomes less clear. Also, the discovery of scale invariance in deep inelastic electron scattering (Chapter

9) where very short wavelength photons interact with the quarks, suggested that the quarks indeed had very small masses and moreover had pointlike interactions with photons, i.e. were Dirac fermions.

7.1 Quark spin flip matrix elements

7.1.1 STATIC MAGNETIC MOMENTS

The quark magnetic moments were originally assumed to be described by a single parameter μ with a normalisation such that the magnetic moment operator for the quarks has the form (Becchi and Morpurgo, 1965b; Dalitz, 1965)

$$\mathbf{M} = \mu\left(\tfrac{2}{3}\boldsymbol{\sigma}_\mathrm{u} - \tfrac{1}{3}\boldsymbol{\sigma}_\mathrm{d} - \tfrac{1}{3}\boldsymbol{\sigma}_\mathrm{s}\right) \tag{7.1}$$

The proportionality of M_i and e_i is a consequence of SU(3) invariance. In particular, the d and s are in the same U-spin multiplet (having the same charge) and hence have the same electromagnetic properties if the electromagnetic current is a U-spin scalar; consequently

$$\mathbf{M}_\mathrm{d} = \mathbf{M}_\mathrm{s} \tag{7.2}$$

Furthermore, if

$$j_{\mathrm{e.m.}} = AI_3 + BY \tag{7.3}$$

then for arbitrary A, B we have

$$\mathbf{M}_\mathrm{u} + \mathbf{M}_\mathrm{d} + \mathbf{M}_\mathrm{s} = 0 \tag{7.4}$$

after summing over the triplet since

$$\sum_{i=1}^{3} I_3 = \sum_{i=1}^{3} Y = 0 \tag{7.5}$$

Hence the single scale parameter μ related the M_i to e_i.

Since SU(3) is not exact then one might question the validity of the assumption in equation (7.1). The effect of mass breaking upon the magnetic moment operator and ensuing calculations is discussed later.

Morpurgo (1965) calculated the proton magnetic moment by taking the expectation value of M between proton wavefunctions (Tables 3.7 and 4.3). The result is obtained by exploiting the overall symmetry of the wavefunction so that

$$\sum_{i=1}^{3} \langle p|M_z^i|p \rangle \equiv 3\mu \langle p|e^{(3)}\sigma_z^{(3)}|p \rangle \tag{7.6}$$

The calculation is now straightforward as follows.

For the **56**, $L = 0$ state

$$p \equiv \frac{1}{\sqrt{2}} \psi_S(\phi_{M,S}\chi_{M,S} + \phi_{M,A}\chi_{M,A}) \tag{7.7}$$

Since $\langle \psi_S | \psi_S \rangle = 1$ and the expectation values of $e^{(3)}$ and $\sigma_z^{(3)}$ are

$$\langle \phi_{M,A}|e^{(3)}|\phi_{M,A} \rangle = \tfrac{2}{3}(\text{proton}); \ -\tfrac{1}{3}(\text{neutron}) \tag{7.8}$$

$$\langle \phi_{M,S}|e^{(3)}|\phi_{M,S} \rangle = 0(\text{proton}); \ +\tfrac{1}{3}(\text{neutron}) \tag{7.9}$$

$$\langle \chi_{M,A}^{\uparrow}|\sigma_z^{(3)}|\chi_{M,A}^{\uparrow} \rangle = 1 \tag{7.10}$$

$$\langle \chi_{M,S}^{\uparrow}|\sigma_z^{(3)}|\chi_{M,S}^{\uparrow} \rangle = -\tfrac{1}{3} \tag{7.11}$$

then we have

$$\mu_P \equiv 3\langle p|e^{(3)}\sigma_z^{(3)}|p \rangle = \mu\tfrac{3}{2}(\tfrac{2}{3} \times 1 + 0 \times \tfrac{1}{3}) = \mu \tag{7.12}$$

Consequently the quark scale moment is identical to the proton's magnetic moment in this nonrelativistic calculation. This means that

$$\mu = 2 \cdot 79 \frac{e}{2m} (h = c = 1) \tag{7.13}$$

and so

$$\frac{2 \cdot 79}{m_P} = \frac{g_q}{m_q} \tag{7.14}$$

with the consequence that for a massive quark a large g factor or anomalous moment is required. Conversely, a Dirac quark would require a mass of around 340 MeV and the nonrelativistic nature of the problem would be unclear (see also the discussion of section 18.1.2 for the modern viewpoint on this topic).

The corresponding calculation for the neutron yields

$$\mu_N \equiv 3\langle n|e^{(3)}\sigma_z^{(3)}|n \rangle = \mu\tfrac{3}{2}(-\tfrac{1}{3} \times 1 + \tfrac{1}{3} \times (-\tfrac{1}{3})) = -\tfrac{2}{3}\mu \tag{7.15}$$

and so there is the parameter free prediction of the quark model that

$$\frac{\mu_P}{\mu_N} = -\frac{3}{2} \tag{7.16}$$

to be compared with the data,

$$-\frac{2 \cdot 79}{1 \cdot 91} \equiv -1 \cdot 46 \tag{7.17}$$

This result had been found earlier by Beg *et al.* (1964) within the framework of SU(6) symmetry. In the quark model it arises from the fact that the spin–unitary spin wavefunctions of the nucleons are totally symmetric (equation 4.29). There is no need for SU(6) symmetry to be exact in the quark model derivation since the spatial wavefunctions ψ_S for proton and neutron could be quite different.

From the above discussion it is a straightforward exercise to calculate μ_Λ. You should do this and confirm that

$$\mu_\Lambda = -\tfrac{1}{3}\mu \tag{7.18}$$

and hence that

$$\frac{\mu_\Lambda}{\mu_P} = -\frac{1}{3} \tag{7.19}$$

In the early days of the quark model this was compared with the extant data $\mu_\Lambda/\mu_P = -0 \cdot 29 \pm 0 \cdot 05$ and was satisfactory (Combe *et al.*, 1966). Since that time data have improved and today we believe that this ratio is (Particle Data Group, 1976)

$$\frac{\mu_\Lambda}{\mu_P} \simeq -0 \cdot 24 \pm 0 \cdot 02 \tag{7.20}$$

and a less satisfactory situation has arisen. This is a reflection of SU(3) symmetry breaking and it seems possible to systematically accommodate this.

If the quark masses $m_{u,d}$ are light, around $340\,\text{MeV}$, then the separation in mass between $\Delta, \Sigma^*, \Xi^*, \Omega^-$ suggests that the strange quark is somewhat more massive, $m_s \simeq 500\,\text{MeV}$ say. Because the quark magnetic moment operators are inversely proportional to their masses

$$\frac{\mu_d}{\mu_s} = \frac{m_s}{m_d} \tag{7.21}$$

This enables the smaller value of μ_Λ to be "understood". This use of light masses is discussed in sections 15.2.1, 17.2.3 and 18.1.

7.1.2 TRANSITION MAGNETIC MOMENTS

Returning to the case of nonstrange baryons we study the transition $\gamma P \to \Delta^+$ (1236). The transition of $\frac{1}{2}^+$ to $\frac{3}{2}^+$ can take place by interaction with either the M1 or E2 multipoles (the magnetic dipole or electric quadrupole respectively).

Becchi and Morpurgo (1965a) noted that in the quark model the E2 transition is forbidden, in line with the data. This is essentially because the E2 transition is proportional to the charge operator which cannot cause transitions between the quark spin $\frac{1}{2}$ and $\frac{3}{2}$ states and hence the matrix element vanishes by orthogonality of the quark spin wavefunctions. Furthermore the E2 transition involves the spherical haromic Y_2 which cannot lead to transitions between $L = 0$ spatial wavefunctions (recall that the proton and $\Delta(1236)$ are both $L = 0$ in the quark model).

The M1 transition involves the quark magnetic moments—hence the spin operator in equation (7.1)—and this can lead to transitions between $S = \frac{1}{2}$ and $S = \frac{3}{2}$ (Dalitz and Sutherland, 1966). The matrix element is

$$\mu_{P\Delta}^* = \langle p, m = +\tfrac{1}{2}|\mu_z|\Delta, m = +\tfrac{1}{2}\rangle \qquad (7.22)$$

The wavefunctions for Δ are given in Tables 3.4 and 4.3 (equation 4.26) and a straightforward exercise will show that

$$\mu_{P\Delta}^* = \frac{2\sqrt{2}}{3}\mu_P \qquad (7.23)$$

This calculation would only be realistic in a world where the proton and Δ were degenerate. For nondegenerate particles the effects of recoil in transitions like $\gamma N \to N^*$ have to be taken into account, and this brings us to the next stage in the development of transition studies which is the incorporation of recoil in a systematic fashion. Before entering the arena we should note that there exists a selection rule which may be hoped to be independent of this question.

7.1.3 MOORHOUSE'S SELECTION RULE

Moorhouse (1966) noticed that photoexcitation of *protons* to states N^* in octets with quark spin $\frac{3}{2}$ is predicted to vanish. A particular example of such a state is the D_{15} (1690) which belongs to $^4\mathbf{8}$ in the 70plet of SU(6).

The selection rule is a direct result of the vanishing matrix element in equation (7.9) for *protons*.

The matrix element for $\gamma p \to N^*$ ($S = \frac{3}{2}$) necessarily involves the spin flip, magnetic interaction in order to give a transition from the nucleon (quark spin $= \frac{1}{2}$) to the $^4\mathbf{8}$ with quark spin total of $\frac{3}{2}$. The wavefunction for the intital 56plet is

$$|56, {}^2\mathbf{8}\rangle = \frac{1}{\sqrt{2}} \psi_{00}^S (\phi_{M,S}\chi_{M,S} + \phi_{M,A}\chi_{M,A}) \tag{7.24}$$

while the final state is

$$|70, {}^4\mathbf{8}\rangle = \frac{1}{\sqrt{2}} (\psi_{M,S}\phi_{M,S} + \psi_{M,A}\phi_{M,A})\chi_S \tag{7.25}$$

The spin flip can only reach the χ_S from the initial $\chi_{M,S}$ since the matrix element $\langle \chi_S | \sigma_+^{(3)} | \chi_{M,A} \rangle$ vanishes identically. Hence the transition is proportional to the expectation value of $\langle \phi_{M,S} | e^{(3)} | \phi_{M,S} \rangle$. This is finite for neutron targets but vanishes for protons. Hence the whole matrix element vanishes and the Moorhouse selection rule follows.

7.2 Quark model amplitudes for resonance photoexcitation

The basic assumption is that a single quark absorbs the photon and leads to excitation of the system. The nonrelativistic form of the interaction is written by analogy with nuclear physics (Becchi and Morpurgo, 1965a,b; Faiman and Hendry, 1969; Copley, Karl and Obryk, 1969)

$$\mathcal{H} = \sum_{j=1}^{3} \mathbf{J}_j . \mathbf{A}(\mathbf{r}_j) = \sum_{j=1}^{3} e^{(j)} [-2ig\mathbf{s}^j . \mathbf{k} \times \mathbf{A} + (\mathbf{p}^j + \mathbf{p}'^j) . \mathbf{A}] \frac{e}{2m_q} \tag{7.26}$$

where $ee^{(j)}$, $s^{(j)}$ are the charge and spin operators of the j-th quark ($\mathbf{s} = \frac{1}{2}\boldsymbol{\sigma}$ where $\boldsymbol{\sigma}$ are the Pauli matrices of section 2.2), $\mathbf{p}(\mathbf{p}')$ are the initial and final momenta of the quark that has absorbed the photon of momentum $\mathbf{k}$, $ge e_j/2m$ is the magnetic moment of the bound quark. The quantity $eg/2m$ is equal to μ, the quark scale magnetic moment which is taken to be equal to the proton's magnetic moment, and hence $\mu = 0.13 \text{ GeV}^{-1}$.

The electromagnetic field of a photon of momentum **k** and polarisation **ε** is given by

$$\mathbf{A}(\mathbf{r}_j) = \sqrt{4\pi} \frac{1}{\sqrt{2k_0}} \boldsymbol{\varepsilon} [a_k^+ \exp(i\mathbf{k} \cdot \mathbf{r}_j) + a_k \exp(-i\mathbf{k} \cdot \mathbf{r}_j)] \quad (7.27)$$

where **ε** is a unit vector of polarisation. The computation is greatly simplified if the photon momentum **k** is chosen as the quantisation axis (z-axis). A real photon is polarised transverse to this axis since **k** . **A** = 0. Without any loss of generality one can focus attention on right-handed photons (photons with helicity +1) for which

$$\boldsymbol{\varepsilon} = -\frac{1}{\sqrt{2}}(1, i, 0) \quad (7.28)$$

As usual we take advantage of the overall symmetry of the SU(6)× 0(3) baryon wavefunction to write

$$\mathcal{H}' = \sum_{j=1}^{3} \mathbf{J}_j \cdot \mathbf{A}(\mathbf{r}_j) = 3\mathbf{J}_{(3)} \cdot \mathbf{A}(\mathbf{r}_3) \quad (7.29)$$

and hence

$$\mathcal{H}' = 3\sqrt{2} \sqrt{\frac{\pi}{k}} \mu [e^{(3)} \exp(-ikz^{(3)})]$$

$$\times \left\{ 2i(s_x \varepsilon_y - s_y \varepsilon_x)k + \frac{2}{g}\left(-\frac{1}{\sqrt{2}}\right)(p_x + ip_y)^{(3)} \right\} \quad (7.30)$$

or

$$\mathcal{H}' = 6\sqrt{\frac{\pi}{k}} \mu [e^{(3)} \exp(-ikz^{(3)})] \left\{ k(s_x + is_y)^{(3)} - \frac{1}{g}(p_x + ip_y)^{(3)} \right\} \quad (7.31)$$

The first term is the magnetic interaction with a quark and flips the quark's spin projection , s_z, by one unit whereas the second term is the interaction of the field with the quark's convection current and raises the L_z of the system by one unit. For the reader already familiar with some of the literature on the Melosh transformation, these terms are the explicit model realisation of the general structure BS_+ and AL_+ respectively (see also section 7.3 and equation 7.112 et seq.).

One can now compute the matrix elements of $\mathcal{H}'$ between nucleon states with $J_z = \pm\frac{1}{2}$ and resonance states with $J_z = \frac{3}{2}$ and $\frac{1}{2}$. These are the helicity amplitudes $A_{3/2}$ and $A_{1/2}$ respectively, in terms of which the radiative width of the resonance can be calculated

$$\Gamma_\gamma = \frac{k^2}{4\pi} \frac{m_N}{m_R} \frac{8}{2J+1}\{|A_{1/2}|^2 + |A_{3/2}|^2\} \qquad (7.32)$$

The contribution of a resonance to the total cross-section for single pion photoproduction is given by

$$\sigma_T(\gamma p \to N^* \to \pi^0 p) = \left\{\begin{matrix}\frac{1}{3}\\\frac{2}{3}\end{matrix}\right\} \frac{m_N}{m_R} \frac{x}{\Gamma} 2\{|A_{1/2}|^2 + |A_{3/2}|^2\} \qquad (7.33)$$

where x and Γ are elasticity and total resonance width and the $\frac{1}{3}, \frac{2}{3}$ factors correspond to $I = \frac{1}{2}, \frac{3}{2}$ respectively.

The data on pion photoproduction have been analysed and $A_{1/2}, A_{3/2}$ for several resonances extracted with both proton and neutron targets. A detailed calculation of these amplitudes has been made by Copley *et al.* (1969a,b) and of Γ_γ by Faiman and Hendry (1969). To illustrate the method we will consider two interesting examples, namely photo-production of $D_{13}(1520)\{^28$ of **70** with $L = 1\}$ and $F_{15}(1690)\{^28$ of **56** with $L = 2\}$ which are the most prominently photoproduced resonances apart from the $P_{33}(1235)$. These are some interesting properties in the photoproduction of these states which have not been understood outside of the quark model discussion which follows. These phenomena are seen in Figs 1, 9, 11 of Walker (1969).

In $\gamma p \to \pi^+ n$ the total cross-section shows three clear peaks which are predominantly $P_{33}(1235)$, $D_{13}(1520)$, $F_{15}(1690)$ respectively. However, for the colinear production $(\theta_{\pi^+} = 0^0)$ the first peak is seen but the second and third have vanished. The total cross-section is proportional to

$$\sigma_T \sim |A_{1/2}|^2 + |A_{3/2}|^2 \qquad (7.34)$$

whereas

$$\sigma(\theta = 0^0) \sim |A_{1/2}|^2 \qquad (7.35)$$

because the $A_{3/2}$ cannot contribute for kinematic reasons in this configuration (the pion has $J = 0$ and so only $J_z = \frac{1}{2}$ can appear in the final colinear system). Consequently one infers that the excitation of the D_{13} and F_{15} off protons is entirely in the $A_{3/2}$ mode, i.e.

$$A_{1/2}^P(D_{13}, F_{15}) \simeq 0 \qquad (7.36)$$

In photoproduction off neutrons, on the other hand, one finds a suppression of the F_{15} peak at $90°$, i.e. the excitation is dominantly $A_{1/2}$. In fact detailed analysis suggests

$$A^N_{3/2}(F_{15}) \simeq 0 \tag{7.37}$$

This result turns out to be a selection rule flowing from the interaction equation (7.31). The results $A^P_{1/2} \simeq 0$ for D_{13} and F_{15} are more subtle and will be dealt with later.

7.2.1   WHY IS $A_{3/2}(F_{15}) \cong 0$?

The F_{15} has $L = 2$ and $S = \frac{1}{2}$: consequently to have $J_z = \frac{3}{2}$ it must have $L_z \neq 0$. The nucleon has $L_z = 0$ and so the transition can only proceed by L_z flip, i.e. by the $p_x + ip_y$ term in $\mathcal{H}'$

$$\mathcal{H}' = 6\sqrt{\frac{\pi}{k}} \mu [e^{(3)} \exp (-ikz^{(3)})] \left\{ k(s_x + is_y)^{(3)} - \frac{1}{g}(p_x + ip_y)^{(3)} \right\} \tag{7.38}$$

This interaction in spin and unitary space is therefore proportional to the charge operator. The matrix element of the charge operator between neutron and F_{15} is zero because the F_{15} is in $^2\mathbf{8}$ of a 56plet and so has the same behaviour under $SU(6)$ as does the neutron. The interaction is therefore proportional to the neutron's charge and hence zero;

$$\langle F^0_{15} | e_{(3)} | n \rangle \equiv \langle n | e_{(3)} | n \rangle = 0 \tag{7.39}$$

Notice that if there had been a spin–orbit interaction in $\mathcal{H}'$ then there could be an L_z flip accompanied by a spin operator (the C and D terms in the general formulation in equation 7.68). In this case the transition would no longer be proportional to the neutron charge due to the presence of the spin operator, and so the matrix element would not necessarily vanish. There is some evidence to suggest that $A^N_{3/2}$ is small but nonzero in which case such spin–orbit terms may be necessary. This question is of interest in connection with higher-order relativistic effects and the Melosh transformation.

7.2.2 WHY ARE $A^P_{1/2}(D_{13}, F_{15}) \simeq 0$?

The vanishing of these amplitudes does not follow from any selection rule and has been one of the most interesting and controversial topics in the quark model applied to resonance production. That the quark model gave a possible explanation of these vanishing amplitudes was first noticed by Copley et al. (1969a,b). Indeed, this work stimulated Walker to devote his rapporteur talk at the 1969 International Conference on Electron and Photon Interactions to the quark model and to place particular emphasis on this result (Walker, 1969). In turn the work of Feynman et al. (1971) was stimulated and with the appearance of the latter work, a rejuvenation of interest in the quark model took place.

The production amplitudes for the D_{13} in particular are very instructive for appreciating much of the subsequent development of the subject. We will therefore study them in some detail. This will also act as an explicit model example of how to calculate matrix elements in the quark model.

The intention is to calculate matrix elements of $\mathcal{H}'$ (equation 7.31) between quark wavefunctions for **56**, $L = 0$ with $J_z = \pm\tfrac{1}{2}$ and **70**, L, $J_z = \tfrac{3}{2}, \tfrac{1}{2}$. For the **56** the wavefunction reads

$$({}^2\mathbf{8}, \mathbf{56}) = \frac{1}{\sqrt{2}} \psi^S_{00}(\phi^{M,S}\chi^{M,S}_{\pm 1/2} + \phi^{M,A}\chi^{M,A}_{\pm 1/2}) \qquad (7.40)$$

where the subscripts on ψ are LL_z and the subscript on χ is S_z. ϕ is the unitary wavefunction for proton or neutron and is exhibited in Table 3.7. For the $({}^2\mathbf{8}, \mathbf{70})$ we have

$$({}^2\mathbf{8}, \mathbf{70})\psi_{JJ_z} = \sum_{L_z + S_z = J_z} \langle JJ_z | LL_z, SS_z \rangle \frac{1}{2} \Big\{ \psi^{M,S}_{LL_z}(\phi^{M,S}\chi^{M,S} - \phi^{M,A}\chi^{M,A})$$

$$+ \psi^{M,A}_{LL_z}(\phi^{M,S}\chi^{M,A} + \phi^{M,A}\chi^{M,S}) \Big\} \qquad (7.41)$$

where the coupling of the quarks orbital and spin angular momenta to give the total J are explicitly taken account of by the Clebsch–Gordan coefficient (our convention is that of the Particle Data Group, 1976).

We have already exploited the overall symmetry of the wavefunction which has the consequence that, by definition, only quark number three is participating. Hence the excitation of ψ_{00}^S to $\psi^{M,A}$ does not take place since

$$\mathbf{M}_A \equiv \boldsymbol{\rho} = \frac{1}{\sqrt{2}}(\mathbf{r}_1 - \mathbf{r}_2) \tag{7.42}$$

This is a great simplification as now only the

$$\psi^{M,S}(\phi^{M,S}\chi^{M,S} - \phi^{M,A}\chi^{M,A}) \tag{7.43}$$

piece of the **70** wavefunction will be operative.

First of all we will calculate the amplitude $A_{3/2}$. To reach $J_z = \frac{3}{2}$ for the states in $^2\mathbf{8}$ necessitates $L_z \neq 0$ since the maximum projection of the quarks intrinsic spin is $+\frac{1}{2}$. For the D_{13} with $L = 1$, the $L_z = 1$ is the only possibility available. Consequently only the $p_x + ip_y$ piece of the interaction contributes to this $A_{3/2}$ amplitude. Explicitly

$$A_{3/2}^{P,N} = 6\sqrt{\frac{\pi}{k}}\mu\frac{1}{2\sqrt{2}}\langle J\tfrac{3}{2}|11, \tfrac{11}{22}\rangle(\psi_{11}^{M,S}|\exp(-ikz^{(3)})|\psi_{00}^S)\left(-\frac{1}{g}\right)$$
$$\times [\langle\phi^{M,S}|e^{(3)}|\phi^{M,S}\rangle - \langle\phi^{M,A}|e^{(3)}|\phi^{M,A}\rangle] \tag{7.44}$$

where we have used the orthonormality of the quark spinor wavefunctions $\chi^{M,A;M,S}$. For proton and neutron we have from equations (7.8), (7.9) that

$$\langle\phi^{M,S}|e^{(3)}|\phi^{M,S}\rangle = 0(P), \tfrac{1}{3}(N), \tag{7.45}$$

$$(\phi^{M,A}|e^{(3)}|\phi^{M,A}\rangle = \tfrac{2}{3}(P), -\tfrac{1}{3}(N). \tag{7.46}$$

If we define

$$R_{11}^{M,S} \equiv \langle\psi_{11}^{M,S}|\exp(-ikz^{(3)})(p_x + ip_y)^{(3)}|\psi_{00}^S\rangle \tag{7.47}$$

which is a quantity that will depend upon the specific spatial wavefunctions chosen, then

$$A_{3/2}^P = \sqrt{2}\mu g^{-1}\sqrt{\frac{\pi}{k}}R_{11}^{M,S} \tag{7.48}$$

$$A_{3/2}^N = -\sqrt{2}\mu g^{-1}\sqrt{\frac{\pi}{k}}R_{11}^{M,S} \tag{7.49}$$

Notice that there is a relation independent of the particular choice of wavefunction, namely

$$A_{3/2}^P = -A_{3/2}^N \tag{7.50}$$

This follows from the SU(6) states (^{28} 70 and ^{28} 56) together with the transformation of the photon interaction ($\mathscr{H}'$) being $(p_x + ip_y)^{(3)}$ (like L_+). Compare this with the general approach (equation 7.68 and Table 7.1) setting the spin–orbit interaction, C, to zero.

TABLE 7.1

Helicity amplitudes for photoexcitation of the $L = 1$, **70** by the general current $J_+ \sim AL_+ + BS_+ + CS_zL_+$. For SU(3) **10** states the proton and neutron amplitudes are equal. The $(ABC)_{01}$ refer to the expectation values of the J_+ operator ABC between $L = 0$ and $L = 1$ orbital wavefunctions.

	$A^P_{1/2}$	$A^P_{3/2}$	$A^N_{1/2}$	$A^N_{3/2}$
$S_{11}(^28)$	$\frac{1}{\sqrt{3}}(A_{01} - C_{01} + B_{01})$	—	$-\frac{1}{\sqrt{3}}\left(A_{01} - \frac{C_{01}}{3} + \frac{B_{01}}{3}\right)$	
$D_{13}(^28)$	$\frac{1}{\sqrt{6}}(A_{01} - C_{01} - 2B_{01})$	$\frac{1}{\sqrt{2}}(A_{01} + C_{01})$	$-\frac{1}{\sqrt{6}}\left(A_{01} - \frac{C_{01}}{3} - \frac{2B_{01}}{3}\right)$	$-\frac{1}{\sqrt{2}}\left(A_{01} + \frac{C_{01}}{3}\right)$
$S_{11}(^48)$	0	—	$\frac{1}{3\sqrt{3}}(B_{01} - C_{01})$	—
$D_{13}(^48)$	0	0	$\frac{1}{3\sqrt{15}}(B_{01} - 4C_{01})$	$\frac{1}{\sqrt{5}}\left(B_{01} - \frac{2}{3}C_{01}\right)$
$D_{15}(^48)$	0	0	$-\frac{1}{\sqrt{15}}(B_{01} + C_{01})$	$-\sqrt{\frac{2}{15}}(B_{01} + C_{01})$
$S_{31}(^210)$	$\frac{1}{\sqrt{3}}\left(A_{01} + \frac{C_{01}}{3} - \frac{B_{01}}{3}\right)$	—		
$D_{33}(^210)$	$\frac{1}{\sqrt{6}}\left(A_{01} + \frac{C_{01}}{3} + \frac{2B_{01}}{3}\right)$	$\frac{1}{\sqrt{2}}\left(A_{01} - \frac{C_{01}}{3}\right)$		

If the photon had transformed as a vector meson then it would have had a $^3S_1$ $q\bar{q}$ structure. The symmetry group $SU(6)_W \otimes 0(2)_{L,z}$ applied to this transition would forbid the photoproduction of D_{13} with helicity $\frac{3}{2}$. This result is an immediate consequence of the L_z conservation of the above group symmetry, as discussed in section 6.3. In the explicit quark model, the amplitudes $A_{3/2}$ are not zero (equations 7.48 and 7.49). The reason of course is that the second term in the interaction operator equations (7.26) and (7.31), flips L_z and violates $SU(6)_W$ vertex symmetry.

We now turn to study the $A_{1/2}$ amplitude. This can proceed by either the $p_x + ip_y$ term or by the S_+ term in equation (7.31). If the initial state has $J_z = S_z = -\frac{1}{2}$ then the former term leads to a final state with $J_z = +\frac{1}{2}$ comprising $L_z = -1$ (orbital flip) while the latter produced the $J_z = +\frac{1}{2}$ state with $L_z = 0$ and $S_z = +\frac{1}{2}$ (spin flip). If we define

$$R_{10}^{\mathrm{M,S}} \equiv \langle \psi_{10}^{\mathrm{M,S}} | \exp(-ikz^{(3)}) | \psi_{00}^{\mathrm{S}} \rangle \tag{7.51}$$

then the matrix element becomes

$$A_{1/2}^{\mathrm{P,N}} = 6\sqrt{\frac{\pi}{k}}\mu \frac{1}{2\sqrt{2}} \Big\{ \langle J\tfrac{1}{2} | 10, \tfrac{1}{2}\tfrac{1}{2} \rangle R_{10}^{\mathrm{M,S}} k [\langle \phi^{\mathrm{M,S}} | e^{(3)} | \phi^{\mathrm{M,S}} \rangle \langle \chi_{1/2}^{\mathrm{M,S}} | S_+^{(3)} | \chi_{-1/2}^{\mathrm{M,S}} \rangle$$

$$- \langle \phi^{\mathrm{M,A}} | e^{(3)} | \phi^{\mathrm{M,A}} \rangle \langle \chi_{1/2}^{\mathrm{M,A}} | S_+^{(3)} | \chi_{-1/2}^{\mathrm{M,A}} \rangle]$$

$$+ \langle J\tfrac{1}{2} | 11, \tfrac{1}{2}-\tfrac{1}{2} \rangle R_{11}^{\mathrm{M,S}} \left(-\frac{1}{g} \right)$$

$$\times [\langle \phi^{\mathrm{M,S}} | e^{(3)} | \phi^{\mathrm{M,S}} \rangle - \langle \phi^{\mathrm{M,A}} | e^{(3)} | \phi^{\mathrm{M,A}} \rangle] \Big\} \tag{7.52}$$

where we have again used the fact that

$$\langle \chi^{\mathrm{M,A}} | S_+ | \chi^{\mathrm{M,S}} \rangle \equiv \langle \chi^{\mathrm{M,S}} | S_+ | \chi^{\mathrm{M,A}} \rangle = 0 \tag{7.53}$$

Since

$$\langle \chi_{1/2}^{\mathrm{M,A}} | S_+^{(3)} | \chi_{-1/2}^{\mathrm{M,A}} \rangle = 1 \tag{7.54}$$

and

$$\langle \chi_{1/2}^{\mathrm{M,S}} | S_+^{(3)} | \chi_{-1/2}^{\mathrm{M,S}} \rangle = -\frac{1}{3} \tag{7.55}$$

we have that

$$A_{1/2}\binom{\mathrm{P}}{\mathrm{N}} = \sqrt{2}\sqrt{\frac{\pi}{k}}\mu \Big\{ g^{-1} R_{11}^{\mathrm{M,S}} \langle J\tfrac{1}{2} | 11, \tfrac{1}{2}-\tfrac{1}{2} \rangle \binom{+1}{-1}$$

$$+ R_{10}^{\mathrm{M,S}} k \langle J\tfrac{1}{2} | 10, \tfrac{1}{2}\tfrac{1}{2} \rangle \binom{-1}{+\frac{1}{3}} \Big\} \tag{7.56}$$

From this one can choose $J = \frac{1}{2}(S_{11})$ or $J = \frac{3}{2}(D_{13})$. For the latter we have finally

$$D_{13} \begin{cases} A_{1/2}^{\mathrm{P}} = \sqrt{\dfrac{2}{3}}\sqrt{\dfrac{\pi}{k}}\mu [g^{-1} R_{11}^{\mathrm{M,S}} - \sqrt{2}k R_{10}^{\mathrm{M,S}}] \\[4mm] A_{1/2}^{\mathrm{N}} = \sqrt{\dfrac{2}{3}}\sqrt{\dfrac{\pi}{k}}\mu \left[-g^{-1} R_{11}^{\mathrm{M,S}} + \dfrac{\sqrt{2}}{3} k R_{10}^{\mathrm{M,S}} \right] \end{cases} \tag{7.57}$$

Notice that for $J = \frac{1}{2}(S_{11})$ one would have

$$S_{11} = \begin{cases} A_{1/2}^P = \sqrt{\frac{2}{3}}\sqrt{\frac{\pi}{k}}\mu[\sqrt{2}g^{-1}R_{11}^{M,S} + kR_{10}^{M,S}] \\[3mm] A_{1/2}^N = \sqrt{\frac{2}{3}}\sqrt{\frac{\pi}{k}}\mu[-\sqrt{2}g^{-1}R_{11}^{M,S} - \frac{1}{3}kR_{10}^{M,S}] \end{cases} \quad (7.58)$$

Hence there is the possibility of the $A_{1/2}$ amplitudes vanishing by cancellation of the orbital flip (R_{11}) and spin flip (R_{10}) for *either* the D_{13} or the S_{11} *but not both*.

Copley *et al* (1969a,b) explicitly evaluated the $R_{10}^{M,S}$, $R_{11}^{M,S}$ using harmonic oscillator wavefunctions. They obtained

$$R_{10}^{M,S} = \frac{ik}{\alpha\sqrt{3}}\exp\left(\frac{-k^2}{6\alpha^2}\right); \quad R_{11}^{M,S} = i\sqrt{\frac{2}{3}}\alpha\exp\left(\frac{-k^2}{6\alpha^2}\right) \quad (7.59)$$

The relevant experimental datum is the relative importance of $A_{1/2}$ and $A_{3/2}$. In the model calculation this becomes

$$\left(\frac{A_{1/2}}{A_{3/2}}\right)^P = -\frac{1}{\sqrt{3}}\left(1 - gk\sqrt{2}\frac{R_{10}}{R_{11}}\right) = -\frac{1}{\sqrt{3}}\left(1 - \frac{gk^2}{\alpha^2}\right) \quad (7.60)$$

This bracket vanishes immediately if $k^2 = \alpha^2/g$.

As an exercise it is useful to calculate the $A_{1/2}$ for $F_{15}(1690)$ which is $^2\mathbf{8}$ in $\mathbf{56}$, $L = 2$. Verify that

$$A_{1/2}^P = \frac{2\sqrt{2}}{\sqrt{5}}\sqrt{\frac{\pi}{k}}\mu\left[\sqrt{\frac{3}{2}}kR_{20} - g^{-1}R_{21}\right]$$

Using the harmonic oscillator wavefunctions one finds that

$$R_{20} = -\frac{1}{6}\sqrt{\frac{2}{3}}\left(\frac{k}{\alpha}\right)^2\exp\left(\frac{-k^2}{6\alpha^2}\right); \quad R_{21} = -\frac{k}{3}\exp\left(\frac{-k^2}{6\alpha^2}\right)$$

Compute also the $A_{3/2}$ and verify that the crucial ratio of amplitudes in this case is

$$\left(\frac{A_{1/2}}{A_{3/2}}\right)^P = -\frac{1}{\sqrt{2}}\left(1 - gk\sqrt{\frac{3}{2}}\frac{R_{20}}{R_{21}}\right) = -\frac{1}{\sqrt{2}}\left(1 - \frac{gk^2}{2\alpha^2}\right)$$

Comparing equation (7.63) with equation (7.60) we see that the $A_{1/2}$ can vanish for *both* D_{13} and F_{15} if

$$k_{F_{15}}^2 \simeq 2k_{D_{13}}^2$$

In the laboratory frame this is indeed approximately true,

$$k^2_{D_{13}} \simeq 0\cdot 17 \,\mathrm{GeV^2}, \qquad k^2_{F_{15}} \simeq 0\cdot 34 \,\mathrm{GeV^2}$$

Copley *et al.* (1969a,b) chose $g = 1$ and $0\cdot 17 \,\mathrm{GeV^2}$ to force this vanishing; these values of parameters also gave good fits to the photo-production couplings of several other resonances they examined. Hence the vanishing of the helicity amplitude $A_{1/2}$ for F_{15} and D_{13} off protons could be accommodated. However, two worrying questions are left unanswered.

i. What is so special about the laboratory frame?

ii. If $g = 1$, then the quark mass is around 340 MeV since $\mu_q = \mu_P$. Hence the use of the nonrelativistic interaction in equation (7.26) is unjustified. In particular, spin–orbit terms could be important in the interaction (compare our discussion of the $A^N_{3/2} \simeq 0$ for the F_{15}).

7.2.3 HELICITY STRUCTURE OF RESONANCE ELECTROPRODUCTION

Notwithstanding these questions, two interesting properties of these results were noted by Close and Gilman (1972). The ratio $A_{1/2}/A_{3/2}$ is a function of $|\mathbf{k}|$ which in a given frame is a function of the resonance mass and also the photon mass. If one now considers electroproduction of the resonance instead of photoproduction then $|\mathbf{k}|$ will change (in whatever frame the calculation was performed) and so a cancellation between $|\mathbf{k}|^2$

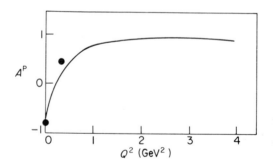

FIG. 7.1. Change in helicity structure of D_{13} photo and electroproduction in the oscillator quark model and data. $A^P \equiv (|A_{1/2}|^2 - |A_{3/2}|^2)/(|A_{1/2}|^2 + |A_{3/2}|^2)$. See also equation (13.12).

and α^2/g in photoproduction would not arise in electroproduction. In fact with the parameterisation of Copley et al. a very dramatic effect is to be expected, namely that $A^P_{1/2}(D_{13})$ which is nearly zero in photoproduction becomes the dominant amplitude in electroproduction. Indeed, this amplitude is predicted to dominate even by $Q^2 = 0.5 \text{ GeV}^2$. A similar behaviour was also found to be present in the quark model of Feynman et al. (section 7.4).

In the days when this behaviour was noticed in the model, there was no indication in the data supporting it. More recently, however, improved data do appear to be consistent with a dramatic change in the helicity structure as Q^2 varies in line with the model predictions (Evangelides et al., 1974; Devenish and Lyth, 1975) (Fig. 7.1). This phenomenon is also of interest in connection with deep inelastic scattering of polarised electrons on polarised protons (section 13.2).

7.2.4 ALGEBRAIC APPROACH TO PHOTO- AND ELECTROPRODUCTION

The second feature of the helicity amplitudes noted in Close and Gilman (1972) was that, independent of the spatial wavefunctions and $|\mathbf{k}|^2$, the assumed transformation property of equation (7.26) or equation (7.31) as having two pieces, that is (i) spin flip or magnetic term ($k(s_x + is_y)$ in equation 7.31) and (ii) orbital flip ($g^{-1}(p_x + ip_y)$ in equation 7.31), enabled two algebraic relations to be written between combinations of $A^{P,N}_{1/2}$, $A^{P,N}_{3/2}$ for the D_{13}. These relations are independent of the relative size of the two terms S_+, L_+ and hence of R_{11} and R_{10}. We have already noted one of these,

$$A^N_{3/2} = -A^P_{3/2} \tag{7.62}$$

The second one is

$$-A^N_{1/2} = \frac{2}{3\sqrt{3}} A^P_{3/2} + \frac{1}{3} A^P_{1/2} \tag{7.63}$$

Experimental verification or refutation of these is a direct test for the assumed transformation property of $\mathcal{H}'$ (equations 7.26 and 7.31) and the **70**, **56** assignments of the D_{13} and proton states, independent of spatial wavefunctions, choice of frames etc.

As we shall see in section 7.3, the well-determined amplitudes $A_{3/2}^{N,P}$ suggest that the interaction $\mathcal{H}'$ contains more than just these two terms and that spin–orbit interactions should be included. One can proceed piecemeal and include further terms as suggested in analogous atomic or nuclear interactions. Such a spin–orbit term reads (Bowler, 1970; Kubota and Ohta, 1976).

$$\mathcal{H}'_{S,O} = \sum_{j=1}^{3} e_j \left[\frac{k}{4m_q} (2g-1) i \mathbf{s} \cdot \mathbf{p} \times \mathbf{A} \right] \frac{e}{2m_q} \tag{7.64}$$

and in the notation analogous to $\mathcal{H}'$ in equation (7.31) reads

$$\mathcal{H}'_{S,O} = -6\sqrt{\pi k}\mu \left[e^{(3)} \exp\left(-ikz^{(3)}\right) \right]$$

$$\times \left(\frac{1}{2m_q} \left(2 - \frac{1}{g} \right) \right) (S_z^{(3)} (p_x + ip_y)^{(3)} - S_+^{(3)} p_z^{(3)}) \tag{7.65}$$

By including this interaction in $\mathcal{H}'$ one can break the relation equating the magnitudes of $A_{3/2}^N$ and $-A_{3/2}^P$. It will also lead to a violation of the neutron target selection rule

$$A_{3/2}^N(F_{15}) = 0 \tag{7.66}$$

One could then force

$$A_{1/2}^P(D_{13}) \simeq 0, \; A_{1/2}^P(F_{15}) \simeq 0 \tag{7.67}$$

in photoproduction and see what Q^2 dependence results for $|A_{1/2}/A_{3/2}|$ in electroproduction.

More efficiently we can ask what is the most general algebraic structure of $\mathcal{H}'$ that can arise if we assume only that the photon interacts with a single quark in the hadron and that the total amplitude is additive in the quarks. This question will be investigated in section 7.3 and the answer is

$$\mathcal{H} = \mathbf{J} \cdot \mathbf{A}; \; J_+ = AL_+ + BS_+ + CS_z L_+ + DS_- L_+ L_+ \tag{7.68}$$

Hence there are four possible combinations of orbital or spin projection flips with A, B, C, D unknowns. The matrix elements for photoproduction of resonances are quoted as a function of these quantities in Table 7.1. Comparison with the results of Copley et al. shows that their work is a particular case of the general structure ($C = D = O$).

7.3 Current matrix elements—a general approach

A common feature of nearly all quark models is that $A \to B\gamma$ is triggered by a single quark $q \to q\gamma$ and the full amplitude is obtained by summing over the constituent quarks in A, B. The decay $A \to B\gamma$ defines a colinear axis z; the hadrons AB are classified in irreducible representations of $SU(6)_W \otimes 0(2)_{L,z}$. The $q \to q\gamma$ transition defines an axis z'. Along this axis the real photon can only flip the quark's spin and the decay transforms simply under $SU(6)_W \otimes 0(2)_{L,z'}$.

If the quarks had no momentum transverse to the hadron decay axis (z) then the z-axis and z'-axis ($q \to q\gamma$) would be coincident. The decay $A \to B\gamma$ would then involve only quark spin flips along the hadron decay axis and so the hadron decay would be $SU(6)_W \times 0(2)_{L,z}$ conserving (i.e. $\Delta L_z = 0$).

In general the quarks have momenta transverse to the hadron decay axis with the consequence that $z' \neq z$. Hence the groups $SU(6)_W \times 0(2)_{L,z}$ and $SU(6)_W \times 0(2)_{L,z'}$ differ. We can see the physical realisation of this in two equivalent ways. First we shall show how the simple classification of states AB under $SU(6)_W \times 0(2)_{L,z}$ leads to configuration mixing under $SU(6)_W \times 0(2)_{L,z'}$; secondly we can see how the simple $SU(6)_W \times 0(2)_{L,z'}$ structure of the $q \to q\gamma$ transition is complicated under $SU(6)_W \times 0(2)_{L,z}$.

7.3.1 CONFIGURATION MIXING

A quark at rest is described by a two-component spinor $\chi^\uparrow$ or $\chi^\downarrow$ (the two spin orientations along the z-axis) and is in a basic representation of $SU(2)_z$ or, with three flavours, of $SU(6)$ (technically $SU(6)_{\text{static}}$). If the quark is boosted by P_z then the spinor $\chi^\uparrow$ becomes (equation 6.11)

$$q^\uparrow(p_z) = \sqrt{\left(\frac{E+m}{2m}\right)} \begin{pmatrix} \chi^\uparrow \\ \dfrac{\sigma_z p_z}{E+m} \chi^\uparrow \end{pmatrix} \tag{7.69}$$

and since $\sigma_z \chi^\uparrow = \chi^\uparrow$ we have that

$$q^\uparrow(p_z) = \sqrt{\left(\frac{E+m}{2m}\right)} \begin{pmatrix} \chi^\uparrow \\ \dfrac{|p|}{E+m} \chi^\uparrow \end{pmatrix} \tag{7.70}$$

A similar procedure for $\chi^\downarrow$ yields

$$q^\downarrow(p_z) = \sqrt{\left(\frac{E+m}{2m}\right)} \left(\frac{\chi^\downarrow}{\frac{-|p|\chi^\downarrow}{E+m}}\right)$$

Hence the quarks have preserved their characteristic representations of SU(2) as the spin orientations are unchanged. However if the boost had been in the x or y direction then the $\chi^\uparrow$ becomes

$$q^\uparrow(p_x) = \sqrt{\left(\frac{E+m}{2m}\right)} \left(\frac{\chi^\uparrow}{\frac{\sigma_x p_x \chi^\uparrow}{E+m}}\right) \qquad (7.71)$$

and since $\sigma_x p_x \equiv \sigma_+ p_- + \sigma_- p_+$ (where $\sigma_\pm \equiv (\sigma_x \pm i\sigma_y)/2$ etc.) then

$$q^\uparrow(p_x) = \sqrt{\left(\frac{E+m}{2m}\right)} \left(\frac{\chi^\uparrow}{\frac{p_+ \chi^\downarrow}{E+m}}\right) \qquad (7.72)$$

and the spin has flipped in the lower components. The quark still has $J_z = +\frac{1}{2}$ of course; the upper components have $S_z = +\frac{1}{2}$ and the lower have $S_z = -\frac{1}{2}$ and $L_z = +1$ (see also the discussion in section 18.1).

If a baryon contains three quarks at rest in S-waves then it forms an irreducible representation of SU(6). If the quarks are in motion along the z-axis the hadron is in an irreducible representation of SU(6)$_W$. Quark momenta transverse to the z-axis yield components in the hadron wavefunction which are not in S-wave, have $L_z = \pm 1$ and $S_z = \mp\frac{1}{2}$ (equation 7.72). Hence the hadron will be in a mixture of SU(6) configurations. Therefore we see that a nucleon at rest or in motion along the z-axis is only classified in a 56plet of SU(6) to the approximation that quark (transverse) momenta are negligible.

If we know the probabilities for the quarks to have momenta $p_{x,y,z}$ in the nucleon, then we can determine the mixture of SU(6)$_W \times$ O(2)$_{L,z}$ configurations to which the nucleon belongs. The wavefunction for the nucleon will thereby be written in terms of $\chi^\uparrow$ and $\chi^\downarrow$ states. A decay $A \to B\gamma$ along the z-axis will then be described by a sum over $q(\chi^{\uparrow \text{or} \downarrow}) \to q(\chi^{\downarrow \text{or} \uparrow}) + \gamma$ transitions along the z-axis, and hence the quark's spin is flipped since $L_z = 0$ along the z-axis. Hence the current is simple and the nucleon configuration is complicated. Radiative matrix elements computed this way are sometimes referred to as calculated in the "current basis".

7.3.2 CURRENT MIXING

In a frame where the $q \to q\gamma$ decay is along some axis z' then the quark spin is flipped along this axis. One can boost to a frame where the decay $A \to B\gamma$ is collinear. This decay will then define an axis $z \neq z'$ in general. Along the z-axis the γq interaction is not collinear and so the interaction can now flip L_z as well as S_z. If the momentum distribution of the quarks is known then the relative importance of $L_z \neq 0$ and $S_z \neq 0$ interactions can be determined. The former term(s) will be SU(6)$_W \times$ $0(2)_{L,z}$ violating.

The photoabsorption by a single quark can transform in any of four ways in general. With total $J_z = 1$ one can flip the L_z with the γ-quark vertex transforming as $S = 0$. This term will be written L_+ to denote that it raises the orbital projection by one unit. The γ-quark vertex can transform as $S = 1$ yielding three possibilities corresponding to the cases where the quark's spin is flipped up (S_+) or down (in which case L_z must be flipped by two) or remains unaltered (and hence L_z is flipped by one).

Hence the most general form of the interaction operator for the current absorption by a single quark contains four pieces whose relative importance is *a priori* arbitrary (Melosh, 1974; Lipkin, 1974; Hey and Weyers, 1974; Gilman and Karliner, 1974; Close *et al.*, 1974). We will write

$$J_+ = AL_+ + BS_+ + CS_z L_+ + DS_- L_+ L_+$$

The A, B, C, D are SU(6) singlet operators whose relative magnitudes are arbitrary and in general their expectation values will depend upon the particular orbital wavefunctions of the initial and final state hadrons. In explicit models these expectation values can be computed. This is illustrated in sections 7.2 and 7.4 for various models.

Given the classification of hadrons in SU(6)$\times$0(3) multiplets it is straightforward to compute matrix elements for use as Clebsch–Gordan coefficients multiplying the $(A, B, C, D)_{i,f}$ expectation values (i, f being the 0(3) states of N^* and N). Hence an algebraic pattern will emerge (Table 7.1) among the matrix elements *which will be common to any explicit model having* $q \to q\gamma$ *as the basic mechanism*. We will illustrate this in sections 7.4.1 and 7.4.2; equation 7.108 *et seq.*

To illustrate this approach we return to the case of photoexcitation of the D$_{13}$(1520), the prominent resonance in **70**, $L = 1^-$. The idea is (i) test if the general algebraic structure is satisfactory by looking for relations which are testable independent of $ABCD$, and (ii) having been satisfied by its validity (at least to a first approximation) pursue explicit models to see if their predictions are realised—note that these will now depend upon explicit values for $ABCD$.

Since the photon is a mixture of $I = 0$ and $I = 1$, then both neutron and proton targets give interesting structure. The results of typical data analyses which extract the magnitudes of the various helicity amplitudes for the production of this state are shown in Table 7.2 in unspecified units (Moorhouse and Oberlack, 1973; Moorhouse et al., 1974; Metcalf and Walker, 1974; Knies et al., 1974).

TABLE 7.2
Helicity amplitudes for $D_{13}(1520)$ photoproduction

	$A^P_{1/2}$	$A^P_{3/2}$	$A^N_{1/2}$	$A^N_{3/2}$
MO	-26 ± 15	194 ± 31	-85 ± 14	-124 ± 13
KMO	-19 ± 8	169 ± 12	-77 ± 5	-120 ± 10
MOR	0 ± 6	174 ± 6	-88 ± 7	-119 ± 25
MW	-6 ± 6	165 ± 11	-66 ± 10	-118 ± 13

These analyses have the following common features:

i. $A^P_{1/2} \lesssim 0, \qquad A^N_{1/2} < 0$

ii. $|A^{P,N}_{3/2}| > |A^{P,N}_{1/2}|$ (7.74)

iii. $A^P_{3/2} > -A^N_{3/2}$

and furthermore the ratios $A^N_{1/2} : A^P_{1/2} : A^N_{3/2} : A^P_{3/2}$ are the same (within the errors) for each analysis.

From the general structure of the $J^{e.m.}$ matrix elements we can form one relation independent of ABC. This reads

$$A^N_{1/2} + \tfrac{1}{3}A^P_{1/2} = \sqrt{\tfrac{1}{3}}(A^N_{3/2} + \tfrac{1}{3}A^P_{3/2}) \qquad (7.75)$$

which is in good agreement with the data and so the general structure appears to be a reasonable first approximation. If we define $R = C/A$ then we find two relations

$$A^N_{3/2} = -\frac{1 + \dfrac{R}{3}}{1 + R} A^P_{3/2} \qquad (7.76)$$

$$A_{1/2}^{N} + \frac{1}{3}A_{1/2}^{P} = -\frac{2}{3\sqrt{3}}\frac{A_{3/2}^{P}}{1+R} \qquad (7.77)$$

and elimination of R between these would recover the single relation in equation (7.75).

In *any* explicit model with C absent, then R is zero and hence

$$A_{3/2}^{N} = -A_{3/2}^{P} \qquad (7.78)$$

The relations with zero value for R were first noticed by Close and Gilman (1972). If R is unity then one finds the relations of the $^3P_0$ model (Peterson and Rosner, 1972, 1973), in particular

$$A_{3/2}^{N} = -\tfrac{2}{3}A_{3/2}^{P} \qquad (7.79)$$

From these relations (equations 7.76 and 7.77) which only involve the largest couplings of this prominent resonance, it seems that

$$R \neq 0 \qquad (7.80)$$

If we also define

$$\tilde{\mu} = B_{01}/A_{01} \qquad (7.81)$$

then we would find for $A_{1/2}^{P}$ that

$$A_{1/2}^{P} = \frac{A_{01}}{\sqrt{6}}(1 - 2\tilde{\mu} - R) \qquad (7.82)$$

and since $A_{1/2}^{P} \simeq 0$ then we have roughly that

$$A_{01} : B_{01} : C_{01} \simeq 4 : 1 : 2 \qquad (7.83)$$

A detailed fit to the full 70plet of resonances only changes this result by less than 5 per cent (Cashmore *et al.*, 1975b). It seems therefore that quark models will be inadequate unless a spin–orbit term (C) is included. We shall consider this question after returning to the development of some dynamical quark models with respect to electromagnetic interactions.

7.4   Relativistic models

The main problem of the nonrelativistic model is that the excitation energies of the states are comparable to the masses themselves and so the nonrelativistic principles are not warranted. However, the nonrelativistic model does appear to have some phenomenological success. Stimulated by this, Feynman *et al.* (1971) attempted to create a relativistic model of quarks with harmonic interaction. The model uses a simple relativistic four-dimensional generalisation of the nonrelativistic three-dimensional oscillator and enables definite calculations to be performed. This idea seems to have been first discussed by Fujimura *et al.* (1970).

The model is not a complete relativistic theory (pair creation and other similar effects that would arise from a relativistic quantum field theory are not included) and hence certain general features of complete relativistic field theories are violated, in particular unitarity. This arises from the four-dimensional nature of the oscillator which means that excitations of the timelike mode will exist in addition to the spatial excitations of the nonrelativistic three-dimensional oscillator.

Explicitly the three-dimensional Hamiltonian of the non-relativistic model

$$2mH_{3D} = \mathbf{p}^2 + m^2\omega_0^2\mathbf{X}^2$$

$$\equiv \mathbf{p}^2 + \Omega^2\mathbf{X}^2 \tag{7.84}$$

is generalised to

$$K_{4D} \equiv (\mathbf{p}^2 - p_0^2) + \Omega^2(\mathbf{X}^2 - t^2) \equiv p^2 + \Omega^2 X^2 \tag{7.85}$$

where $p^2 \equiv p_\mu p^\mu$ and the conjugate position is X_μ so that $p_\mu \equiv i(\partial/\partial X_\mu)$. Feynman *et al.* write for three quarks

$$K = 3(p_a^2 + p_b^2 + p_c^2) + \frac{\Omega^2}{36}((X_a - X_b)^2 + (X_b - X_c)^2 + (X_c - X_a)^2) + C \tag{7.86}$$

the 3 and 36 being simply useful normalisations for their calculations, and C is a constant.

The timelike degree of freedom is a problem due to it having the opposite sign relative to the space components. For example the ground state wavefunction for any pair of quarks is proportional to $\exp [(t_1 - t_2)^2 - (\mathbf{X}_1 - \mathbf{X}_2)^2]$. This tends to zero as their spatial separation increases which is physically sensible and is familiar from the nonrelativistic model. However the wavefunction and probability tend to infinity as their time separation increases, which is nonsense. Feynman *et al.* also illustrate the problems by showing that the time modes have *negative* norm and positive energy. This means that the mass spectrum

$$p_0^2 = \Omega(\sum n_{\text{space}} - \sum n_{\text{time}}) \tag{7.87}$$

has states with imaginary mass if timelike modes are excited.

To avoid these problems FKR suppose that only spatial excitations exist. The model therefore begins to look very much like the earlier nonrelativistic works.

To satisfy unitarity both spacelike and timelike modes would be required to be excitable. By restricting attention only to the spacelikes, FKR argue that their matrix elements will be too large (the timelike have negative norm) and empirically this indeed turns out to be the case.

To overcome this difficulty they suppress the matrix elements by multiplying each one by an *arbitrary ad hoc adjustment factor*

$$F = \exp\left(\frac{-M_1^2 Q^2}{\Omega(M_1^2 + M_2^2)}\right) \tag{7.88}$$

where $M_1^2 = P_1^2$ etc. for three lines of four momentum $P_1 + P_2 + P_3 = 0$ and with Q the spatial momentum of one of the particles 2 or 3 in the rest system of particle P_1. This adjustment factor is analogous to the $\exp(-\text{const } \mathbf{Q}^2)$ of the nonrelativistic three-dimensional oscillator model. The four-dimensional model would have given $\exp[\text{const }(q_0^2 - \mathbf{Q}^2)]$. The *ad hoc* function F replaces this. At this point it is not clear whether any advance has been made over the nonrelativistic three-dimensional model.

7.4.1 ELECTROMAGNETIC INTERACTIONS IN THE FKR MODEL

The expression for K (equation 7.85) does not appear to involve the spin $\frac{1}{2}$ nature of the quarks. The simplest way to incorporate this is to interpret

$$p^2 \equiv \not{p}\not{p} \equiv (p_\mu \gamma^\mu)(p_\nu \gamma^\nu) \tag{7.89}$$

This leaves K unchanged and enables us to calculate the effect of perturbing by an electromagnetic field. Replacing $\not{p}_a$ by $\not{p}_a - e\not{A}(u_a)$ in the presence of an electromagnetic potential, the first-order perturbation is

$$\delta K = 3 \sum_j e_j(\not{p}\not{A}(u_j) + \not{A}(u_j)\not{p}) \tag{7.90}$$

with j summed over the quarks as usual. For interaction with a plane wave with momentum q_μ the electromagnetic potential at quark j is

$$A_\mu \propto \varepsilon_\mu \exp(ik \cdot u_j) \tag{7.91}$$

where ε_μ is the wave's polarisation vector. The interaction is $j_\mu^v \varepsilon^\mu$ where

$$j_\mu^v = 3 \sum_j e_j(\not{p}\gamma_\mu e^{ik.u} + \gamma_\mu e^{ik.u}\not{p}) \tag{7.92}$$

It is instructive to compare this current with the more familiar $\bar{u}\gamma_\mu u$ of electrodynamics. This latter current arises when one describes the free quark by a Dirac Hamiltonian which is linear in mass and the electromagnetic perturbations are on the operator

$$m = \not{p} - e\not{A} \tag{7.93}$$

The present model however has perturbed in

$$m^2 = (\not{p} - e\not{A})(\not{p} - e\not{A}) \tag{7.94}$$

and so $\not{A}$ is always accompanied by a $\not{p}$, hence the $\bar{u}\not{p}\gamma_\mu u$ structure of the current.

The current (equation 7.92) can be written out explicitly using the following representation of the Dirac matrices (equation 6.15)

$$\gamma_t = \begin{pmatrix} 1 & 0 \\ 0 & -1 \end{pmatrix}, \qquad \boldsymbol{\alpha} = \gamma_t\boldsymbol{\gamma} = \begin{pmatrix} 0 & \boldsymbol{\sigma} \\ \boldsymbol{\sigma} & 0 \end{pmatrix}, \qquad \boldsymbol{\sigma} \equiv \begin{pmatrix} \boldsymbol{\sigma} & 0 \\ 0 & \boldsymbol{\sigma} \end{pmatrix} \tag{7.95}$$

(where $\boldsymbol{\sigma}$ are the Pauli matrices). Set

$$p_a = (\varepsilon_a, \mathbf{p}_a), \qquad k = (\nu, \mathbf{k}) \tag{7.96}$$

then moving the plane wave from the right-hand side to the left

$$\not{p}_a e^{ik.u_a} = e^{ik.u_a}(\not{p}_a + \not{k}) \tag{7.97}$$

yields for $\mu = x$ or y (i.e. interaction with a real photon)

$$\begin{aligned} j_\mu^v &= 3e_a e^{ik.u_a}((\not{p} + \not{k})\gamma_\mu + \gamma_\mu\not{p}) \\ &= 3e_a e^{ik.u_a}((\nu + \varepsilon)\boldsymbol{\alpha} - p_i\gamma_i\gamma_\mu - k_i\gamma_i\gamma_\mu + \gamma_\mu\gamma_t\varepsilon - \gamma_\mu\gamma_i p_i) \\ &\equiv 3e_a e^{ik.u_a}(\nu\boldsymbol{\alpha} \cdot \boldsymbol{\varepsilon} - p_i(\gamma_i\gamma_j + \gamma_j\gamma_i)\varepsilon_j - k_i\gamma_i\gamma_j\varepsilon_j) \end{aligned} \tag{7.98}$$

Now

$$\gamma_i \gamma_j \equiv \sigma_i \sigma_j = g_{ij} + i\varepsilon_{ijk}\sigma_k \tag{7.99}$$

hence

$$j_\mu^v = 3e_a \, e^{ik.u_a}(v\boldsymbol{\alpha} \cdot \boldsymbol{\varepsilon} - 2\mathbf{p} \cdot \mathbf{e} - \cancel{\mathbf{k} \times \boldsymbol{\varepsilon}} - i\boldsymbol{\sigma} \cdot \mathbf{k} \times \boldsymbol{\varepsilon}) \tag{7.100}$$

which is exact between four spinors and we have chosen the gauge $\mathbf{k} \cdot \boldsymbol{\varepsilon} = 0$. Only the $\boldsymbol{\alpha} \cdot \boldsymbol{\varepsilon}$ term can mix the upper (large) and lower (small) components of the four spinors. For the initial state with

$$p_1 = (E_1, \mathbf{p}_1) \tag{7.101}$$

and final state with

$$p_2 = (E_2, \mathbf{p}_2) \equiv (E_1 - v, \mathbf{p}_1 - \mathbf{k}) \tag{7.102}$$

then the four spinors in the representation of equation (6.17) become the familiar

$$\chi_1 = \sqrt{\left(\frac{E_1 + m_1}{2m_1}\right)}\left(\begin{array}{c} \chi \\ \dfrac{\boldsymbol{\sigma} \cdot \mathbf{p}_1}{E_1 + m_1}\chi \end{array}\right) \tag{7.103}$$

and

$$\bar{\chi}_2 = \sqrt{\left(\frac{E_2 + m_2}{2m_2}\right)}\left(\chi^+, -\frac{\boldsymbol{\sigma} \cdot \mathbf{p}_2}{E_2 + m_2}\chi^+\right) \tag{7.104}$$

Here χ are two-component rest state or Pauli spinors.

The exact expression in equation (7.100) can be placed now between the Dirac explicit spinors of equations (7.103) and (7.104) and expanded out to obtain the form of the interaction involving two-component or Pauli spinors. This becomes

$$-2\mathbf{p}_1 \cdot \boldsymbol{\varepsilon} + i\boldsymbol{\sigma} \cdot \mathbf{k} \times \boldsymbol{\varepsilon} \qquad \text{(large–large)} \tag{7.105}$$

$$+v\left(\frac{\boldsymbol{\sigma} \cdot \boldsymbol{\varepsilon}\boldsymbol{\sigma} \cdot \mathbf{p}_1}{E_1 + m_1} - \frac{\boldsymbol{\sigma} \cdot \mathbf{p}_2\boldsymbol{\sigma} \cdot \boldsymbol{\varepsilon}}{E_2 + m_2}\right) \qquad \text{(large–small)} \tag{7.106}$$

$$+\frac{2\mathbf{p} \cdot \boldsymbol{\varepsilon}\boldsymbol{\sigma} \cdot \mathbf{p}_2\boldsymbol{\sigma} \cdot \mathbf{p}_1 - \boldsymbol{\sigma} \cdot \mathbf{p}_2\boldsymbol{\sigma} \cdot \mathbf{k} \times \boldsymbol{\varepsilon}\boldsymbol{\sigma} \cdot \mathbf{p}_1}{(E_1 + m_1)(E_2 + m_2)} \qquad \text{(small–small)}$$

$$\tag{7.107}$$

where $\mathbf{k} \equiv \mathbf{p}_2 - \mathbf{p}_1$ and the origin of each of these terms is also shown (large–large meaning that the upper components were involved etc).

Notice that the transformation properties of the first line are $L_z = +1(L_+)$ and $S_z = +1(S_+)$ respectively, i.e. like the A and B terms in the general transformation in equation (7.68). If p_2 and p_1 are non-zero then the second line and third line contain $S_z L_+$ structure (the C term in equation 7.68) while the fourth line contains $S_- L_+ L_+$ (the D term).

In the operator equations (7.106) and (7.107) the dependence on quark number 3 is explicit, an implicit unit operator acts on quarks 1 and 2 as in the traditional formalism (section 7.2). Hence this interaction falls within the general

$$AL_+ + BS_+ + CS_z L_+ + DS_- L_+ L_+ \tag{7.108}$$

structure of section 7.3. The reason for this is that the assumption of single quark interaction has been made in the model.[1]

Note that if either p_1 or $p_2 = 0$ then the small–small components vanish and hence in turn the D term is absent. Furthermore the large–small structure takes on $\boldsymbol{\sigma} \cdot \mathbf{k} \times \boldsymbol{\varepsilon}$ form and hence the C term vanishes. FKR chose to specialise to a coordinate system where $\mathbf{p}_1 = 0$, hence $\mathbf{p}_2 = -\mathbf{k}$ and the interaction collapses to

$$-2\mathbf{p} \cdot \boldsymbol{\varepsilon} + i\boldsymbol{\sigma} \cdot (\mathbf{k} \times \boldsymbol{\varepsilon})\left(1 + \frac{\nu}{E_2 + m_2}\right) \tag{7.109}$$

FKR then define

$$g^2 \equiv \frac{(m_1 + m_2)^2 - k^2}{4 m_1 m_2} \equiv \frac{E_2 + m_2}{2 m_2} \tag{7.110}$$

and so finally the interaction becomes

$$-2\mathbf{p} \cdot \boldsymbol{\varepsilon} + i\boldsymbol{\sigma} \cdot (\mathbf{k} \times \boldsymbol{\varepsilon})\left(1 + \frac{\nu}{2 m_2 g^2}\right) \tag{7.111}$$

which is equation (32) of their paper. Notice that the structure is simply

$$AL_+ + BS_+ \tag{7.112}$$

which, as stated above, is due to the $\mathbf{p}_1 = 0$ constraint.

Hence after all the discussion of relativistic effects and so forth, the model has ended up with identical transformation properties for the electromagnetic current as in the three-dimensional oscillator model of

---

1 For a relativistic treatment to be fully consistent, "non-additive" interactions of quarks are necessary (see, e.g. Kellett, 1974a,b).

section 7.2. This is indeed the case as can be seen by taking the general matrix elements in Table 7.1 and setting $C = D = 0$. The explicit values for A_{01} and B_{01} turn out to be[1]

$$A_{01} \equiv \sqrt{2\Omega} \tag{7.113}$$

$$B_{01} \equiv \frac{3}{\sqrt{2}} \lambda \rho \tag{7.114}$$

where

$$\lambda = \sqrt{\frac{2}{\Omega}} k, \qquad \rho = \sqrt{\Omega} \left(1 + \frac{\nu}{2m_2 g^2} \right) \lambda \tag{7.115}$$

and

$$\mathbf{k}^2 = -k^2 + \frac{(M^2 - m^2 + k^2)^2}{4M^2} \tag{7.116}$$

For $k^2 = 0$ these expressions simplify to read

$$\rho = \sqrt{2}(m_1 - m_2), \qquad k \equiv |\mathbf{k}| = \frac{M^2 - m^2}{2M} \tag{7.117}$$

The correspondence with the three-dimensional model is exact and for $\mathbf{70}$, $L = 1$ states reads

$$\lambda \rho \equiv \tfrac{2}{3} \mu R_{10}^{M,S} \sqrt{\pi k} \tag{7.118}$$

$$\sqrt{\Omega} \equiv \sqrt{2} \sqrt{\frac{\pi}{k} \frac{\mu}{g}} R_{11}^{M,S} \tag{7.119}$$

At $L = 2$ it is

$$\lambda(\lambda \rho) \equiv -2\sqrt{\tfrac{2}{3}} \mu \sqrt{\pi k} R_{20} \tag{7.120}$$

$$\lambda \sqrt{\Omega} \equiv -2\sqrt{\frac{\pi}{k} \frac{\mu}{g}} R_{21} \tag{7.121}$$

7.4.2 RELATED OSCILLATOR MODELS

Other models that have made a four-dimensional generalisation of the three-dimensional oscillator Hamiltonian include those of Fujimura *et al.* (1970) and Lipes (1972). These authors choose as solutions of the

---

[1] See Table 1 of Feynman *et al.* (1971).

wave equation wavefunctions which are normalisable (noninfinite) in all four variables and hence in the rest frame

$$\psi_0(\rho, \xi, \eta; \mathbf{0}) = \exp\left[-\frac{iM_0\rho}{\sqrt{3}}\right]\left(\frac{\alpha}{3\pi}\right)^2 \exp\left[-\frac{\alpha}{6}(\eta_t^2 + \boldsymbol{\eta}^2 + \xi_t^2 + \boldsymbol{\xi}^2)\right]$$

(7.122)

where

$$\rho \equiv \frac{1}{\sqrt{3}}(X_1 + X_2 + X_3)$$

$$\xi \equiv \frac{1}{\sqrt{2}}(X_1 - X_2)$$

(7.123)

$$\eta \equiv \frac{1}{\sqrt{6}}(X_1 + X_2 - 2X_3)$$

are four vectors which separate the c.m. from the internal motion and furthermore separates the internal motion into two independent oscillators. Notice that the time component enters with the same sign as the spatial in contrast to FKR where the amplitude grew with the time separation. This can be written covariantly as

$$\psi_0(\rho, \xi, \eta; \mathbf{p}) = \exp\left[-\frac{iP \cdot \rho}{\sqrt{3}}\right]$$
$$\times \left(\frac{\alpha}{3\pi}\right)^2 \exp\left[-\frac{\alpha}{6}\left(2\left(\frac{P \cdot \eta}{M_0}\right)^2 - \eta^2 + 2\left(\frac{P \cdot \xi}{M_0}\right)^2 - \xi^2\right)\right] \quad (7.124)$$

where P is the c.m. momentum of the composite system.

Fujimura *et al.* also consider a more general wavefunction

$$\psi_0^{(n)} = \left(\frac{\alpha}{3\pi}\right)^2 \sqrt{(2n-1)} \exp\left[-\frac{\alpha}{6}\left(2n\left(\frac{P \cdot \eta}{M_0}\right)^2 - \eta^2 + 2n\left(\frac{P \cdot \xi}{M_0}\right)^2 - \xi^2\right)\right]$$

(7.125)

the parameter n distinguishing the timelike extension from the spatial. The $n = 0$ case is essentially the (not normalisable in t) FKR solution. The $n = 1$ is that of Lipes.

The electromagnetic current again has an $AL_+ + BS_+$ structure $C, D = 0$ as can be seen by inspecting equation (13a) of Lipes (1972). Consequently there is an exact correspondence between the results of

Lipes Table 3 for the matrix elements and the analogous Table 1 of FKR and those of CKO. This correspondence becomes

$$A^{(01)}_{\text{Lipes}} \equiv \sqrt{\tfrac{2}{3}}\sqrt{\Omega}_{\text{FKR}}; \ B^{(01)}_{\text{Lipes}} \equiv \sqrt{\tfrac{3}{2}}\rho_{\text{FKR}} \qquad (7.126)$$

for the **70**, $L = 1$ states.

Lipes' solution also has problems with the timelike excitations, states of imaginary mass being present if the number of timelike excitations is sufficiently large. To eliminate this he restricts attention to spacelike excitations in the rest frame of the particle, very analogous to the FKR approach.

However Lipes appears to calculate explicitly the "form factor" analogue of the FKR adjustment factor F at equation (7.88). In comparison with electroproduction the shape of his form factor agrees nicely with the data. A careful reading of his section 3a will reveal that one power of $(1 + q^2/4m^2)$ has been arbitrarily omitted in order to attain this agreement. A possible excuse for this is offered.

7.5   Electromagnetic interactions in the $^3P_0$ model

In the general approach to pion emission at equation (6.79) we argued that the matrix element contained two pieces transforming as W_z and $(W_+L_- - W_-L_+)$ respectively. The emission of a transversely polarised photon involved four independent contributions (equation 7.73).

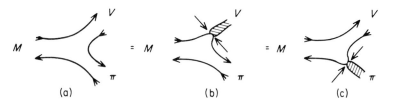

FIG. 7.2. $M \to V\pi$ in the $^3P_0$ model where (b) V or (c) π is regarded as elementary.

Imagine for a moment that this photon is the result of an elementary vector meson emission in a decay $M \to V\pi$ (Fig. 7.2). We would infer that this process contains four degrees of freedom. However, in the $^3P_0$ model, we could instead regard the π as elementary and hence infer that only two degrees of freedom characterise the problem (equation 6.79).

This suggests that in the $^3P_0$ model not all of $ABCD$ in equation (7.73) are independent. The $^3P_0$ model in fact constrains

$$A_{if} = C_{if}; \quad D_{if} = 0$$

and from Table 7.2 we indeed see that data are consistent with $A_{01} = C_{01}$. The origin of this result can be heuristically seen as follows. (For more details see Petersen and Rosner, 1973, and Le Youanc *et al.*, 1973.)

The presence of a $^3P_0$ in a matrix element for $M \to BC$ gives rise to χ^m spin and $Y_1^{-m}(\mathbf{k} - \mathbf{k}_B)$ orbital pieces (equations 6.99 and 6.100). Rewriting χ^m with the aid of the Pauli $\boldsymbol{\sigma}$ matrices, then this coupling of spin and orbital angular momentum generates a $\boldsymbol{\sigma} \cdot (\mathbf{k} - \mathbf{k}_B)\sqrt{(3/4\pi)}$ contribution to the matrix element. We may regard this as the intrinsic $^3P_0$ operator. Now consider its effect upon amplitudes for $M \to BC$ where B is a π or a ρ.

From equations (6.99) and (6.100) for $B \equiv \pi$ we find that (compare equation 6.102)

$$\chi^{-m}I_m(M; BC)$$

$$= \sqrt{\frac{3}{4\pi}} \frac{1}{8} \delta(\mathbf{k}_B + \mathbf{k}_C) \int d^3k \, \boldsymbol{\sigma} \cdot (\mathbf{k} - \mathbf{k}_B)\psi_M(\mathbf{k}_B + \mathbf{k})\psi_C(\mathbf{k}) \tag{7.127}$$

where we have approximated $\psi_B(\mathbf{k}) \simeq \psi_B(0)$ (compare equation 6.103 for an elementary meson where $R \to 0$.) We explicitly see the independent σ_z and $\sigma_\pm$ structures (compare equation 6.79).

$$\langle \sigma_z \rangle \propto \int d^3k (k_z - k_B)\psi_M(\mathbf{k}_B + \mathbf{k})\psi_C(\mathbf{k}) \tag{7.128}$$

$$\langle \sigma_\pm \rangle \propto \int d^3k (k_x \pm ik_y)\psi_M(\mathbf{k}_B + \mathbf{k})\psi_C(\mathbf{k}) \tag{7.129}$$

With particular wavefunctions (e.g. harmonic oscillator) the relative importance of these terms can be computed (c.f. equation 6.119 for $B \to \omega\pi$).

If $B \equiv \rho$ then the $^3S_1$ vector meson will give a contribution χ^n to the matrix element (like the $^3P_0$) but there is no orbital angular momentum contribution from the S-wave ρ. Coupling to the emitted photon with polarisation vector $\boldsymbol{\varepsilon}$ yields a $\boldsymbol{\sigma} \cdot \boldsymbol{\varepsilon}$ contribution in the matrix element which will multiply the $^3P_0$ operator. Using the identity of Pauli matrices (equation 7.99) we can rewrite this:

$$\boldsymbol{\sigma} \cdot \boldsymbol{\varepsilon}\boldsymbol{\sigma} \cdot (\mathbf{k} - \mathbf{k}_V) = \mathbf{k} \cdot \boldsymbol{\varepsilon} + i\boldsymbol{\sigma} \cdot (\mathbf{k} - \mathbf{k}_V) \times \boldsymbol{\varepsilon} \qquad (7.130)$$

(the $\mathbf{k}_V \cdot \boldsymbol{\varepsilon}$ vanishes for transversely polarised photons). Hence we see the $B\sigma_\pm$ term at $\boldsymbol{\sigma} \cdot \mathbf{k}_V \times \boldsymbol{\varepsilon}$

$$\langle \sigma_\pm \rangle \propto \int d^3\mathbf{k}(k_z - k_B)\psi_M(\mathbf{k}_B + \mathbf{k})\psi_C(\mathbf{k}) \qquad (7.131)$$

and the $AL_\pm$ and $C\sigma_z L_\pm$ pieces are equal

$$\langle L_\pm; \sigma_z L_\pm \rangle \propto \int d^3\mathbf{k}(k_x \pm ik_y)\psi_M(\mathbf{k}_B + \mathbf{k})\psi_C(\mathbf{k}) \qquad (7.132)$$

The π or ρ emission are therefore both described by *two* independent amplitudes in the $^3P_0$ model; one conserving $SU(6)_W$ and one violating it.

7.6 Massive quark models

A realistic dynamical quark model should have a relativistically invariant basic formulation and also take into account that free quarks have not been observed. One way to realise this is to have the quark mass much larger than the hadron masses. If heavy quarks exist with masses greater than, say, 20 GeV, then mesons will arise as poles in $q\bar{q}$ scattering and hence can be described by the Bethe–Salpeter equation in the $q - \bar{q}$ channel (Bethe and Salpeter, 1951).

The dynamical input is in the Bethe–Salpeter kernel, namely the form of the $q\bar{q}$ interaction. Llewellyn-Smith (1969) noted that if the quarks have small relative momenta ($|\mathbf{p}|^2 \ll M_Q^2$) (Morpurgo, 1965) then an approximate SU(6) symmetry can hold if they interact through a Dirac scalar potential. However, there are some interesting effects, e.g. the low lying vectors mesons contain $^3D_1$ as well as the usual $^3S_1$ component.

For a single free quark the spinor has a "large" and "small" component (equation 6.17). The latter is of the order $\boldsymbol{\sigma} \cdot \mathbf{p}/M$ times the former. In general a four-component spinor may be separated into upper and lower (χ_+, χ_-) two component spinors. A $q\bar{q}$ state at rest may be described by a wavefunction

$$\chi(\mathbf{P} = 0) = \begin{pmatrix} \chi_{+-} & \chi_{++} \\ \chi_{--} & \chi_{-+} \end{pmatrix}$$

where $\chi_{\alpha\beta}$ transforms like $q_\alpha \otimes \bar{q}_\beta$. In the weakly bound nonrelativistic pictures the "small" component is indeed small and

$$\chi_{+-} \sim \chi_{-+} \sim \frac{\boldsymbol{\sigma} \cdot \mathbf{p}}{M} \chi_{++}; \qquad \chi_{--} \sim \left(\frac{\mathbf{p}^2}{M^2}\right) \chi_{++}$$

In the nonrelativistic quark model discussed elsewhere in this chapter it is implicitly assumed that such relations hold true even in tightly bound situations with the consequence that $\chi_{\pm\mp}, \chi_{--}$ can be neglected if $\mathbf{p}^2 \ll M^2$.

For a pseudoscalar meson in the approximation that the mass is zero

$$\chi \sim \gamma_5 \equiv \begin{pmatrix} 0 & 1 \\ 1 & 0 \end{pmatrix}$$

Hence χ_{+-} and $\chi_{-+} = 0$ but $\chi_{++} = \chi_{--}$ (plus corrections proportional to the bound state mass). Detailed study of the coupled Bethe–Salpeter equations shows that the usual nonrelativistic picture will emerge only if the interquark potential is a Dirac scalar, since only for this case is

$$\chi_{--} \ll \chi_{++}$$

(Llewellyn-Smith, 1969).

Developments in the Bethe–Salpeter studies include that of Sundaresan and Watson (1970) who use a relativistic generalisation of the three-dimensional harmonic oscillator interaction between q and $\bar{q}$ and generate a meson spectrum with linearly rising Regge trajectories. These ideas have been intensively studied by Bohm et al. (1972, 1973, 1974). These authors use a Bethe–Salpeter kernel which simulates a smooth harmonic potential well for small separations of q and $\bar{q}$ but which vanishes asymptotically. In their 1972 paper they use scalar "quarks" and obtain meson Regge trajectories together with daughters. Their 1973 paper incorporates spin $\frac{1}{2}$ quarks.

Electromagnetic interactions in a massive quark model and, in particular, the deep inelastic electron-scattering phenomenology (where the quarks appear to be light—Chapters 9 to 14) have been investigated by Preparata (section 10.4).

8 Coloured Quarks

8.1 Symmetric quark model

In Chapter 5 we saw that a nice description of the baryon spectrum emerged if one demanded that the baryons were built from three quarks and that the total $SU(6) \otimes 0(3)$ three-quark wavefunction was *symmetric* under the interchange of any pair of quarks (the "symmetric quark model"). Yet baryons are *antisymmetrised* with respect to one another since they are fermions and obey the Pauli principle. Consequently quarks are funny: they are symmetrised in sets of three but one set of three is antisymmetrised with another set.

Greenberg (1964) suggested that quarks are parafermions of order 3, as a result of which the above paradox is avoided. Mathematically this means that the creation (or annihilation) operator for a quark with given flavour, spin, momentum quantum numbers (generically labelled λ) becomes

$$a^+ = \sum_{i=1}^{3} a^+_{(i)\lambda}, \qquad a = \sum_{i=1}^{3} a_{(i)\lambda} \qquad (8.1)$$

where i is some new label which we might call "colour". These operators commute for the same colour but anticommute otherwise

$$\{a_{\lambda(i)}, a^+_{\mu(i)}\} = \delta_{\lambda\mu}, \{a^+_{\lambda(i)}, a^+_{\mu(i)}\} = 0 \qquad (8.2)$$

$$[a_{\lambda(i)}, a^+_{\mu(j)}] = 0 \; (i \neq j)$$

$$[a^+_{\lambda(i)}, a^+_{\mu(j)}] = 0 \; (i \neq j) \qquad (8.3)$$

We can now create a three-quark state by acting on the vacuum Φ_0 with the operator $f^+_{\lambda\mu\nu}$ where

$$f^+_{\lambda\mu\nu} \equiv \{\{a^+_\lambda, a^+_\mu\}, a^+_\nu\}$$

$$\equiv 4 \sum_{\substack{ijk=1 \\ i\neq j\neq k\neq i}}^{3} a^+_{\lambda(i)} a^+_{\mu(j)} a^+_{\nu(k)} \tag{8.4}$$

The state $f^+_{\lambda\mu\nu}\Phi_0$ is therefore symmetric under any permutation of the $\lambda\mu\nu$ labels as are the three-quark states of Chapter 5. However, this system is a fermion since $\{f^+_{\lambda\mu\nu}, a^+_\sigma\} = 0$ and hence

$$\{f^+_{\lambda\mu\nu}, f^+_{\alpha\beta\gamma}\} = 0 \tag{8.5}$$

which implies that the creation operators for the three body systems anticommute (hence they are fermions satisfying Pauli statistics).

The price we have paid is to introduce a new threefold degree of freedom for each quark flavour.[1] If we call this "colour", then the quarks can have three primary colours, red, yellow and blue, and the proton requires each of its quarks to have a different colour.

8.2   Charges of coloured quarks

The Gell-Mann–Nishijima relation

$$Q = T_3 + \frac{Y}{2} \equiv F_3 + \frac{1}{\sqrt{3}} F_8 \tag{8.6}$$

defines the charge operator as a function of the $SU(3)$ generators. From the commutation relations in equation (2.47) it follows that

$$[Q, U_\pm] = [Q, U_3] = 0 \tag{8.7}$$

and hence the charge operator is a U-spin scalar. This is exemplified by the fundamental triplet representation where the d and s quarks form a U-spin doublet and have the same charge of $-\frac{1}{3}$.

If equation (8.6) is generalized to

$$Q = \left(T_3 + \frac{Y}{2}\right) + \frac{1}{3} t \tag{8.8}$$

[1] Compare with the model of Freund and Lee (1964).

where t is an arbitrary c-number, then the commutation relations in equation (8.2) are preserved but a triplet will now have charges $(z, z-1, z-1)$ where

$$z \equiv \tfrac{1}{3}(t+2) \tag{8.9}$$

The uds triplets in colour states RBY will have charges

	u	d	s
R	z_R	$z_R - 1$	$z_R - 1$
B	z_B	$z_B - 1$	$z_B - 1$
Y	z_Y	$z_Y - 1$	$z_Y - 1$

$$\tag{8.10}$$

subject to the constraint that

$$z_R + z_B + z_Y = 2 \tag{8.11}$$

(which follows from the requirement that $\Delta^{++}(u_R u_B u_Y)$ have charge $+2$). Equivalently, the $z_R z_B z_Y$ correspond to $t_R t_B t_Y$ via equation (8.9) and the constraint in equation (8.11) corresponds to

$$t_R + t_B + t_Y = 0 \tag{8.12}$$

hence familiar baryon charges satisfy the standard Gell-Mann–Nishijima relation in equation (8.6).

Notice that the "average charge" of uds is therefore

$$e_u = \tfrac{1}{3}(z_R + z_B + z_Y) = \tfrac{2}{3}$$
$$e_d = e_s = \tfrac{1}{3}[(z_R - 1) + (z_B - 1) + (z_Y - 1)] = -\tfrac{1}{3} \tag{8.13}$$

and so all baryons made of one R, one B and one Y quark will have the same charges as in the familiar uncoloured quark model since the average quark charges are the same as for "uncoloured" quarks.

Two particular models are one where electromagnetism is colourblind ($z_R = z_B = z_Y = \tfrac{2}{3}$) (Greenberg and Zwanziger, 1966; Gell-Mann, 1972) and one where electromagnetism can "spectrum analyse" the quarks ($z_Y = z_B = 1$, $z_R = 0$) (Han and Nambu, 1965). The former therefore contains three identical triplets and if the RBY generate a symmetry, $SU(3)_{colour}$, then $SU(3)_{colour}$ can be an exact symmetry of Nature. In the Han–Nambu model the quarks have integer charges.

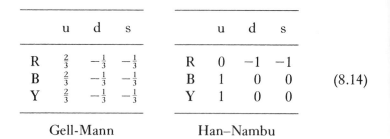

	u	d	s		u	d	s	
R	$\frac{2}{3}$	$-\frac{1}{3}$	$-\frac{1}{3}$	R	0	-1	-1	
B	$\frac{2}{3}$	$-\frac{1}{3}$	$-\frac{1}{3}$	B	1	0	0	(8.14)
Y	$\frac{2}{3}$	$-\frac{1}{3}$	$-\frac{1}{3}$	Y	1	0	0	
	Gell-Mann			Han–Nambu				

8.3 Colour and $\pi^0 \to 2\gamma$ decay

Colour is also useful in connection with the $\pi^0 \to \gamma\gamma$ rate for which Adler (1969) and Bell and Jackiw (1969) have given an exact formula in a field theory of quarks and gluons. The amplitude is a known constant times

$$\left(\sum_i e_i^2\right)_{I_3=+1/2} - \left(\sum_i e_i^2\right)_{I_3=-1/2} \tag{8.15}$$

with e_i the quark charges for $I_3 = \pm\frac{1}{2}$ quarks (Fig. 8.1). To agree with

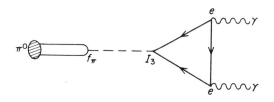

FIG. 8.1. The $\pi^0 \to \gamma\gamma$ decay is related to the triangular coupling of two photons and the isotriplet axial-vector current; hence the structure of equation (8.15).

experiment within the errors requires that $\sum_i I_3^i e_i^2 = \frac{1}{2}$. Fermi–Dirac (uncoloured) quarks would yield

$$\tfrac{1}{2}(e_u^2 - e_d^2) = \tfrac{1}{6} \tag{8.16}$$

a factor of three too small in amplitude (nine in rate). With the three colour degrees of freedom the Gell-Mann colour scheme immediately rectifies this. In fact any three-colour model with the constraint in equation (8.11) will be satisfactory since

$$\sum_i I_3^i e_i^2 = \tfrac{1}{2} \tag{8.17}$$

requires that

$$(z_R^2 + z_B^2 + z_Y^2) - [(z_R - 1)^2 + (z_B - 1)^2 + (z_Y - 1)^2] = 1 \qquad (8.18)$$

and hence

$$z_R + z_B + z_Y = 2 \qquad (8.11 \text{ bis})$$

8.4 Colour as a symmetry

Suppose that the RYB degree of freedom generates an $[SU(3)]_{\text{colour}}$ group (see also Han and Nambu, 1974; Close, 1975). Baryons are therefore $[SU(6) \otimes 0(3) \otimes SU(3)_{\text{colour}}]_{\text{antisymmetric}}$ if the Pauli principle is to be satisfied. The familiar baryons of the symmetric quark model (section 5.1) are therefore

$$[SU(6) \otimes 0(3)]_{\text{symmetric}} \otimes [SU(3)_{\text{colour}}]_{\text{antisymmetric}} \qquad (8.19)$$

which requires that they be *colour singlets* (since a totally antisymmetric three-body state in $SU(3)$ is a singlet). Note that the antisymmetric colour state requires the three quarks to be one red, one blue and one yellow.

The quarks form a representation of $SU(3)_{\text{flavour}} \otimes SU(3)_{\text{colour}}$. In place of equation (8.8) we may write now

$$Q = \left(I_3 + \frac{Y}{2}\right) + \left(\alpha \tilde{I}_3 + \beta \frac{\tilde{Y}}{2}\right) \qquad (8.20)$$

where the tildas refer to the $SU(3)_{\text{colour}}$ group and α, β are arbitrary constants. We will define RB to be the $\tilde{I}_3 = +\frac{1}{2}, -\frac{1}{2}$ states respectively. The charges of the u quarks will be

$$z_R = \frac{2}{3} + \frac{\alpha}{2} + \frac{\beta}{6}$$

$$z_B = \frac{2}{3} - \frac{\alpha}{2} + \frac{\beta}{6} \qquad (8.21)$$

$$z_Y = \frac{2}{3} \qquad - \frac{\beta}{3}$$

and of course we recover equation (8.11).

The Gell-Mann coloured quarks have $\alpha = \beta = 0$ and so their charges transform as a singlet under $(SU(3)_{colour})$. The Han–Nambu charges arise if $\alpha = \beta = -1$ and hence are 3^* under $[SU(3)]_{colour}$ and $[3, 3^*]$ under $(SU(3)_f \otimes SU(3)_c)$. The weight diagram of the Han–Nambu quarks is shown in Fig. 8.2. Note the inverted triangle (triplet) of uds and the triangles (antitriplet) that show the spectrum analysis of each quark into RBY of $SU(3)_c$. If $\alpha = \beta = +1$ one obtains charges transforming as $(3, 3)$ (Greenberg, 1975).

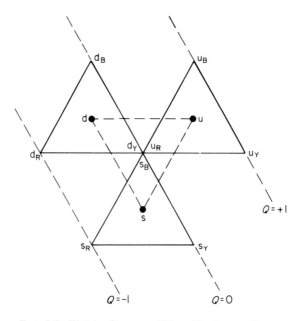

FIG. 8.2. Weight diagram of Han–Nambu quarks.

The currently observed meson spectroscopy places no constraints whatsoever on $z_R z_B z_Y$. First of all consider uncoloured quarks where $\pi^+ \equiv u\bar{d}$. The charge is given by

$$\langle u\bar{d} | e_q + e_{\bar{q}} | u\bar{d} \rangle = e_u + e_{\bar{d}} = \tfrac{2}{3} + \tfrac{1}{3} = 1 \qquad (8.22)$$

in an obvious notation. Now consider coloured quarks. The colour singlet π^+ can be written

$$\left(u\bar{d}, \frac{1}{\sqrt{3}} (R\bar{R} + B\bar{B} + Y\bar{Y}) \right) \equiv \frac{1}{\sqrt{3}} (u_R \bar{d}_R + u_B \bar{d}_B + u_Y \bar{d}_Y) \qquad (8.23)$$

The charge calculation now becomes

$$\langle \pi^+, \tilde{\mathbf{1}}|\hat{Q}|\pi^+, \tilde{\mathbf{1}}\rangle = \tfrac{1}{3}\langle u_R\bar{d}_R + u_B\bar{d}_B + u_Y\bar{d}_Y|e_q + e_{\bar{q}}|u_R\bar{d}_R + u_B\bar{d}_B + u_Y\bar{d}_Y\rangle$$

$$= \tfrac{1}{3}(z_R + z_B + z_Y) + \tfrac{1}{3}((1 - z_R) + (1 - z_B) + (1 - z_Y))$$

$$= 1 \qquad (8.24)$$

Notice that this result was obtained *independently* of any constraint on $z_{R,B,Y}$. Meson spectroscopy requires only that uds form a triplet and that in $SU(3)_c$ the mesons are colour neutral ($R\bar{R}, B\bar{B}, Y\bar{Y}$ in arbitrary weights but not $R\bar{B}$ etc.). The failure to see nine coloured pions suggests the colour singlet assignment for the conventional states.

In electron–positron annihilation below the threshold for producing colour nonsinglet states,

$$R \equiv \frac{\sigma(e^+e^- \to \text{hadrons})}{\sigma(e^+e^- \to \mu^+\mu^-)} = \sum_i e_i^2 = 2 \qquad (8.25)$$

(section 11.5) independent of the explicit $z_{R,B,Y}$. Intuitively this is because only the "average" quark charges are being seen, hence effectively $\tfrac{2}{3}, -\tfrac{1}{3}, -\tfrac{1}{3}$ with three freedoms. This can be seen explicitly by considering the photon in colour models

$$\gamma \sim z_R(u_R\bar{u}_R + d_R\bar{d}_R + s_R\bar{s}_R) - (d_R\bar{d}_R + s_R\bar{s}_R) + z_B(u_B\bar{u}_B + d_B\bar{d}_B + s_B\bar{s}_B)$$

$$- (d_B\bar{d}_B + s_B\bar{s}_B) + z_Y(u_Y\bar{u}_Y + d_Y\bar{d}_Y + s_Y\bar{s}_Y) - (d_Y\bar{d}_Y + s_Y\bar{s}_Y) \qquad (8.26)$$

Some algebraic manipulation separates this into a colour $\tilde{\mathbf{1}}$ and colour $\tilde{\mathbf{8}}$ piece of form

$$(\tfrac{2}{3}u\bar{u}, (R\bar{R} + B\bar{B} + Y\bar{Y})) + ((z_R - \tfrac{2}{3})u\bar{u}, (R\bar{R} - Y\bar{Y}))$$

$$+ ((z_B - \tfrac{2}{3})u\bar{u}, (B\bar{B} - Y\bar{Y})) \qquad (8.27)$$

with similar structure for the quarks d and s.

The colour singlet piece of the photon is therefore independent of the particular $z_{R,B,Y}$. To obtain information on these quantities requires experiments receiving contributions from above colour threshold. One example would be the crossing of colour threshold in e^+e^- annihilation whence R (equation 8.25) would rise to a new value $3(z_R^2 + z_B^2 + z_Y^2) - 2$ (or with α, β defined in equation 8.20, $R = 2 + \tfrac{1}{2}(3\alpha^2 + \beta^2)$).

In the particular case of Han–Nambu charges the photon is written

$$\gamma_{HN} \sim u_Y\bar{u}_Y + u_B\bar{u}_B - d_R\bar{d}_R - s_R\bar{s}_R \to u\bar{u}(Y\bar{Y} + B\bar{B}) - (d\bar{d} + s\bar{s})R\bar{R}$$

$$(8.28)$$

Trivial algebra enables this to be rewritten as

$$\gamma_{HN} = (\tfrac{2}{3}u\bar{u} - \tfrac{1}{3}d\bar{d} - \tfrac{1}{3}s\bar{s})(R\bar{R} + B\bar{B} + Y\bar{Y})$$
$$- (u\bar{u} + d\bar{d} + s\bar{s})(\tfrac{2}{3}R\bar{R} - \tfrac{1}{3}B\bar{B} - \tfrac{1}{3}Y\bar{Y}) \tag{8.29}$$

and hence

$$\gamma_{HN} = (\mathbf{8}, \tilde{\mathbf{1}}) - (\mathbf{1}, \tilde{\mathbf{8}}) \tag{8.30}$$

The $(\mathbf{8}, \tilde{\mathbf{1}})$ contains the familiar ρ, ω, ϕ vector mesons which are colour singlets and are produced by the photon in e^+e^- annihilation. The $(\mathbf{1}, \tilde{\mathbf{8}})$ piece of the photon can excite vector mesons which are singlets of $SU(3)_f$ and octets of $\overline{SU(3)}_c$. If colour $SU(3)$ is a good symmetry of the strong interactions then these $\tilde{\mathbf{8}}$ states will not decay by strong interactions into $\tilde{\mathbf{1}}$ hadrons. Consequently they will be metastable. However a state $(\mathbf{1}, \tilde{\mathbf{8}})$ can decay into conventional hadrons $(\mathbf{1}, \tilde{\mathbf{1}})$ and a photon due to the $(\mathbf{1}, \tilde{\mathbf{8}})$ piece of the photon. It was suggested that the metastable J/ψ vector meson with mass 3.1 GeV might be such a state. However it does not have dominantly radiative decays and it now appears more likely that it is the first manifestation of a fourth flavour of quark (charmed quark) and hence the $^3S_1$ $c\bar{c}$ brother of $\phi(s\bar{s})$ and ρ, ω (Chapter 16).

No evidence for colour octet states has yet been demonstrated. If such states exist with masses of the order of M GeV (where M is some number larger than about 5) then in e^+e^- annihilation above M GeV one can produce coloured neutral $\tilde{\mathbf{8}}$ states along with a photon

$$e^+e^- \to \gamma(8, \tilde{1}) \to M(\mathbf{8}, \tilde{\mathbf{8}}) + \gamma(1, \tilde{\mathbf{8}}) \tag{8.31}$$

and above $2M$ GeV one can pair produce the whole octet of coloured states. At energies $E < M$ GeV, only colour singlet hadrons are produced and

$$R \equiv \frac{\sigma(e^+e^- \to hadrons)}{\sigma(e^+e^- \to \mu^+\mu^-)} \equiv \sum_i e_i^2 = 2$$

whereas for $E > 2M$ GeV the colour degrees of freedom are all unfrozen and R will rise to a value of 4 (since in the Han–Nambu model $4 \equiv \sum_i e_i^2$). Although the data do appear to undergo this sort of behaviour around 4 GeV, it is more likely that this is due to a fourth flavour of quark (charm-c) being unfrozen. It seems that there is no compelling reason to believe that colour is unfrozen at energies so far probed in the e^+e^- annihilation.

If colour does ever reveal itself, then it will be interesting to look for "exotic" states like mesons with charge 2 (e.g. $u_Y\bar{d}_R$) or charge three baryons. Also, once the colour is revealed, there is no reason why coloured quarks cannot be produced. This sort of possibility has been actively investigated by Pati and Salam (1973a,b; 1974; 1975; it is described in detail by Pati, 1977).

The idea that we have been carrying in the back of our minds in all of this discussion is that colour singlets lie lowest in mass and that colour nonsinglets (e.g. $\tilde{8}, \tilde{3}, \ldots$) have much higher masses. Imagine what would happen if the colour nonsinglets were pushed up to infinite masses. Clearly only colour $\tilde{1}$ would exist as physically observable states and quarks would in consequence be permanently confined. At any finite energy we would see only the "average" quark charges and phenomenologically we could not distinguish this from the Gell-Mann model where the quarks form three identical triplets. The photon here is

$$\gamma_{GM} \sim (\tfrac{2}{3}u\bar{u} - \tfrac{1}{3}d\bar{d} - \tfrac{1}{3}s\bar{s})(R\bar{R} + B\bar{B} + Y\bar{Y}) = (8, \tilde{1}) \qquad (8.32)$$

and colour singlet states are all that can be produced. In e^+e^- annihilation, $R = 2$ at all energies unless further flavours of quarks reveal themselves. This model has been used as the basis for a possible field theory of strong interactions (quantum chromodynamics) discussed in section 15.2 (Fritzsch and Gell-Mann, 1972; Weinberg, 1973; Gross and Wilczek, 1973b; Fritzsch et al., 1974).

8.5   Further reading

After the discovery of the heavy metastable J/ψ meson in 1974 there was speculation that it might be the first evidence for a nonsinglet coloured hadron (see, e.g. Krammer et al., 1974; Sanda and Terezawa, 1975; Kenny et al., 1975a,b; Mathews, 1975; Yamaguchi, 1975; Close, 1975; Marinescu and Stech, 1975; Greenberg, 1975). Subsequent data have ruled out this interpretation of J/ψ but, even so, these papers may give one a feeling for the phenomena to be expected when (if) colour threshold is crossed.

The possibility that quarks and leptons might be unified in the Han–Nambu colour scheme is described in lectures by Pati (1977). These lectures also discuss the possibility that quarks may be liberated and may already have been produced as free particles in e^+e^- annihilation.

PART 2
DEEP INELASTIC PHENOMENA
AND PARTONS

9   Lepton Scattering and Partons

The discovery that high-energy electrons have a significant probability to scatter from a proton with large transfers of energy and momentum (Panofsky, 1968) suggests that the proton's charge is localised on a few scattering centres (analogous to Rutherford's inference from α-particle scattering that the atom contained a nucleus). The energy and angular distributions of the scattered electrons exhibit a correlation known as "scaling" which suggests that the scattering centres are structureless spin $\frac{1}{2}$ Dirac particles. This result combined with data on neutrino interactions indicates that these constituents have the same quantum numbers as the quarks of Table 3.1.

The scaling phenomenon is often referred to by saying that the quarks are "pointlike" (Bjorken, 1967; Feynman, 1969; Bjorken and Paschos, 1969). It is not known whether they are indeed pointlike or whether high resolution experiments will reveal a deeper substructure. However, just as in the atomic case the deeper structure of the nucleus is at first order phenomenologically irrelevant in describing atomic phenomena, so it appears that we can to a good first approximation ignore any quark substructure.

Hence we will investigate the deep inelastic scattering of electrons and neutrinos on protons and neutrons assuming that the beams scatter incoherently on the constituent quarks. We will look for evidence in the data supporting the notion that these quarks carry the quantum numbers that are expected from the spectroscopic phenomenology described in previous chapters.

9.1 Electron scattering

The cross-section for the exclusive process $1 + 2 \rightarrow 3 + 4$ may be written (appendix B of Bjorken and Drell)

$$d\sigma = \frac{1}{|\mathbf{v}_1 - \mathbf{v}_2|} \frac{1}{2E_1 2E_2} |A|^2 \frac{d^3 p_3}{(2\pi)^3 2E_3} \frac{d^3 p_4}{(2\pi)^3 2E_4} (2\pi)^4 \delta^{(4)}$$
$$\times (p_1 + p_2 - p_3 - p_4) \quad (9.1)$$

where E_1, $\mathbf{v}_1$, are the energy and velocity of particle 1, etc. The phase space and kinematic factors have been made explicit and the physics is contained in the Lorentz scalar invariant amplitude $|A|^2$. In particular this amplitude contains a sum over the spins of particles 3 and 4 and an average for 1 and 2. If the particles are fermions then equation (9.1) requires that the Dirac spinors be normalised such that

$$\bar{u}(p, s) u(p, s) = 2m \quad (9.2)$$

and hence

$$\sum_{\pm s} u_\alpha(p, s) \bar{u}_\beta(p, s) = (\not{p} + m)_{\alpha\beta} \quad (9.3)$$

(that equation 9.3 follows from 9.2 may be verified explicitly using the spinors in equation 6.17, normalised by equation 9.2 in place of equation 6.18).

We will study $e\mu \rightarrow e\mu$ and $ep \rightarrow ep$ scattering concentrating on $|A|^2$. The main difference between electron–proton and electron–muon scattering is that whereas the muon and electron interact by exchanging a photon (Fig. 9.1) in a known and calculable fashion (the electron–photon and muon–photon couplings are believed to be known exactly, viz. "pointlike coupling") the proton is a more complicated object containing an unknown structure. The purpose of electron–proton scattering experiments is to explore that structure. In order to see the way in which the proton's structure manifests itself in the cross-section it will be useful to compare the form of the electron–proton cross-section with the form of the analogous cross-section with a muon target. Apart from the fact that the muon and proton have different masses the only differences in the cross-sections will come from the corrections to the "pointlike coupling" which arise due to the proton's structure (e.g. its anomalous magnetic moment).

This detailed derivation of the kinematics and formalism of elastic and inelastic electron scattering can be omitted if desired. A heuristic derivation of the essential formalism is presented in section 9.2.2 and a discussion of the relation between scale invariance and the scattering from the pointlike objects is given in sections 9.4 and 11.1.

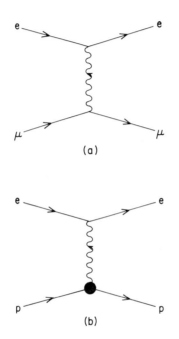

FIG. 9.1. (a) Electron–muon and (b) electron–proton scattering in the one photon exchange approximation.

9.1.1   ELECTRON–MUON SCATTERING

The electromagnetic current of the electron or muon is formally $\bar{u}(p')\gamma_\mu u(p)$. For $e\mu \to e\mu$ the invariant amplitude $|A|^2$ in equation (9.1) is

$$|A|^2 = \tfrac{1}{4} \sum_{s_1 s_2 s_3 s_4} \left| \bar{u}(k', s_3)\gamma^\mu u(k, s_1) \frac{e^2}{q^2} \bar{u}(p', s_4)\gamma_\mu u(p, s_2) \right|^2 \qquad (9.4)$$

where the $\tfrac{1}{4}$ comes from averaging over the initial spins. Exploiting equation (9.3) we can rewrite equation (9.4) as

$$|A|^2 = \frac{e^4}{q^4} L^{(e)}_{\mu\nu} L^{\mu\nu}_{(\mu)} \tag{9.5}$$

$$L^{(e)}_{\mu\nu} = \tfrac{1}{2} \text{Trace} \, (\not{k}' + m)\gamma_\mu (\not{k} + m)\gamma_\nu \tag{9.6}$$

$$= 2[k'_\mu k_\nu + k_\mu k'_\nu - g_{\mu\nu}(k \cdot k' - m^2)] \tag{9.7}$$

and analogously for $L^{\mu\nu}_{(\mu)}$. Here m is the lepton mass and the rules for evaluating traces in proceeding from equation (9.6) to (9.7) are given in appendix A of Bjorken and Drell. Contracting together the tensors $L^{(e)}_{\mu\nu} L^{\mu\nu}_{(\mu)}$ yields

$$L^{(e)}_{\mu\nu} L^{\mu\nu}_{(\mu)} = 8[k' \cdot p'k \cdot p + p \cdot k'k \cdot p' - m^2 p \cdot p' - M^2 k \cdot k' + 2m^2 M^2] \tag{9.8}$$

In the laboratory frame where the muon is initially at rest

$$p = (M, \mathbf{0}), \qquad k = (E, \mathbf{k}), \qquad k' = (E', \mathbf{k}'), \qquad q^2 = (k - k')^2$$

and if we neglect terms proportional to m^2 and M^2 then we see that $q^2 = -2k \cdot k' = -2p \cdot p'$ and hence

$$L^{(e)}_{\mu\nu} L^{\mu\nu}_{(\mu)} = 8\left[2M^2 EE' + \frac{q^2}{2} M(E' - E) + \frac{M^2 q^2}{2} \right] \tag{9.9}$$

Using

$$q^2 = -4EE' \sin^2 \frac{\theta}{2}$$

$$E' - E = -\nu = \frac{q^2}{2M}$$

$$M(E' - E) = EE'(\cos \theta - 1)$$

(see Fig. 9.2 for the definition of these quantities) then

$$L^{(e)}_{\mu\nu} L^{\mu\nu}_{(\mu)} = 16M^2 EE'\left\{ \cos^2 \frac{\theta}{2} - \frac{q^2}{2M^2} \sin^2 \frac{\theta}{2} \right\} \tag{9.10}$$

Now insert $|A|^2$ (equation 9.5 with 9.10) into the cross-section (equation 9.1). This yields

$$\frac{d^2\sigma}{dE' \, d\Omega_e} = \frac{4\alpha^2 (E')^2}{q^4}\left\{ \cos^2 \frac{\theta}{2} - \frac{q^2}{2M^2} \sin^2 \frac{\theta}{2} \right\} 2M \frac{d^3 p_4}{2E_4} \delta^{(4)}(q + p - p_4) \tag{9.11}$$

If the muon is not observed then we may integrate over d^3p_4. Since

$$\int \frac{d^3p_4}{2E_4} \delta^{(4)}(q+p-p_4) \to \delta[(p+q)^2 - M^2] \qquad (9.12)$$

then with $(p)^2 = M^2$ and $p \cdot q \equiv M\nu$ we find $(Q^2 \equiv -q^2)$

$$\frac{d^2\sigma}{dE'\,d\Omega_e} = \frac{4\alpha^2(E')^2}{Q^4}\left\{\cos^2\frac{\theta}{2} + \frac{Q^2}{2M^2}\sin^2\frac{\theta}{2}\right\}\delta\left(\nu - \frac{Q^2}{2M}\right) \qquad (9.13)$$

This double differential cross-section is in a form which will compare most immediately with the inelastic electron scattering (equation 9.35).

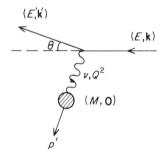

FIG. 9.2. Electron–proton interactions in the laboratory frame.

To show how it relates to a form frequently met in the literature we integrate over dE' using the identity $\int dx\,\delta(f(x)) \equiv [df/dx]^{-1}$ and obtain

$$\frac{d\sigma}{d\Omega} = \frac{4\alpha^2(E')^2\cos^2\dfrac{\theta}{2}}{Q^4}\left[1 + \frac{2E}{M}\sin^2\frac{\theta}{2}\right]^{-1}\left\{1 + \frac{Q^2}{2M^2}\sin^2\frac{\theta}{2}\right\} \qquad (9.14)$$

By energy conservation

$$1 + \frac{2E}{M}\sin^2\frac{\theta}{2} \equiv \frac{1}{E'}\left(E' + \frac{Q^2}{2M}\right) = \frac{E}{E'} \qquad (9.15)$$

and hence

$$\frac{d\sigma}{d\Omega} = \left(\frac{d\sigma}{d\Omega}\right)_{\text{Mott}}\frac{E'}{E}\left(1 + \frac{Q^2}{2M^2}\sin^2\frac{\theta}{2}\right) \qquad (9.16)$$

which has factored out the Mott cross-section for relativistic electron scattering in a Coulomb field (Mott, 1929; Chapter 7 of Bjorken and Drell, 1964). The multiplicative factor in equation (9.16) arises from the recoil ($E' \neq E$) and magnetic interaction. We can manipulate the Mott cross-section into the form

$$\left(\frac{d\sigma}{dQ^2}\right)_{Mott} = \frac{4\pi\alpha^2}{Q^4}\left(1 - \frac{\nu}{E} - \frac{Q^2}{4E^2}\right) \xrightarrow{\nu^2 \gg Q^2, E \to \infty} \frac{4\pi\alpha^2}{Q^4} \tag{9.17}$$

One sometimes meets the "no structure" cross-section (Rutherglen, 1969) defined by

$$\left(\frac{d\sigma}{d\Omega}\right)_{NS} \equiv \left(\frac{d\sigma}{d\Omega}\right)_{Mott} \frac{E'}{E} \tag{9.18}$$

which physically is Coulomb scattering with recoil.

9.1.2 ELASTIC ELECTRON–PROTON SCATTERING

Electron–proton scattering takes place by photon exchange just like the previous example but now the photon–proton coupling is not known (contrast the γ_ρ coupling for the muon). If account is taken of the fact that the proton spinors $\bar{u}(p)$ and $u(p)$ obey the Dirac equation then the most general form of the proton's electromagnetic current $J_\rho = \bar{u}(p)\Gamma_\rho u(p)$ can only involve γ_ρ, q_ρ and $i\sigma_{\rho\alpha}q^\alpha$ multiplied by general functions of $p^2(\equiv M^2)$, $p'^2(\equiv M^2)$ and q^2 standing between the spinors, viz.

$$J_\rho \sim \bar{u}(p')\{\Gamma_1(q^2)\gamma_\rho + \Gamma_2(q^2)i\sigma_{\rho\alpha}q^\alpha + \Gamma_3(q^2)q_\rho\}u(p) \tag{9.19}$$

where $\sigma_{\rho\alpha} = (i/2)[\gamma_\rho, \gamma_\alpha]$. Current conservation, $q^\rho J_\rho = 0$, forces $\Gamma_3(q^2) \equiv 0$ since the terms multiplying $\Gamma_{1,2}$ vanish identically and hence we have

$$J_\rho = \bar{u}(p')\left\{F_1(q^2)\gamma_\rho + \kappa\frac{F_2(q^2)}{2M}i\sigma_{\rho\alpha}q^\alpha\right\}u(p) \tag{9.20}$$

where $F_1(0) = 1$ (proton charge) and $F_2(0) = 1$. The quantity κ is the anomalous magnetic moment of the proton ($=1\cdot79$ Bohr magnetons). Using equation (9.20) for the proton's electromagnetic current, we can calculate the electron–proton scattering cross-section in direct analogy to the electron–muon case. The only essential difference comes in the invariant amplitude $|A|^2$ where $L_{(\mu)}^{\mu\nu}$ is replaced by

$$L_{(P)}^{\mu\nu} = \tfrac{1}{2}\, \text{Trace} \, (\not{p}' + M)\Gamma^\mu (\not{p} + M)\Gamma^\nu \qquad (9.21)$$

where

$$\Gamma^\mu \equiv \gamma^\mu F_1(q^2) + \frac{\kappa F_2(q^2)}{2M} i\sigma^{\mu\alpha} q_\alpha$$

in contrast to the muon target where $\Gamma^\mu \equiv \gamma^\mu$. In place of equation (9.14) one finds for the cross-section

$$\frac{d\sigma}{d\Omega} = \left(\frac{d\sigma}{d\Omega}\right)_{\text{Mott}} \frac{E'}{E}\left[\left(F_1^2 + \frac{\kappa^2 Q^2}{4M^2} F_2^2\right) + \frac{Q^2}{2M^2}(F_1 + \kappa F_2)^2 \tan^2 \frac{\theta}{2}\right]$$

$$(9.22)$$

which reduces to the muon example if $F_1 \equiv 1$, $F_2 = 0$.

Alternatively one can define electric and magnetic form factors

$$G_{\text{E}} \equiv F_1 - \frac{\kappa Q^2}{4M^2} F_2$$

$$G_{\text{M}} \equiv F_1 + \kappa F_2 \qquad (9.23)$$

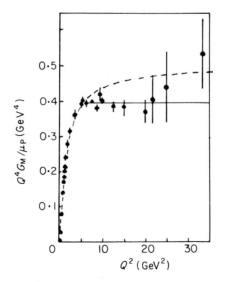

FIG. 9.3. The proton form factor G_{M}. The dotted curve is the dipole form in equation (9.25). The solid line indicates that $G_{\text{M}} \propto 1/Q^4$ for $Q^2 > 4\ \text{GeV}^2$.

in terms of which (compare equation 9.13)

$$\frac{d^2\sigma}{d\Omega\, dE'} = \frac{4\alpha^2 (E')^2}{Q^4}\, \delta\left(\nu - \frac{Q^2}{2M}\right)$$

$$\times \left\{ \cos^2\frac{\theta}{2}\left[\frac{G_E^2 + (Q^2/4M^2)G_M^2}{1 + Q^2/4M^2}\right] + \frac{Q^2 G_M^2}{2M^2}\sin^2\frac{\theta}{2}\right\} \quad (9.24)$$

The data (Taylor, 1975; and Fig. 9.3) suggest that $G_{E,M}(Q^2)$ decrease as Q^2 increases. This decrease is consistent with a dipole behaviour

$$\frac{G_M(Q^2)}{G_M(0)} = \left(1 + \frac{Q^2}{0\cdot 7 \text{ GeV}^2}\right)^{-2} \quad (9.25)$$

and causes the elastic cross-section to die out rapidly as Q^2 increases, e.g.

$$\frac{d\sigma(ep \to ep)/dQ^2}{d\sigma(e\mu \to e\mu)/dQ^2} \propto (Q^2)^{-4} \quad (9.26)$$

9.1.3 INELASTIC ELECTRON SCATTERING

We come now to the focal point of the discussion, namely the process $e + p \to e + $ hadrons where the energy of the final electron and its angle of scatter are all that one measures. A spectrometer sits at some angle θ to the incident electron beam and records the energies of the scattered electrons and counts events. We assume that the one photon exchange mechanism is dominant (Fig. 9.4).

The differential cross-section may be written

$$\frac{d^2\sigma}{d\Omega\, dE'} = \frac{\alpha^2}{Q^4}\frac{E'}{E}|A|^2 \equiv \frac{\alpha^2}{Q^4}\frac{E'}{E}L_{\mu\nu}^{(e)}W^{\mu\nu} \quad (9.27)$$

where (equations 9.4 to 9.7)

$$L_{\mu\nu}^{(e)} = \tfrac{1}{2}\sum_{s'} \bar{u}(k's')\gamma_\mu u(ks)\bar{u}(ks)\gamma_\nu u(k's') \quad (9.28)$$

and the hadronic tensor is

$$W_{\mu\nu} = \tfrac{1}{2}\sum_n \langle p|J_\mu^+|n\rangle\langle n|J_\nu|p\rangle (2\pi)^3\delta^4(p + q - p_n) \quad (9.29)$$

Compare equation (9.27) with Gilman (1968); our $L_{\mu\nu}$ is four times his, and all other quantities are identical with his (this is useful in comparing with the conventions of Bjorken and Walecka, 1966, and Drell and Walecka, 1964, and is discussed in Gilman, 1968).

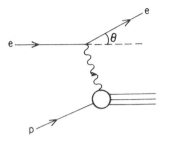

FIG. 9.4. Electron–proton inelastic inclusive scattering in the one photon exchange approximation.

If the initial proton is unpolarised then the most general form for $W^{\mu\nu}$ which is symmetric in $\mu\nu$ is

$$W^{\mu\nu} = W_1 g^{\mu\nu} + \frac{W_2}{M^2} p^{\mu} p^{\nu} + \frac{W_4}{M^2} q^{\mu} q^{\nu} + \frac{W_5}{M^2}(p^{\mu} q^{\nu} + q^{\mu} p^{\nu}) \qquad (9.30)$$

($L_{\mu\nu}$ is symmetric in $\mu\nu$ and hence only $W^{\mu\nu(\text{sym})}$ contributes to the unpolarised cross-section.) The W_i are in general functions of ν, q^2. Gauge invariance constrains $q_{\mu} W^{\mu\nu} = 0$ for any ν, q^2. Applying this to equation (9.30) requires

$$W_5 = -W_2(p \cdot q/q^2) \qquad (9.31)$$

$$W_4 = W_2(p \cdot q/q^2)^2 - W_1 M^2/q^2 \qquad (9.32)$$

and so

$$W^{\mu\nu} = W_1(\nu, q^2)\left(-g^{\mu\nu} + \frac{q^{\mu} q^{\nu}}{q^2}\right)$$

$$+ \frac{W_2(\nu, q^2)}{M^2}\left[\left(p^{\mu} - \frac{p \cdot q}{q^2} q^{\mu}\right)\left(p^{\nu} - \frac{p \cdot q}{q^2} q^{\nu}\right)\right] \qquad (9.33)$$

Contracting with $L_{\mu\nu}$ (equation 9.7) yields

$$L_{\mu\nu} W^{\mu\nu} = 4 W_1 k \cdot k' + \frac{W_2}{M^2} \{4p \cdot kp \cdot k' - 2k \cdot k' M^2\} \quad (9.34)$$

In the laboratory frame (compare equation 9.8 *et seq.*), equation (9.34) substituted into equation (9.27) yields

$$\frac{d^2\sigma}{dE' \, d\Omega} = \frac{4\alpha^2 (E')^2}{Q^4} \left\{ \cos^2 \frac{\theta}{2} W_2(\nu, q^2) + 2 W_1(\nu, q^2) \sin^2 \frac{\theta}{2} \right\} \quad (9.35)$$

Comparing this with equations (9.24) and (9.13) we can see the form that $W_{1,2}$ have for elastic scattering $(Q^2 \equiv -q^2)$

$$W_1^{el}(\nu, Q^2) = \frac{Q^2}{4M^2} G_M^2(Q^2) \delta\left(\nu - \frac{Q^2}{2M}\right) \quad (9.36)$$

$$W_2^{el}(\nu, Q^2) = \frac{G_E^2(Q^2) + (Q^2/4M^2) G_M^2(Q^2)}{1 + Q^2/4M^2} \delta\left(\nu - \frac{Q^2}{2M}\right) \quad (9.37)$$

(which are identical to equation 8 in Gilman, 1968). Notice the presence of the form factors $G_{E,M}(Q^2)$. For a Dirac particle with "pointlike" coupling (equation 9.13) we have for the dimensionless combinations

$$2MW_1^{pt}(\nu, Q^2) = \frac{Q^2}{2M\nu} \delta\left(\frac{Q^2}{2M\nu} - 1\right) \quad (9.38)$$

$$\nu W_2^{pt}(\nu, Q^2) = \delta\left(\frac{Q^2}{2M\nu} - 1\right) \quad (9.39)$$

which are functions only of the dimensionless ratio $\omega = 2M\nu/Q^2$, no scale of mass being present (contrast equations 9.36 and 9.37 where the $G_{E,M}(Q^2)$ contain an explicit mass scale in equation 9.25).

The two structure functions $W_{1,2}(\nu, q^2)$ are related to the photoabsorbtion cross-sections for transverse photons (helicity ± 1) and longitudinal-scalar photons (helicity 0). It is because the photoabsorbtion has these two independent cross-sections σ_S, σ_T that two independent structure functions $W_{1,2}$ are present in the electron-scattering cross-section. A photon moving along the z-axis with energy ν and squared mass Q^2 can have polarisation vectors

$$\varepsilon(\pm 1) = \mp \frac{1}{\sqrt{2}}(0; 1, \pm i, 0) \quad (9.40)$$

$$\varepsilon(0) = \frac{1}{\sqrt{Q^2}} [\sqrt{(Q^2 + \nu^2)}; 0, 0, \nu] \qquad (9.41)$$

(gauge invariance demands that $q \cdot \varepsilon = 0$). If the incident flux of photons is K then defining

$$\sigma_{\pm 1,0} = \frac{4\pi^2 \alpha}{K} \varepsilon_{\pm 1,0}^{+\mu} W_{\mu\nu} \varepsilon_{\pm 1,0}^{\nu} \qquad (9.42)$$

yields

$$W_1 = \frac{K}{4\pi^2 \alpha} \sigma_T (\sigma_T \equiv \tfrac{1}{2}(\sigma_+ + \sigma_-)) \qquad (9.43)$$

$$W_2 = \frac{K}{4\pi^2 \alpha} (\sigma_T + \sigma_S) \frac{Q^2}{Q^2 + \nu^2} \qquad (9.44)$$

A useful quantity is

$$R \equiv \frac{\sigma_S}{\sigma_T} = \frac{W_2}{W_1} \left(1 + \frac{\nu^2}{Q^2}\right) - 1 \qquad (9.45)$$

When $Q^2 = 0$ the photon flux is $K = \nu$. When $Q^2 \neq 0$ there is some arbitrariness about the definition. Gilman's convention (1968) is to choose $K = |\mathbf{q}| = \sqrt{(\nu^2 + Q^2)} \approx \nu + Q^2/2\nu$. Hand (1963), however, takes K to be the "equivalent photon energy"; namely that energy which would produce a final hadron state of the same mass as would have been created by a real photon with energy ν. Hence, since the final state mass M^* satisfies $M^{*2} = M^2 + 2M\nu - Q^2$ then $K^{\text{Hand}} = \nu - Q^2/2M$. The structure functions $W_{1,2}$ are well defined in terms of the experimentally measured $d^2\sigma/d\Omega\, dE'$ (equation 9.35). The photoabsorbtion cross-sections are then defined modulo this arbitrary flux factor K.

Knowing the form for $d^2\sigma/d\Omega\, dE'$ in terms of $W_{1,2}$ and also our definitions of $W_{1,2}$ in terms of $\sigma_{T,S}$, we can substitute these latter in and obtain $d^2\sigma/d\Omega\, dE'$ in terms of the photoabsorption cross-sections. This form is widely met in the literature and completes our kinematic study. The reader is invited to perform the algebra as an exercise and verify that

$$\frac{d^2\sigma}{d\Omega\, dE'} = \Gamma(\sigma_T + \varepsilon\sigma_S) \qquad (9.46)$$

where

$$\varepsilon = \left[1 + 2\frac{Q^2 + \nu^2}{Q^2}\tan^2\frac{\theta}{2}\right]^{-1}$$

$$\Gamma = \frac{K\alpha}{2\pi^2 Q^2}\frac{E'}{E}\frac{1}{1-\varepsilon}$$

(9.47)

9.2 Deep inelastic scattering and partons

The structure functions $W_{1,2}$ are functions of both ν and q^2 which can be independently varied (since the former depends only on the electron's energy loss while the latter depends on the scattering angle). The squared mass of the unobserved hadron system W^2 is

$$W^2 = (p + q)^2 = M^2 + 2p \cdot q + q^2$$

(9.48)

which in the laboratory frame reads $(Q^2 \equiv -q^2)$.

$$W^2 = M^2 + 2M\nu - Q^2$$

(9.49)

The kinematic region probed ($\nu - Q^2$ plane) is shown in Fig. 9.5.

The data may be summarised as follows. For fixed W (e.g. elastic scattering or resonance production)

$$\left.\begin{array}{l}MW_1(W, Q^2)\\\nu W_2(W, Q^2)\end{array}\right\} \xrightarrow{Q^2 \to \infty} 0$$

(9.50)

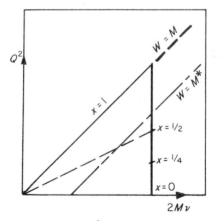

FIG. 9.5. The $\nu - Q^2$ plane probed in $ep \to eX$.

due to the resonance form factors killing the cross-section at large Q^2 (compare equations 9.36 and 9.37). However for fixed $\omega = 2M\nu/Q^2$ and $Q^2 \gtrsim 1$ GeV^2 there is the remarkable phenomenon (Fig. 9.6) of Bjorken scaling (Bjorken, 1969), namely that

$$MW_1(\omega, Q^2) \to F_1(\omega)$$
$$\nu W_2(\omega, Q^2) \to F_2(\omega)$$

(9.51)

In addition it appears that $\sigma_L \ll \sigma_T$ and hence that

$$\omega F_2(\omega) \simeq 2F_1(\omega)$$

(9.52)

The Q^2 independence of dimensionless $F_{1,2}$ for fixed values of the dimensionless ω implies that the structure functions are independent of

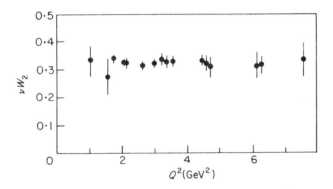

FIG. 9.6. Approximate Q^2 independence of νW_2 at fixed $\omega \simeq 4$.

any mass scale (scale invariant). This phenomenon arises naturally if the inelastic electron–proton scattering is due to incoherent elastic scattering from pointlike constituents in the proton since for this latter process no scale of mass is present (equation 9.39).

If the basic scattering of the electron is incoherently on a parton carrying fraction x of the target's four momentum (Fig. 9.7) and if the parton mass and transverse momentum are negligible then

$$\nu W_2(\nu, Q^2) \to F_2(\omega) = \sum_i \int dx\, e_i^2 x f_i(x) \delta\left(x - \frac{1}{\omega}\right)$$

(9.53)

where the sum is over the various species of parton (u, d, s, c . . .), of charges e_i and $f_i(x)$ is the probability that the parton has momentum in the interval $x \rightarrow x + dx$. The important feature here is the $xf(x)$ structure from which many relations will be seen to flow.

We will see later that the phenomenology is consistent with the charged pointlike constituents having the quantum numbers of quarks.

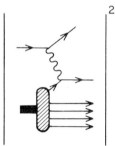

FIG. 9.7. Inclusive $ep \rightarrow eX$ viewed as incoherent elastic scattering of electrons from partons.

That the partons have spin $\frac{1}{2}$ is supported by the observation of a small value of $R = \sigma_L/\sigma_T$. We can see that this quantity will yield information on the parton's spin as follows.

If one sits in a frame where the photon and parton momenta are collinear then a spin 0 parton could not absorb a photon with helicity

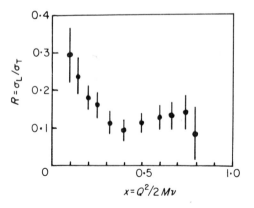

FIG. 9.8. $R \equiv \sigma_L/\sigma_T$ as a function of x.

±1. Hence for the spin zero partons with negligible transverse momenta,

$$\frac{\sigma_L}{\sigma_T} \to \infty \qquad (9.54)$$

This is not at all like the data (Atwood *et al.*, 1976), so very little if any charge of the proton is carried by spinless constituents (at least for the range of $x \gtrsim 0.1$ so far studied). Spin $\frac{1}{2}$ partons on the other hand give $\sigma_L/\sigma_T \to 0$ which qualitatively is like Fig. 9.8. These results will be derived in the following paragraphs.

9.2.1 *stu* APPROACH TO ELECTRON SCATTERING AND SCALING

If the inelastic electron scattering from the proton is due to the incoherent elastic scatterings of the electron on the Dirac quarks, then the inelastic cross-section will be given by the sum of elastic electron–quark cross-sections. These latter are essentially the same as the electron muon cross-sections.

We have calculated the e–μ cross-section (section 9.1.1). By comparing with the inelastic cross-section (equation 9.35) we will show how MW_1 and νW_2 are predicted to scale. If one rewrites these equations in terms of the Mandelstam invariants s, t, u

$$s = (p + k)^2, \qquad t = (p - p')^2, \qquad u = (p - k')^2 \qquad (9.55)$$

then the results are almost immediate. Moreover, it is then very easy to make contact with hadronic phenomenology where these variables are widely employed. Finally, one can gain an intuitive feeling for the physics behind the angular dependences in the scattering cross-sections.

9.2.1.1 *Electron–muon scattering*

First we look at the $e^-\mu^+ \to e^-\mu^+$ cross-section. Neglecting masses at large s, t, u then the $L_{\mu\nu}W^{\mu\nu}$ (equation 9.8) becomes

$$L_{\mu\nu}W^{\mu\nu} \to 2(s^2 + u^2) \qquad (9.56)$$

Then, noting that

$$\frac{d^2\sigma}{dQ^2\,d\nu} \equiv \frac{\pi}{EE'}\frac{d^2\sigma}{d\Omega\,dE'}$$

and that

$$s = 2ME, \qquad u = -2ME', \qquad t = -Q^2 \qquad (9.57)$$

yields (equations 9.14 and 9.15),

$$\frac{d\sigma}{dt} = \frac{4\pi\alpha^2}{t^2}\frac{1}{2}\left(\frac{s^2+u^2}{s^2}\right) \qquad (9.58)$$

In order to make contact with the inelastic formalism we should make explicit the energy-momentum conservation. We do this by writing (compare equation 9.13)

$$\frac{d^2\sigma}{dt\,du} = \frac{4\pi\alpha^2}{t^2}\frac{1}{2}\left(\frac{s^2+u^2}{s^2}\right)\delta(s+t+u) \qquad (9.59)$$

(recall that $s+t+u = \Sigma m^2$ and so at high energies, neglecting the masses, we have $u = -(s+t)$). Now recalling equation (9.57) we see that

$$s+t+u = 2m(E-E') - Q^2 = 2M\nu - Q^2 \qquad (9.60)$$

It will also be useful, later, to notice that

$$\frac{-t}{s+u} = \frac{Q^2}{2M\nu} \equiv \frac{1}{\omega} \qquad (9.61)$$

9.2.1.2 Electron–parton scattering

In the parton model, the *inelastic* electron-target scattering is hypothesised to arise from the sum of incoherent *elastic* scattering of electrons on the partons in the target. If the partons have spin $\frac{1}{2}$ and couple to the photon just as does the μ^+ of the previous example ("pointlike coupling") then we can easily obtain an expression for the cross-section.

Let us neglect any parton momentum transverse to the target so that

$$p_{parton} = x p_{target} \qquad (9.62)$$

Then from the previous example we can write the cross-section for elastic scattering on a muon(parton) with momentum xp. To do this we note that

$$s \to xs, \qquad u \to xu \qquad (9.63)$$

but t remains untouched since this can be defined involving the electron vertex alone. Then (compare equation 9.59)

$$\left(\frac{d^2\sigma}{dt\,du}\right)_{e\mu(x)\to e\mu(x)} = \frac{4\pi\alpha^2}{t^2}\frac{1}{2}\left(\frac{s^2+u^2}{s^2}\right)x\delta(t+x(s+u)) \qquad (9.64)$$

If the target is built from partons of types (flavours) i and the probability for a parton i to have momentum fraction x to $x+dx$ is $f_i(x)$ then the inelastic electron-target cross-section will be (after summing over all the elastic parton contributions)

$$\left(\frac{d^2\sigma}{dt\,du}\right)_{eN\to eX} = \frac{4\pi\alpha^2}{t^2}\frac{1}{2}\frac{s^2+u^2}{s^2}\int dx \sum_i e_i^2 x f_i(x)\frac{1}{s+u}\delta\left(x-\frac{1}{\omega}\right) \qquad (9.65)$$

where we used equation (9.61) in rewriting the delta function.

We already see the appearance of the structure in equation (9.53). To obtain that expression explicitly we must compare the equation (9.65) with the expression for $eN \to eX$ which involves $W_{1,2}$ (equation 9.35). This can be immediately recast as

$$\frac{d^2\sigma}{dt\,du} = 2M\frac{u}{s}\frac{4\pi\alpha^2}{t^2}\left\{\cos^2\frac{\theta}{2}W_2(\nu, Q^2) + 2\sin^2\frac{\theta}{2}W_1(\nu, Q^2)\right\} \qquad (9.66)$$

If the hadronic final state has mass W then

$$s + t + u = M^2 + W^2 \qquad (9.67)$$

and if we are in the deep inelastic region $s, t, u, W^2 \to \infty$ then

$$\sin^2\frac{\theta}{2} = -\frac{tM^2}{su} \qquad \left(\text{since } Q^2 = +4EE'\sin^2\frac{\theta}{2}\right) \qquad (9.68)$$

$$\nu = \frac{s+u}{2M} \qquad (9.69)$$

$$x \equiv \frac{Q^2}{2M\nu} = -\frac{t}{s+u} \qquad (9.70)$$

and so equation (9.66) becomes

$$\left(\frac{d^2\sigma}{dt\,du}\right)_{eN\to eX} = \frac{4\pi\alpha^2}{t^2}\frac{1}{2}\frac{1}{s^2(s+u)}[2xF_1(s+u)^2 - 2usF_2] \quad (9.71)$$

where $F_1 \equiv MW_1$, $F_2 \equiv \nu W_2$ are in principle functions of x, t (this follows most rapidly from equation 9.34).

Since s and u can be independently varied we compare coefficients in equations (9.65) and (9.71). This shows immediately that $F_{1,2}$ are only functions of x

$$2xF_1(x) = F_2(x) = \sum_i e_i^2 x f_i(x) \quad (x \equiv 1/\omega) \quad (9.72)$$

which is the master formula of the spin $\frac{1}{2}$ parton model (Bjorken and Paschos 1964; Callan and Gross 1969).

9.2.2   HEURISTIC APPROACH TO ELECTRON–PARTON SCATTERING

The relation between $2xF_1(x)$ and $F_2(x)$ corresponds to

$$\frac{\sigma_L}{\sigma_T} \to 0 \quad (9.73)$$

and is a consequence of the spin $\frac{1}{2}$ nature of the quarks. We have already noted that spin 0 constituents would contribute only in σ_L and that this was readily understood physically by simple helicity considerations. The above result for the spin $\frac{1}{2}$ is ultimately related to the helicity conserving nature of the electromagnetic current. It is this helicity conservation that is at the root of the angular (s, u) distributions in equation (9.71) and hence of the $F_{1,2}$ interrelation.

To illustrate this we can make a heuristic derivation of the formulae for elastic electron–parton scattering.

The cross-section for high-energy electron scattering in a Coulomb field (e.g. the field of an infinitely massive object like a nucleus) is

$$\frac{d\sigma}{dt} = \frac{4\pi\alpha^2}{t^2}\left(\frac{1+\cos\theta}{2}\right)^2 \quad (9.74)$$

where $t \equiv -Q^2$ and θ is the scattering angle in the centre of mass (which is the same as the laboratory in this case). One can qualitatively understand this result:

i. The dimensions of $d\sigma/dt$ are [energy]$^{-4}$. Since the photon propagator provides t^{-2} in $d\sigma/dt$ then no further dimensional quantities occur.

ii. The high-energy electron–photon vertex conserves helicity. Hence $180°$ scattering is forbidden and in turn this is the origin of the angular dependence. Note that as $E \to \infty$ and $\nu^2 \gg Q^2$ then $\theta \to 0$ and so equation (9.17) is recovered.

The cross-section for $e^-\mu^+ \to e^-\mu^+$ is (equation 9.58)

$$\frac{d\sigma}{dt} = \frac{1}{2}\frac{4\pi\alpha^2}{t^2}\left(\frac{u^2}{s^2}+1\right) \tag{9.75}$$

In the centre of mass $-u/s \simeq (1+\cos\theta)/2$ and so the first term is the same as in the Coulomb example. The difference is in the extra presence of an isotropic term and an overall factor of $\frac{1}{2}$.

The factor $\frac{1}{2}$ arises due to the averaging over the two spin states of the "target" muon (contrast the previous example).

When e^- and μ^+ have net $J_z = \pm 1$ the $180°$ scattering is angular momentum forbidden as before—hence the u^2/s^2. When $J_z = 0$ the $180°$ scattering can occur (contrast the previous example)—hence the presence of an isotropic term.

The resulting cross-section is symmetric in s^2 and u^2. Hence the electron–proton inelastic scattering cross-section will be symmetric in s^2 and u^2 if it is the sum of such pointlike elastic contributions. Hence the equal weightings of the coefficients of $(s+u)^2$ and $-2us$ emerge with the result

$$2xF_1 = F_2 \leftrightarrow \frac{\sigma_L}{\sigma_T} \to 0 \tag{9.76}$$

9.3 Scaling, the impulse approximation and confinement

The naive quark–parton model is the subnucleon analogue of the familiar impulse approximation of nuclear physics (Drell, 1970; Drell et al., 1970; Drell and Yan, 1971). The current–target interaction in the impulse approximation may be represented by Fig. 9.9, the mathematical formulation having been described in the preceding section. The implicit assumptions that must be satisfied if the incoherent impulse approximation is valid are:

1. During the time of the current interaction one can neglect interactions between the partons (Fig. 9.9(a)).
2. Final state interactions can be ignored (Fig. 9.9(b)).

Intuitively one parton has been struck so violently that it has recoiled from its fellow partons and so can be regarded as quasi-free, independent of their influence. In the case of a nucleus this is reasonable since the nucleus shatters as a result of the struck parton (a nucleon in this example) being truly removed from its friends and so the final state interactions are indeed absent.

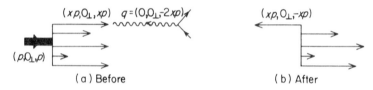

FIG. 9.9. Current interaction in the parton Breit frame: $-q^2/2P \cdot q \equiv x$. The parton recoils from its fellows as if quasi-free.

If the target is a proton then the partons are quarks. The quarks appear to be permanently confined within the proton (in contrast to a nucleus whose partons—the nucleons—can be removed). Hence an implicit assumption is being made in the quark–parton model, namely that the final state interactions which confine the quarks act at large space-time distances of the order of the proton size, much larger than the parton size and the timescale of the current–parton interaction. Then during the time of interaction the parton can justifiably be regarded as quasi-free and the cross-section calculated; the subsequent final state interactions do not affect the calculation. A widely quoted example is that of a particle attached to a slack elastic band. If the particle is struck by an impulse then it moves off as if free. The cross-section for another particle to interact with it via an impulse is correctly calculated by pretending that the elastic band is not there. The confining effect of the elastic band acts later.

One attempt to formulate a quark–parton model, including the confining effect of final state interactions, while still retaining the quasi-free aspect of the impulse approximation has been that of Preparata (1975) (section 10.4). An alternative has been that of bag models (e.g. Chapter 18) where the confinement arises due to an external pressure, the deep inelastic scaling behaviour arising from impulse interaction with quasi-free quarks in the bag (Jaffe, 1975; Jaffe and Patrascioiu, 1975).

Two-dimensional quantum chromodynamics is a mathematical laboratory where quarks can be confined ('t Hooft, 1974; for an introduction see Ellis, 1977). In this model the scaling phenomena also arise, and the quark confinement is found not to affect the scaling properties of the model (Einhorn, 1976; Callan *et al.*, 1976). If this model is a guide to the real four-dimensional world then we may hope that some *a posteriori* justification will be given to the naive parton model (where the scaling phenomena and other properties of interactions are discussed while the final state interactions and confining properties are ignored).

At the present time the only real justification for the naive quark–parton model is the extent to which it is empirically successful (Chapter 11).

9.4 Scaling and pointlike structure

Before we investigate the consequences of the master formula, equation (9.72), it is instructive to ponder on the relation between pointlike substructure and the existence of scaling phenomena.

In any electron scattering process, the mass W of the produced system at the hadronic vertex is given by

$$W^2 \equiv (p+q)^2 = M^2 + 2p \cdot q + q^2$$

$$\underset{\text{(lab)}}{=} M^2 + 2M\nu - Q^2 \tag{9.77}$$

where $Q^2 \equiv -q^2 > 0$ is the squared mass and ν is the laboratory energy of the exchanged photon. Consider first the case of elastic scattering. Here $W^2 = M^2$ so that

$$2M\nu = Q^2 \tag{9.78}$$

or equivalently

$$x \equiv \frac{Q^2}{2M\nu} = 1 \qquad (9.79)$$

Hence elastic scattering is described by just *one* kinematical variable and so the scattering of an electron elastically on a target proton will involve

$$\delta(x - 1) \qquad (9.80)$$

from the kinematic constraint (more precisely it is the dimensionless quantity νW_2 (equation 9.39) that is proportional to this dimensionless delta function).

There is also a possibility that there will be an additional Q^2 dependence $f(Q^2)$ arising dynamically due to the target having a finite extension in space, hence an internal structure which can be excited in the photoabsorbtion and which responds differently to different values of Q^2 (for example at large values of Q^2 it is more probable that the proton will break up than at the smaller values where less momentum hits it). This $f(Q^2)$ is known as the proton's elastic form factor which typically has a dipole type of dependence $(1 + Q^2/0 \cdot 7 \text{ GeV}^2)^{-2}$. Hence elastic scattering may be summarised by

$$\nu W_2 \propto \quad f(x) \quad \times \quad f(Q^2) \qquad (9.81)$$
$$\updownarrow \qquad\qquad \updownarrow$$
$$\text{kinematic} \quad \text{dynamic-form factor}$$
$$\updownarrow$$
$$\text{finite target size}$$

Actually the proton has spin $\frac{1}{2}$ and hence two form factors (e.g. G_E, G_M in equations 9.36 and 9.37). In the present discussion $f(Q^2)$ refers to the spin averaged combination entering the dimensionless quantity in equation (9.81).

In the form factor, written[1] $(1 + Q^2/\Lambda^2)^{-N}$, the parameter Λ is bigger the more pointlike the target. For a genuine structureless target, like a muon for example, $\Lambda \to \infty$. In such a case the dimensionless quantity νW_2 is controlled only by the dimensionless delta function and no explicit scale of Q^2 is present. Hence we say that νW_2 is scale invariant.

[1] All that concerns us is that the form factor falls as Q^2 increases and Λ^2 sets the scale of Q^2. I do not mean to imply that the form factor necessarily falls as a simple power.

This is intimately related with the structureless or pointlike nature of the scattering centre.

The muon is known to be structureless down to 10^{-15} cm whereas the proton is of order 1 fm (10^{-13} cm) in size and this is reflected in the proton's Q^2 dependent form factor. In elastic scattering if $Q^2 \ll \Lambda^2$ then $f(Q^2) \simeq 1$ and so the target will appear to be pointlike (i.e. no internal structure will be resolved) and νW_2 will scale whereas for $Q^2 \gg \Lambda^2$ the structure is revealed and the form factor causes the cross-section to die with increasing Q^2, breaking the scale invariance of νW_2.

Consider therefore electron scattering from a *nucleus* with mass M and plot the cross-section against $x = Q^2/2M\nu$. At $x = 1$ we will see the

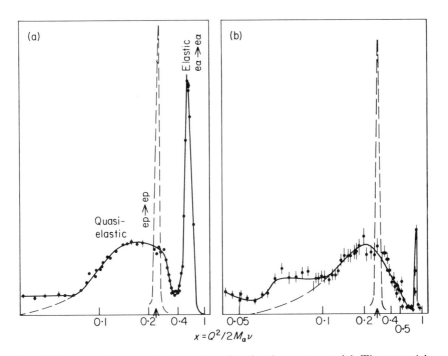

FIG. 9.10. Inelastic e–α scattering as template for the parton model. The α-particle consists of four partons (He$^4 \equiv$ ppnn) and the quasi-elastic peak is seen at $x \simeq \frac{1}{4}$. 400 MeV electrons are detected at (a) $45°(Q^2 \simeq 0.08 \text{ GeV}^2)$ and (b) $60°(Q^2 \simeq 0.1 \text{ GeV}^2)$. The elastic e$\alpha \to$ eα scattering has almost disappeared in (b) whereas the quasi-elastic peak is approximately Q^2 independent and hence scales. The ep $\to$ ep quasi-elastic scattering is shown in the dotted peak centred at $x \simeq 1/N = \frac{1}{4}$ and is smeared out by Fermi momentum. As $x \to 0$ pions are emitted from the nucleon and mask the quasi-elastic peak which is otherwise tending to zero. Compare with Fig. 11.1 for electron–carbon interactions.

coherent elastic scattering from the nucleus. At higher energies we will see coherent excitation of nuclear resonances. If we increase Q^2 so that $Q^2 > \Lambda^2_{\text{nucleus}}$ then the nuclear form factors will kill the elastic nuclear cross-section and also the nuclear resonance production. The cross-section will be dominated by the beam scattering *incoherently* and *elastically* from the nuclear constituents, i.e. the neutrons and protons. This "quasi-elastic scattering" occurs when

$$2m_p \nu = Q^2 \tag{9.82}$$

where m_p is the mass of a *proton or neutron*. Hence this quasi-elastic peak is at $x = m_p/M \simeq 1/N$ where N is the number of constituents (see also West, 1975). If

$$\Lambda^2_{\text{proton}} \gg Q^2 \gg \Lambda^2_{\text{nucleus}} \tag{9.83}$$

then no internal structure of the protons will be resolved (they will appear pointlike) and the scattering will exhibit scaling, i.e. be dependent only on x (specifically $x \equiv 1/N$), there being no effective Q^2 dependence for $Q^2 \ll \Lambda^2_{\text{proton}}$. Data on α-particle targets are shown in Fig. 9.10 and C_{12} in Fig. 11.1. The latter are discussed in section 11.1.

If this was the whole story then the deep inelastic scattering on a nucleus would scale and be a delta function at $x = 1/N$. In practice the nucleons have a Fermi momentum in the nucleus due to their restriction to the finite nuclear size and so the delta function is smeared over a range of x to generate a quasi-elastic peak the area under which is related to the sum of the squared charges of the constituents (compare equation 11.38). This sequence of events is illustrated in Figs 9.10 and 9.11.

In practice we know that the proton is not elementary and so the scale invariance of inelastic scattering on nuclear targets is quickly violated (by $Q^2 \lesssim 0 \cdot 1$ GeV^2). The reason is that the proton form factor introduces a new Q^2 dependence into the cross-section. Equivalently one can think of this dimensionally as being the consequence of the proton's size being revealed and hence a manifest new length scale breaking the scale invariance.

When $Q^2 \gg 0 \cdot 7$ GeV^2 the lepton scattering on the proton dominantly occurs by incoherent elastic scattering from the quarks. Hence on a *nuclear* target there will be a rescaling (Fig. 9.11): at Q^2_1 scaling is due to scattering from the "pointlike" protons while at Q^2_2 there is a new

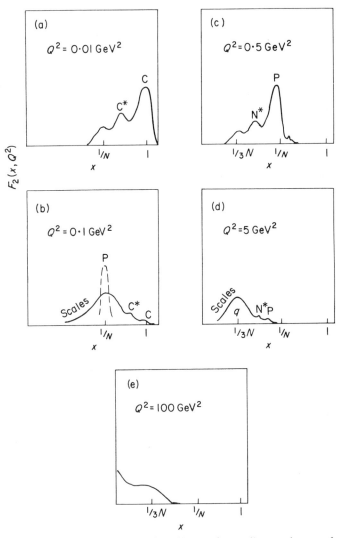

FIG. 9.11. Schematic illustration of scaling and rescaling as layers of matter are unravelled in a carbon nuclear target. (a) eC → eC and eC* (nuclear states excited). (b) $Q_1^2 \simeq 0.1$ GeV^2. Resonances and elastic scattering have been killed. Quasi-elastic ep → ep scales, smeared out by Fermi motion of the protons in the nucleus. At small x pion production begins. (c) For a proton at rest, higher ν and Q^2 produces N* excitations. In real life this whole figure should be smeared by Fermi momentum of the proton in the nucleus. (d) At $Q_2^2 \simeq 5$ GeV^2, C, C* and even N, N* excitations have been killed. Quasi-elastic eq → eq scales. At small x gluon and $q\bar{q}$ sea production begins. (e) $Q^2 \simeq 100$ GeV^2(?). (Valence) quark contributions swamped by continuum (gluons and $q\bar{q}$) or perhaps q^* excitations if quarks are made of prequarks (however even if these q^* exist they will probably be so smeared out that they will not be visible as bumps).

scaling due to the pointlike quarks. Notice the drift of the structure function to smaller x values.

This pattern of scaling, scaling violation and rescaling is quite general if successively more fundamental layers of matter are revealed as Q^2 increases (atomic, nuclear, proton, quark, prequark? . . .). If there is a reasonably sharp transition between length scales (as is certainly the case in the first four cases above) then one can visualise the increasing Q^2 of a photon probe revealing structure at level N but being insensitive to any substructure at level $N + 1$ (Fig. 9.12) (Kogut and Susskind, 1974a,b; Llewellyn Smith, 1975).

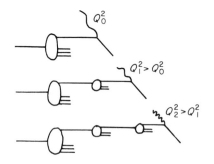

FIG. 9.12. Successive layers revealed as Q^2 resolution improves.

If all of the partons at layer $N + 1$ are charged then they will interact with a photon probe. Since the momentum of a parton at level N must equal that of its $N + 1$ level prepartons then the area under the structure function will be conserved (modulo the squared charges of the prepartons compared to the partons) but the average momentum of a preparton will be less than the parent parton and so the $\langle x \rangle$ at level $N + 1$ will be less than at level N. This results in the pattern exhibited in Fig. 9.13 (compare also Fig. 9.11).

Instead of discrete levels one might have a continuous set of layers as in a field theory for example (see also section 10.6). At level N the proton contains u and d quarks. Level $N + 1$ is where these quarks emit gluons. Level $N + 2$ has the gluons producing further gluons or quark–antiquark pairs. In particular, strange and charmed quarks may appear at this level (Fig. 9.14). As N increases, an equilibrium is approached where the momentum of the proton (originally carried by the three "valence" quarks) is shared among all types of quarks and gluons. Since the number of quarks increases as N increases (or as Q^2 increases) then

$$\langle x_N \rangle \xrightarrow{N \to \infty} 0 \tag{9.84}$$

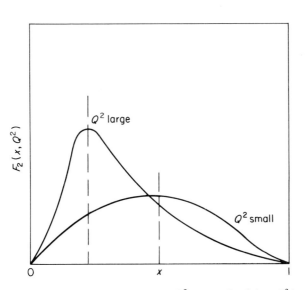

FIG. 9.13. Scaling violation and shift of $F_2(x, Q^2)$ to smaller $\langle x \rangle$ as Q^2 increases. $F_2(x)$ rises at small x and falls at large x as Q^2 increases.

Hence

$$F_2(x, Q^2) \xrightarrow{Q^2 \to \infty} \delta(x) \tag{9.85}$$

and instead of the scaling and rescaling of Fig. 9.11 a continuous violation of scaling will be seen as $F_2(x, Q^2)$ moves to smaller values of x tending ultimately to $\delta(x)$.

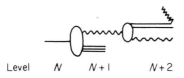

Level N $N+1$ $N+2$

FIG. 9.14. Valence quarks generate gluons and produce a $q\bar{q}$ sea at level $N+2$.

The expected quantitative pattern of scaling violation in a field theory is discussed in section 10.6.2. A detailed discussion of parton models with sequential clusters and their relation with renormalisable field theories is given in Kogut and Susskind (1974a,b) and Llewellyn Smith (1975, 1976).

In the data (Chapter 11) it does appear that a region with (approximate) scaling is seen and the properties of the data do suggest that pointlike quarks are responsible. At the largest values of Q^2 presently attained there are indications of a scaling violation qualitatively similar to that expected in the above discussion (Taylor, 1975; Riordan et al., 1975; Anderson et al., 1976). At large x the $F_2(x, Q^2)$ decreases as Q^2 increases, while at small x it increases. Whether this is evidence for a further layer (prequarks), field theory (QCD?) or is an effect associated with crossing the charm production threshold (section 11.6) is not yet fully clear (see, for example, Llewellyn Smith, 1975).

10 Some Assorted Topics in Parton Models

10.1 Old-fashioned perturbation theory and the infinite momentum frame

In Chapter 9 we made a derivation of the parton model formulae which may have appeared to be Lorentz invariant in that we explicitly formulated the approach in terms of s, t, u. In the literature one often sees the model derived, or applied, in an infinite momentum frame where the target momentum $P \to \infty$ in the z direction (Bjorken and Paschos, 1969; Feynman, 1969, 1972; Drell et al., 1970; Drell and Yan, 1971). In this formalism one makes the dynamical assumption that the constituents of the target also have large p_z but limited p_T and hence the infinite momentum frame is manifestly not a Lorentz invariant concept. Therefore one might ask how this compares with the *stu* approach. In fact, in equation (9.62), we neglected parton momenta transverse to the target. If $s \to \infty$ so that the target has $P \to \infty$ then our neglect of p_T makes the two approaches equivalent. We will now illustrate some of the infinite momentum formalism in order to acquaint the reader with the techniques.

Consider a proton moving in the z direction:

$$p_\mu \equiv [\sqrt{(P^2 + M^2)}; 0_T, P] \xrightarrow{P \to \infty} \left(P + \frac{M^2}{2P}; 0_T, P \right) \qquad (10.1)$$

A parton with mass m and fraction x of the proton's z-momentum and with momentum k_T transverse to the target will have

$$k_\mu = [\sqrt{(x^2 P^2 + m^2 + k_T^2)}, \ k_T, \ xP] \tag{10.2}$$

where $m_T^2 \equiv m^2 + k_T^2$ is sometimes called the "transverse mass". If $x \neq 0$ then

$$k_\mu \simeq \left(xP + \frac{m_T^2}{2xP}, \ k_T, \ xP \right) \tag{10.3}$$

Now consider a photon with momentum q_μ. In the laboratory frame let it have energy ν. The invariant $P \cdot q = M\nu$ can now be exploited to obtain an expression for q_μ in the $P \to \infty$ situation of equation (10.1). Write

$$q_\mu \equiv (q_0, \ q_T, \ q_3) \tag{10.4}$$

then

$$q^2 \equiv q^\mu q_\mu \equiv (q_0 - q_3)(q_0 + q_3) - q_T^2 \tag{10.5}$$

and

$$M\nu = P \cdot q = P(q_0 - q_3) + \frac{M^2}{2P} q_0 \tag{10.6}$$

In order that q^2 and $M\nu$ are P independent as $P \to \infty$ then we require that

$$q_0 - q_3 = A/P$$
$$q_0 = BP \tag{10.7}$$

where A, B are P independent.

There is an infinity of infinite momentum frames corresponding to the infinity of solutions for A and B. Two extremes which are met in the literature are:

i. q_0 and q_3 large of order P:

$$A = 0, \qquad B = \frac{2\nu}{M}, \qquad -q^2 = q_T^2 \tag{10.8}$$

ii. q_0 and q_3 small of order P^{-1}:

$$B = \frac{2M\nu + q^2}{4P^2}, \qquad A = M\nu, \qquad -q^2 = q_T^2 + 0(1/P^2) \tag{10.9}$$

hence

$$q_0 = \left(\frac{2M\nu + q^2}{4P}, q_T, \frac{q^2 - 2M\nu}{4P} \right) \tag{10.10}$$

This frame coincides with the incident electron-target centre of mass frame (the electron being the photon source). Notice that frames with q_0 and q_3 small necessarily require $q^2 \le 0$.

The technical usefulness of the $P \to \infty$ frame approach is if one is using old-fashioned perturbation theory (OFPT) to calculate field theory diagrams.[1] A Feynman diagram is the sum of all possible time-ordered graphs in OFPT. The amplitude for a given Feynman diagram is invariant under frame choice. The *sum* of all the OFPT diagrams will give the Feynman amplitude, independent of frame, but the contribution of each *individual* OFPT diagram depends upon the frame in general. The utility of the $P \to \infty$ frames is that many OFPT diagrams give contributions $0(1/P^N) \to 0$ and hence the calculation is simplified. For example, in the frames where the photon has no energy (equations 10.9 and 10.10) all diagrams vanish where the photon creates a $q\bar{q}$ pair.

In OFPT particles stay on, mass shell and momentum (but not energy) is conserved at any field theory vertex. For example consider the vertex of Fig. 10.1 where a scalar field with momentum p_z fragments into two scalars carrying fractions x and $1-x$ of the initial p_z and in addition $k_T(-k_T)$ transverse momenta. Hence

$$p_1 = \left(P + \frac{M^2}{2P}; 0_T, P \right) \tag{10.11}$$

$$p_2 = \left(|x|P + \frac{m^2 + k_T^2}{2|x|P}; k_T, xP \right) \tag{10.12}$$

$$p_3 = \left(|1-x|P + \frac{\lambda^2 + k_T^2}{2|1-x|P}; -k_T, (1-x)P \right) \tag{10.13}$$

Then

$$E_f \equiv E_2 + E_3 = P(|x| + |1-x|) + \frac{1}{2P} \left(\frac{m^2 + k_T^2}{|x|} + \frac{\lambda^2 + k_T^2}{|1-x|} \right) \tag{10.14}$$

[1] The relation between field theory and parton models is discussed in sections 10.5 and 10.6.

and so

$$\Delta E \equiv E_1 - E_f \rightarrow \begin{cases} P(1 - |1 - x| - |x|) & (x < 0; x > 1) \\ \dfrac{1}{2P}\left(M^2 - \dfrac{m^2 + k_T^2}{x} - \dfrac{\lambda^2 + k_T^2}{1 - x}\right) & (0 < x < 1) \end{cases}$$

(10.15)

For future reference notice that

$$2P \, \Delta E \equiv M^2 - s \qquad (10.16)$$

where $s \equiv (p_2 + p_3)^2$.

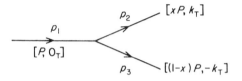

FIG. 10.1. Three-point vertex of scalar fields.

In OFPT the amplitude for such a vertex is inversely proportional to the energy difference ΔE. Hence this amplitude vanishes in the $P \rightarrow \infty$ limit unless $0 < x < 1$, i.e. when the two fields (particles) move forwards (in the same direction as the initial particle). Therefore if a scalar proton showers into a cloud of scalar partons, there will be vanishing probability as $P \rightarrow \infty$ that any parton is moving backwards.

We can illustrate the techniques of OFPT as $P \rightarrow \infty$ by studying the contribution of the Feynman diagram (Fig. 10.2 ($0(g^2)$ in the $g\phi^3$ coupling)) to the elastic form factor of the scalar particle defined by

$$\langle p'|J_\mu(0)|p\rangle = \frac{1}{(2\pi)^3} \frac{1}{2E_p} \frac{1}{2E_{p'}} F(q^2)(p + p')_\mu \qquad (10.17)$$

There are six possible time orderings (Figs 10.2(i-vi)) but only Fig. 10.2(i) survives as $P \rightarrow \infty$ since in each of the other diagrams at least one intermediate particle must be moving backwards due to three-momentum conservation. Hence these diagrams may be neglected as $P \rightarrow \infty$ because they contain energy denominators of $0(P)$ (equation 10.15).

In the surviving diagram $p_{1,2,3}$ have already been parametrised (equations 10.11, 10.12 and 10.13) and ΔE calculated at equation

(10.15). The energy of p_4 is given by

$$E_4 = xP + \frac{m^2 + (k_T + q_T)^2}{2xP} \tag{10.18}$$

where we have chosen to parametrise the photon by

$$q = \left(\frac{M\nu}{2P}; q_T, 0 \right) \tag{10.19}$$

Then $\Delta E' \equiv (E_q + E_1 - E_4 - E_3)$ becomes

$$\Delta E' = \frac{1}{2P} \left(2M\nu + M^2 - \frac{m^2 + (k_T + q_T)^2}{x} - \frac{\lambda^2 + k_T^2}{1-x} \right)$$

Since $W^2 = M^2 + 2M\nu - q_T^2$ (equation 9.49) then

$$\Delta E' = \frac{1}{2P} \left(W^2 - \frac{m^2}{x} - \frac{\lambda^2}{1-x} - \frac{[k_T + (1-x)q_T]^2}{x(1-x)} \right) \tag{10.20}$$

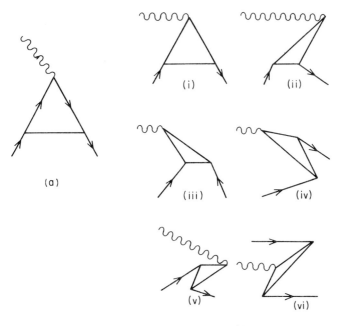

FIG. 10.2. (a) Feynman diagram for elastic form factor at $0(g^2)$ in $g\phi^3$ field theory. (b) (i–vi). Time-ordered diagrams.

Comparing with equation (10.15) we see that ΔE and $\Delta E'$ are intimately related by a shift of momentum and mass

$$\Delta E'(W, k_T + (1-x)q_T) \equiv \Delta E(M, k_T) \tag{10.21}$$

which for elastic scattering ($W \equiv M$) is particularly simple.

The elastic form factor equation (10.17) becomes

$$\langle p'|J_\mu(0)|p\rangle = \frac{1}{2E_p}\frac{1}{2E_{p'}}\frac{g^2}{(2\pi)^3}\int \frac{d^2k_T\, dk_z\,(2k+q)_\mu}{2E_2 2E_3 2E_4(\Delta E)(\Delta E')} \tag{10.22}$$

Examine the $\mu = 0$ component. In equation (10.17) $(p + p')_0$ is $2P$ as $P \to \infty$ while in equation (10.22) $(2k + q)_0$ is $2xP$. Hence in $F(q^2)$ there is a factor x in the numerator due to the current coupling. Explicitly comparing equations (10.17) and (10.22) yields

$$F(q^2) = g^2 \int_0^1 dx \int d^2k_T \frac{2x}{2x \cdot x \cdot (1-x)(2P\,\Delta E)(2P\,\Delta E')} \tag{10.23}$$

where the x factors in the denominator have come from the $2E_{2,3,4}$ energy factors for the internal lines in the loop. With practice one can immediately write down OFPT amplitudes as $P \to \infty$ without all of the above spadework. Integrate over $d^2k_T\, dx$ in each loop; insert the x factors for the energy of each line in the loop, energy denominators become $[2P\,\Delta E(k_T, x)] \equiv D(k_T, x)$ and no P dependence appears. Hence finally

$$F(q^2) = g^2 \int_0^1 \frac{dx}{x(1-x)} \int d^2k_T D^{-1}(k_T, x)D^{-1}(k_T + (1-x)q_T, x) \tag{10.24}$$

The above field theory has been generalised to the case where a composite system is described as a constituent plus a core (Gunion et al., 1973, 1974). The D^{-1} in equation (10.24) is replaced by a two-particle wavefunction ψ

$$\psi(k_T, x) \equiv (M^2 - s + i\varepsilon)^{-1}\phi(s) \tag{10.25}$$

where ϕ is a vertex function which reduces to the coupling constant in the lowest-order Feynman diagram calculation in equation (10.24), since $D \equiv M^2 - s$ (equation 10.16). The generalised form factor is therefore

$$F(q^2) = \int_0^1 \frac{dx}{x(1-x)} \int d^2k_T \psi(k_T, x) \psi[k_T + (1-x)q_T, x]$$

(10.26)

Now let us study the large q^2 dependence of the form factor. Here q_T is large and so s will be large in $\psi[k_T + (1-x)q_T, x]$. Suppose for large s that $\psi \sim s^{-N}F(x)$ (this is discussed in detail in Gunion et al., 1973). Hence

$$\psi(k_T + (1-x)q_T, x) \sim \left[\left(\frac{1-x}{x} q_T^2 \right) \right]^{-N}$$

(10.27)

and the asymptotic behaviour of the form factor becomes

$$F(q^2) \sim (q_T)^{-2N} \int_0^{1-0(m^2/q_T^2)} \frac{dx x^{2N-1}}{1-x} \int \frac{d^2k_T \psi(k_T)}{x^N(1-x)^N}$$

(10.28)

We have separated the x and $1-x$ dependences in this way to exploit the fact that since $\psi \sim [x(1-x)]^N$ as x or $1-x$ tends to zero or 1 then the second integral is a finite nonvanishing function of x. Hence asymptotically

$$F(q^2) \sim (q_T^2)^{-N} \log(q_T^2/m^2)$$

(10.29)

the logarithm arising from the large x behaviour in the integral over dx. For $N = 1$ we recover the $0(g^2)$ field theory example.

This composite formalism (equation 10.25) with large s behaviour in equation (10.27) is at the root of deep elastic and large p_T phenomenology in the works of these authors. This is described in section 14.6. This formalism is also of interest in connection with a relation between the large q^2 behaviour of the elastic form factor (equation 10.29) and the $x \to 1$ behaviour of $\nu W_2(x)$. This is described in section 10.2.

10.1.1 SPIN

The infinite momentum frame is equally useful when discussing time-ordered perturbation theory for particles with spin, e.g. fermions interacting with vector gluons. This is described in detail in Drell et al.

(1970), and Drell (1970). Quantum electrodynamics and renormalisation theory in the infinite momentum frame is described in Brodsky *et al.* (1973).

There is one important feature when spin is included to which attention should be drawn, namely it is no longer necessarily true that vertices are finite only when all particles move forward. Consider a γ_μ vertex. Using the representation of spinors and γ matrices of section 6.2 then immediately for $x_{1,2} > 0$

$$\bar{u}(x_1 P + k_{1\mathrm{T}}) \left\{ \begin{matrix} \gamma_0 \\ \gamma_3 \end{matrix} \right\} u(x_2 P + k_{2\mathrm{T}}) = 0(\sqrt{x_1 x_2} P) \tag{10.30}$$

as in the spinless case. However with γ_T components

$$\bar{u}(x_1 P + k_{1\mathrm{T}}) \gamma_\mathrm{T} u(x_2 P + k_{2\mathrm{T}})$$

$$\simeq \frac{P\sqrt{|x_1 x_2|}}{M} \left(\chi^+, \frac{\chi^+ \boldsymbol{\sigma} \cdot (x_1 P + k_{1\mathrm{T}})}{|x_1| P} \right) \begin{pmatrix} 0 & \boldsymbol{\sigma}_\mathrm{T} \\ \boldsymbol{\sigma}_\mathrm{T} & 0 \end{pmatrix} \left| \begin{matrix} \chi \\ \dfrac{\boldsymbol{\sigma} \cdot (x_2 P + k_{2\mathrm{T}})}{|x_2| P} \chi \end{matrix} \right.$$

$$= \frac{P\sqrt{|x_1 x_2|}}{M} \chi^+ (\sigma_3 \boldsymbol{\sigma}_\mathrm{T}) \chi \left(\frac{x_1}{|x_1|} - \frac{x_2}{|x_2|} \right) \xrightarrow{x_1 : x_2 > 0} 0(1)$$

$$\xrightarrow{x_1 x_2 < 0} 0(P\sqrt{|x_1 x_2|}) \tag{10.31}$$

and hence for γ_T the spin $\frac{1}{2}$ particle prefers to be turned around, the P in the spinor numerator cancelling the P suppression in the energy denominator. The $\gamma_{0,3}$ are called "good" and γ_T "bad" components. Consequently some care is necessary when considering what time orderings survive when spin is present.

As an exercise verify that γ_5 also turns the spinors around. See also p. 1038 of Drell *et al.* (1970).

10.2   The Drell–Yan–West relation

The charge of a system is the integrated probability of finding constituents with fraction x of the target's momentum (as $P \to \infty$) weighted by their charges. Hence

$$F(0) = \sum_i \int_0^1 \mathrm{d}x f_i(x) e_i \tag{10.32}$$

where $f_i(x)$ is identical to that defined in equation (9.53). In the quark-core constituent model we have from equation (10.26) as $q \to 0$ that

$$F(0) = \int_0^1 dx \int \frac{d^2k_T}{x(1-x)} |\psi(k_T)|^2$$

and, comparing with equation (10.32), we can identify

$$f(x) \equiv \int d^2k_T \frac{|\psi(k_T)|^2}{x(1-x)} \tag{10.33}$$

(since in this example one constituent carries unit charge and the core is neutral). From equation (9.53) or (9.72) we see that $\nu W_2(x) \equiv xf(x)$ in this example. Hence

$$\nu W_2(x \to 1) = \int d^2k_T \frac{|\psi(k_T)|^2}{1-x} \tag{10.34}$$

From equations (10.16) and (10.15)

$$s = \frac{m^2 + k_T^2}{x} + \frac{\lambda^2 + k_T^2}{1-x} \tag{10.35}$$

Hence $s \sim (1-x)^{-1}$ as $x \to 1$ and so $\psi(k) \sim s^{-N} \sim (1-x)^N$. Consequently

$$\nu W_2(x \to 1) \sim (1-x)^{2N-1} \tag{10.36}$$

where N is given by the asymptotic behaviour of the elastic form factor

$$F(q^2) \sim (q^2)^{-N} \tag{10.37}$$

This correlation (equations 10.36 and 10.37) is known as the Drell and Yan (1970) and West (1970) relation. For the proton if $F(q^2) \sim q^{-4}$ then $N = 2$ and so $\nu W_2(x \to 1) \sim (1-x)^3$.

The above example is only good for spinless particles and hence is not immediately applicable to the proton or, in general, to systems of quarks. Incorporating spin systematically (section 10.1.7), the Drell–Yan–West relation can be derived by a procedure analogous to the above.

Alternatively the Drell–Yan–West relation incorporating spin $\frac{1}{2}$ quarks can be derived directly from our results of Chapter 9. For elastic scattering at large t (equation 9.58)

$$\frac{d\sigma}{dt}(ep \rightarrow ep) = 2\pi\alpha^2 \frac{s^2 + u^2}{s^2 t^2} G_M^2(t) \qquad (10.38)$$

while inelastic scattering may be described by (equations 9.65 and 9.53, or 9.71 and 9.72)

$$\frac{d^2\sigma}{dt\,dW^2}(ep \rightarrow eX) = 2\pi\alpha^2 \frac{s^2 + u^2}{s^2 t^2} \frac{F_2(\omega \equiv 1 + W^2/-t)}{s + u} \qquad (10.39)$$

For any fixed W^2, in particular $W^2 = M^2$, then $-t \rightarrow \infty$ sends $\omega \rightarrow 1$.

If exclusive scattering connects smoothly on to deep inelastic scattering then the $F_2(\omega) \sim (\omega - 1)^M$ and $G_M^2(t) \sim t^{-2N}$ will be correlated as follows:

$$\frac{d\sigma}{dt}(ep \rightarrow ep) \equiv \int \frac{d^2\sigma\,(ep \rightarrow eX)}{dt\,dW^2} \delta(W^2 - M^2)\,dW^2 \qquad (10.40)$$

and so

$$G_M^2(t) = \int \frac{F_2(\omega - 1 \equiv M^2/-t)}{s + u} \delta(W^2 - M^2)\,dW^2$$

$$= F_2(\omega - 1 \equiv M^2/-t)/(M^2 - t) \qquad (10.41)$$

Then as $-t \rightarrow \infty$ we have

$$t^{-2N} \sim t^{-(M+1)} \qquad (10.42)$$

and hence $2N = M + 1$ obtains.

10.3 A qualitative introduction to the nonperturbative parton model

In renormalisable field theories scaling is broken by logarithms (section 10.6). Landshoff and Polkinghorne (1972) and Landshoff et al. (1971) have formulated a covariant nonperturbative parton model based on field theory. This model gives a finite scaling form for νW_2 due to the dynamical postulate that hadronic amplitudes with external virtual partons tend rapidly to zero as the parton masses become large. We give an outline of their model and in section 10.5 show how it can be obtained from the infinite momentum approach of section 10.1. In particular we shall see how the imposed softening corresponds to a damping of the partons' k_T integration, hence leading to scaling in νW_2.

If we view photoproduction total cross-sections (νW_2) as the imaginary part of the forward Compton amplitude, then scaling arises from the diagrams where a "free" parton propagates between current interactions (Fig. 10.3(a)). If the four momenta are labelled as in Fig. 10.4, then the propagator of the parton (apart from spin dependent

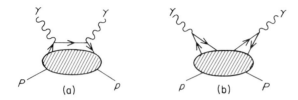

FIG. 10.3. (a) Handbag or box diagram. (b) Cat's ears or six-point diagram.

parts which are not material here, and neglecting any momenta transverse to the proton–photon axis) is $|(xp + q)^2 - m^2 - i\varepsilon|^{-1}$ where m is the parton's mass. The imaginary part of the amplitude includes the delta function $\delta(x^2 M^2 + q^2 + 2xp \cdot q - m^2)$ arising from the imaginary part of the parton propagator. At very large ν, q^2 we may neglect $x^2 M^2$ and m^2 with the result that the amplitude contains the delta function constraint $\delta(2M\nu/q^2 + x)$ and so the structure function is dependent only on the ratio ν/q^2 due to this delta function.

If Fig. 10.3(a) was the *only* diagram that one could draw for Compton scattering in the parton model then the structure functions would scale.

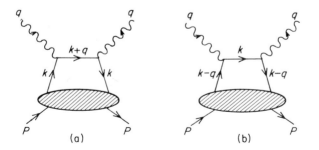

FIG. 10.4. Dominant momentum routing in handbag diagram at large q is shown in (a). In (b) it is suppressed.

However, one can envisage diagrams where the photons couple to different partons, or where the photon couples to the same parton but that parton interacts with other partons between its interactions with the photons (Fig. 10.5). Only if these diagrams are negligible compared to the "incoherent impulse" diagrams (Fig. 10.3(a)) will the scaling automatically obtain.

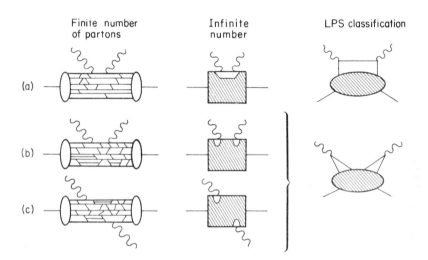

FIG. 10.5. Topological classification of Compton diagrams in the parton model: (a) is distinguished by the absence of interactions on the parton line between the times of its interacting with the photons; (b) contains interactions; and in (c) the photons interact with different partons. The LPS classification corresponds to Fig. 10.3.

In general there could be an arbitrary number N of partons in the proton. The idea of the nonperturbative parton model is to throw $N-1$ partons into a black box and concentrate on the parton(s) that interacts with the external currents. The diagrams come in two types. Figure 10.3(a) we refer to as the box or "handbag" diagram whereas Fig. 10.3(b) is called the "cat's ears". The classification into these two classes will be described later. The blobs contain arbitrary numbers of interacting partons and are treated as parton–proton scattering amplitudes with Regge behaviour. When calculating the Compton amplitude in this model the diagram (amplitude) contains the parton–proton scattering subamplitude. This Regge subamplitude causes the whole Compton amplitude to have Regge behaviour. The important physics, in which we are really interested, comes from the coupling of the photons to the partons and yields the scaling property (which we shall discuss after explaining the classification of the diagrams into "box" or "cat's ears" categories).

For a configuration with any arbitrary number of partons, then if and only if the diagram contains a freely propagating parton between the two photons is it classified in Fig. 10.3(a), the box or "handbag" diagram. Thus the defining feature of this diagram is the freely propagating parton and hence the scale invariance of the structure functions arises from this box diagram. All other diagrams with interactions on the parton which was hit by (or emitted) the photon are classified in the cat's ears; thus the defining feature here is that after the parton has interacted with the photon, the next interaction that that parton undertakes is *not* with the second photon.

As stated above, only the box diagram contains the freely propagating parton required for the scaling of νW_2. Therefore we must invent a rule that will cause this diagram to dominate over the cat's ears in the kinematic realm where this scaling is observed ($Q^2 \gtrsim 1 \, \mathrm{GeV}^2$, $W^2 >$ Resonance masses). The dynamical assumption of this model is that for a parton off its mass shell, then the parton–proton scattering amplitude goes rapidly to zero. If a heavy photon hits a parton then the parton has to transport this mass through the maze until the final photon can take it away. The only way this mass can flow through to the final photon without entering a parton–proton scattering blob is if it flows across the top of Figs 10.3(a) and 10.4. For all other diagrams a heavy parton enters the blob and the amplitude vanishes. Hence the required diagram dominates at large q^2.

At *small* q^2 the partons can stay near their mass shell and so the parton–proton amplitude need not vanish. Hence the scaling does not obtain. Incorporating the cat's ears diagrams one can formulate a model for all q^2 and study $q^2 = 0$ phenomena (Brodsky *et al.*, 1972).

10.4   The massive quark model

In Chapter 7 we mentioned some attempts at formulating massive quark models with particular reference to hadron (in particular meson) spectroscopy. If quarks are indeed very massive then this could be the reason that they are not produced in deep inelastic scattering. The scaling phenomena arise because the quark has a small (~ 300 MeV) effective mass when confined to a space–time volume of 10^{-13} cm. A problem however is that the diagrams which are ultimately responsible for generating the scaling phenomena can only be literally correct if free quarks are produced.

Preparata (1973, 1974) has attempted to incorporate the nonobservability of quarks into the dynamics of a massive quark model (MQM). By formulating the hypothesis that "no quanta are associated with the quark field" he bypasses the dynamical problem of quark confinement (Chapter 15) and formulates an unconventional field theory. A particular feature of his work is that the quark propagator

$$\Delta_{\alpha\beta}(p) = \int \mathrm{d}x \ e^{ip\cdot x}\langle 0|T(q_\alpha(x)\bar{q}_\beta(y))|0\rangle \qquad (10.43)$$

does not have a pole at $p^2 = M_Q^2$ nor a cut at the lowest threshold and hence $\Delta_{\alpha\beta}(p)$ is an entire function in p^2.

Furthermore, since in deep inelastic phenomena the quark mass appears negligible (scale invariance, Chapter 9) he demands that the entire function $\Delta_{\alpha\beta}(p)$ peaks when $p^2 \approx \mu^2$ (with $\mu \approx 300$ MeV an "effective mass" for the quark when confined within a space–time volume of the order of 10^{-13} cm dimension). The consequence of these physical demands is that a Lehmann representation (Chapter 16 of Bjorken and Drell, 1964) cannot hold for the propagator and that the propagator is exponentially damped at high p^2. This unconventional field theory differs from conventional ideas in that the above exponential behaviour prevents the rotation of contours in the complex p_0 plane when performing momentum space integrals. (This is a crucial difference compared to the nonperturbative parton model developed by Landshoff and Polkinghorne (1972) and Landshoff et al. (1971) whose quark propagator dies as $(k^2)^{-1}$ and contour rotations can be performed.)

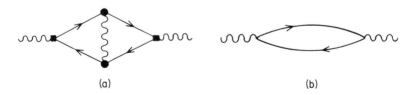

(a)   (b)

FIG. 10.6. Scaling in e^+e^- annihilation: (a) Preparata's model; (b) naive parton model.

At low energies the $q\bar{q}$ scattering amplitude has poles which are the meson states of the Bethe–Salpeter approaches (Chapter 7). At high energy the $q\bar{q}$ amplitude is hypothesised to have Regge behaviour with the $\alpha(0) = 1$ Pomeron the leading intercept. Then in e^+e^- annihilation for example the dominant diagram is in Fig. 10.6. If $\alpha(0) = 1$ for the Pomeron in the loop, then $Q^2\sigma_{total} = $ constant and the canonical free field theory scaling result obtains (Preparata, 1975). This result is formally the same as the free field result (Fig. 10.6(b)) but Preparata's approach does not imply that free quarks appear in the final state.

Similar conclusions emerge in electroproduction. The scaling arises from diagrams like Fig. 10.7 if $\alpha(0) = 1$ (these are topologically of the six-point or cat's ears variety of section 10.3). The four-point handbag diagrams (which gave the scaling in the free field and nonperturbative parton model approaches) are absent in the MQM and so again no free quarks appear in the final state.

Fig. 10.7. Diagram yielding scaling in $ep \rightarrow eX$ in Preparata's model.

10.5 Infinite momentum frame derivation of the covariant nonperturbative parton model

The machinery used in the nonperturbative parton model of section 10.3 exploits Sudhakov variables. Instead of describing this technique we shall show how this model emerges naturally out of the infinite momentum field theory of section 10.1.

We begin by returning to the three-point vertex of Fig. 10.1 and parametrise the momenta as in equations (10.11) to (10.13). The four momentum of particle number 3 calculated with both three momentum and energy conservation is

$$p_1 - p_2 = \left((1-x)P + \frac{M^2x - m^2 - k_T^2}{2xP}; -k_T, (1-x)P \right) \quad (10.44)$$

In time-ordered perturbation theory this quantity squared corresponds to the Feynman off shell variable u. Note that

$$u - \lambda^2 = (1 - x)\left(M^2 - \frac{k_T^2 + m^2}{x} - \frac{k_T^2 + \lambda^2}{1 - x} \right) \qquad (10.45)$$

and, comparing with equation (10.15), we see the similarity with the energy factor appearing in time-ordered perturbation theory. Explicitly

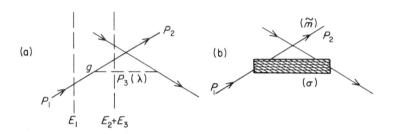

FIG. 10.8. (a) Time-ordered perturbation theory contribution to the parton–proton scattering amplitude. (b) Forward parton–proton amplitude.

the TOPT contribution to the parton–proton scattering amplitude for scalar particles (Fig. 10.8) is

$$\mathcal{M}_u = \frac{g^2}{(2\pi)^3}[2E_3(E_1 - E_2 - E_3 + i\varepsilon)] \xrightarrow{P \to \infty} \frac{g^2}{(2\pi)^3}(u - \lambda^2 + i\varepsilon)^{-1} \qquad (10.46)$$

Now consider the $0(g^2)$ calculation of the elastic form factor (equation 10.24). For $q^2 = 0$ we can extract $f(x)$ (as in equation 10.33). Using (10.45) and (10.46) we see that

$$f(x) = \int d^2k_T \frac{(1 - x)}{2x} \frac{\mathcal{M}_u}{u - \lambda^2} \qquad (10.47)$$

Having established a connection between $f(x)$ and the parton–proton scattering amplitude $\mathcal{M}_u$, we can immediately generalise $\mathcal{M}_u$ to incorporate Regge behaviour. If $\pi\rho(\sigma^2, m^2)$ is the imaginary part of the forward (anti)-parton–proton scattering amplitude (Fig. 10.8(b)) then

$$(2\pi)^3 \mathcal{M}_u = \int \frac{\rho(\sigma^2, m^2) \, d\sigma^2}{u - \sigma^2 + i\varepsilon} \qquad (10.48)$$

where λ^2 of the perturbation result (equation 10.46) has been replaced by a spectral sum variable σ^2. Then

$$\nu W_2(x) = x f(x) = \int d^2 k_T \int d\sigma^2 \tfrac{1}{2}(1-x) \frac{\rho(\sigma^2, \tilde{m}^2)}{(u - \sigma^2 + i\varepsilon)^2} \quad (10.49)$$

We have written $\tilde{m}^2$ here since one must in general take account of the off-shell dependence when the parton–proton amplitude is embedded in the interior of a general amplitude. Since $\tilde{m}^2 \equiv (p_1 - p_3)^2$ then

$$\tilde{m}^2 = \frac{x(1-x)M^2 - x\sigma^2 - k_T^2}{1-x} \quad (10.50)$$

Notice that this coincides with equation (2.25) of Landshoff *et al.* (LPS) (1971) where $\tilde{m}^2$ is their σ^2 (off-shell parton mass) and k_T^2 is the Sudhakov parameter κ^2. By hypothesis in LPS this off-shell dependence damps the parton–proton amplitude. From equation (10.50) we see that in the present approach it corresponds to providing strong convergence of the k_T^2 integrations and leads to Bjorken scaling even in spin $\tfrac{1}{2}$ theories (where scaling is broken by logarithms in undamped perturbation theory).

To complete the correspondence with LPS we absorb two Feynman propagators into $\rho(\sigma^2, \tilde{m}^2)$ and define

$$\mathrm{Im}\, T(\sigma^2, \tilde{m}^2) = \pi \rho(\sigma^2, \tilde{m}^2) / (\tilde{m}^2 - m^2)^2 \quad (10.51)$$

Hence

$$\pi f(x) = \int d^2 k_T \frac{x}{2(1-x)} \int d\tilde{m}^2 \, \mathrm{Im}\, T(\tilde{m}^2, m^2) \quad (10.52)$$

which is the same as equation (2.25) of LPS (1971).

Given this formula for $f(x)$ one can immediately see how Regge behaviour in the parton–proton amplitude drives the Regge behaviour in $\nu W_2(x)$. If $\rho(\sigma^2, \tilde{m}^2) = (\sigma^2)^\alpha \beta(\tilde{m}^2)$ then with $\eta = xm^2$ we have

$$f(x) = \tfrac{1}{2}x \int d^2 k_T \int_0^\infty \frac{d\eta\, \eta^\alpha}{x^{1+\alpha}} \frac{1}{1-x} \frac{\beta(\tilde{m}^2)}{(\tilde{m}^2 - m^2)^2} \xrightarrow{x \to 0} 0(x^{-\alpha}) \quad (10.53)$$

where the latter step follows since the integrals are convergent.

Hence hadronic interactions have been included to all orders and transition from the old-fashioned perturbation theory formulation of the parton model at infinite momentum to the nonperturbative covariant formulation of LPS has been achieved. The application of this formalism is described in detail by Brodsky *et al.* (1973).

Further discussion of the relation between calculational techniques can be found in appendix B of Sivers *et al.* (1977) and Schmidt (1974).

10.6 Parton models and field theory

An essential ingredient of parton models is that at small times and distances (large ν and Q^2) the partons can be regarded as freely moving constituents. Since all other known interactions in Nature appear to be described by field theories then one may suspect that strong interactions are also. In a field theory language the free motion of partons corresponds to demanding that interparton interactions are as soft as in super-renormalisable field theories.

Since in the four-dimensional world there is no super-renormalisable field theory which includes spin $\frac{1}{2}$ fields then the attempts to provide a field-theoretic justification for the parton model have required the imposition of a transverse momentum cut-off (Drell *et al.*, 1970) in order that the scaling obtains. The covariant analogue of this is the softened field theory approach of Landshoff and Polkinghorne (1972) and Landshoff *et al.* (1971). This has been described in section 10.3 and the softening arises by the demand that off-shell parton–hadron scattering amplitudes tend rapidly to zero as the parton mass becomes large. This provides an implicit scale which allows exact scaling. In the absence of *ad hoc* softening or cut-offs renormalisable field theories do not give scale invariance.

Renormalisable field theories (like QED) have dimensionless coupling constants. Physical quantities require infinite renormalisations to define them, i.e. large momenta in loops have infinite amplitude and have to be renormalised order by order in perturbation theory. As the momentum increases in a virtual loop, then intuitively the virtual fluctuation takes place over ever smaller space–time distances. Hence there is never a length or time-scale beyond which interactions can be ignored and so there will always be structure in the interacting fields that is on too small a space–time scale to be resolved however good the resolution of the probe may be. An intuitive picture of the above has been developed in a formulation of the parton model by Kogut and Susskind (1974a,b).

One can proceed beyond perturbation theory by employing the machinery of the renormalisation group.[1] In field theories any function (e.g. a vertex) will depend upon the four momenta p flowing through and on the masses and coupling constants g_i of the theory. If the overall scale of the momenta is changed so that $p_i \to \lambda p_i$ (or in position space $x_i^{-1} \to \lambda x_i^{-1}$) then the renormalisation group analysis shows that the λ dependence can be removed from the p_i and placed instead in dimensionless effective couplings $\bar{g}_i(\lambda)$ which are *scale dependent*. These couplings therefore give a measure of the deviations from free-field behaviour at length scales λ^{-1}.

If $g(\lambda \to \infty) \to g^*$ (a constant) the interacting theory is scale invariant at short distances since all scale dependence has departed from $g(\lambda)$. Such a field theory is known as a "fixed point" theory. In super-renormalisable field theories $g(\lambda \to \infty) \sim 0(\lambda^{-N}) \to 0$. In QED $g(\lambda \to \infty) \to \infty$ and hence interaction can never be ignored. An exciting discovery has been that in non-Abelian gauge field theories (e.g. QCD, section 15.2) $g(\lambda \to \infty) \sim 0(\ln \lambda)^{-1} \to 0$ and hence these theories are asymptotically free ('t Hooft, 1972; Politzer, 1973; Gross and Wilczek, 1973a,b). Since QCD is an example of such a field theory and contains spin $\frac{1}{2}$ fields (quarks?), then it is a natural candidate as a field theory of strong interactions and may provide some rationale for the phenomenological successes of the quark–parton model (Chapter 11).

10.6.1 ASYMPTOTIC FREEDOM

For a gauge field theory involving fermions the renormalisation group analysis shows that for small $\bar{g}$

$$\bar{g}^2(\lambda) = \frac{g^2(\lambda_0)}{[1 + 2bg^2(\lambda_0) \log (\lambda/\lambda_0)]} + 0(\bar{g}^4) \qquad (10.54)$$

where b is a constant. This structure may already be familiar to you for the particular case of QED where (Gell-Mann and Low, 1954)

$$\frac{\bar{e}^2(Q^2)}{Q^2} \equiv e^2 D(Q^2) \qquad (10.55)$$

---

[1] This is described in detail by Politzer (1974) and also by Bogobliubov and Shirkov (1959).

with $D(Q^2)$ the photon propagator and $\bar{e}^2$ the "effective charge". Explicitly

$$\bar{e}^2(Q^2) = e^2 \Big/ \Big\{ 1 - \frac{e^2}{12\pi} \log \frac{Q^2}{m^2} + \cdots \Big\} \qquad (10.56)$$

so that the effective strength $\bar{e}^2$ grows as Q^2 grows and hence QED is not asymptotically free (the coefficient b in equation 10.54 is negative).

If there are F flavours of quark each of which occurs in 3 colours then one can formulate an $SU(3)_{colour}$ non-Abelian gauge field theory of the strong interactions where coloured quarks interact by exchanging an octet of coloured gluons. In such a theory the constant b is found to be[1] ('t Hooft, 1972; Georgi and Politzer, 1974; Gross and Wilczek, 1974)

$$b = \frac{1}{48\pi^2}(33 - 2F) \qquad (10.57)$$

and hence is *positive* if there are less than 17 flavours of quark. In particular, if there are four flavours (u, d, s, c) then

$$\frac{g^2(\lambda_0)}{\bar{g}^2(\lambda)} = 1 + \frac{25}{24\pi^2} g^2(\lambda_0) \log \Big(\frac{\lambda}{\lambda_0}\Big) \qquad (10.58)$$

or with $g^2 = 4\pi\alpha_s$ and $t \equiv \log(\lambda/\lambda_0)$, then

$$\frac{\alpha_s}{\alpha_s(t)} = 1 + \frac{25\alpha_s}{6\pi} \log \Big(\frac{\lambda}{\lambda_0}\Big) \equiv 1 + \frac{25\alpha_s}{12\pi} \log \Big(\frac{\lambda^2}{\lambda_0^2}\Big) \qquad (10.59)$$

Hence in the above field theory the effective coupling constant $\alpha(t)$ indeed tends to zero as $t \to \infty (\lambda^2 \to \infty)$ and so the quarks are asymptotically free.

In the naive quark model if the quarks are massless, coupled to the electromagnetic current in a pointlike manner (Chapter 9) and the quark–gluon coupling $\bar{\alpha} \equiv 0$, then there is no scale of dimension and hence the dimensionless quantities $F_{1,2}$ (equation 9.51) are scale invariant. In the non-Abelian field theory where coloured quarks interact with coloured gluons then at any finite t a *scale dependent* quantity $\bar{\alpha}(t)$ is manifestly present and so $F_{1,2}$ are not scale invariant. Scale invariance is approached asymptotically since $\bar{\alpha}(t \to \infty) \to 0$; at finite t scaling is broken by logarithms of ν, Q^2.

---

[1] A more general formula is given in equation (3.19) of Politzer (1974) for arbitrary $SU(N)$ with fundamental and adjoint representations of coloured quarks.

10.6.2 SCALING VIOLATION WITH INTERACTING QUARKS

A probe with mass $\sqrt{Q_1^2}$ resolves structure on a length scale λ, and sees quarks carrying a fraction x of the target's momentum with probability $q(x)$. When the resolution is improved $(Q_2^2 > Q_1^2)$ then an additional probability $\delta q(x)$ may be revealed which will have arisen as a result of a quark with momentum $y > x$ having radiated a gluon and hence having reduced its momentum from y to x. At Q_1^2 only the parent was resolved; at Q_2^2 this extra contribution at x is resolved (Fig. 10.9). If we write $\tau \equiv \ln Q^2$, then an improvement $\Delta\tau$ in resolution will yield a change $\Delta q(x)$ in the quark probability distribution at x as a result of the quark–gluon vertices of Fig. 10.9.

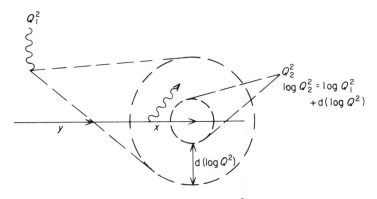

FIG. 10.9. Improving resolution as Q^2 increases.

From the renormalisation group analysis, the quark–gluon vertex effective coupling is scale dependent, $\bar{\alpha}(t)$. The probability at order α that in a change of resolution $d\tau$ the quark will be seen to contain another quark with fraction z of the parent quark's momentum may be written $\alpha(t)P(z)\,dt$. In QED or QCD, for example, the quantity $P(z)$ can be calculated explicitly from the γ_μ quark–gluon vertex and will depend on the quark and gluon helicity states (Bjorken, 1971; Altarelli and Parisi, 1977). Hence in general we have

$$\frac{dq}{d\tau}(x, \tau) = \alpha_s(\tau) \int_x^1 \frac{dy}{y} q(y, \tau) P(x/y) \qquad (10.60)$$

as the driving equation for the change in the quark distribution function as resolution is improved. The convolution integral can be separated into two integrals for $q(y)$ and $P(x/y)$ individually by a Mellin transformation. For arbitrary N we can rewrite equation (10.60) as

$$\frac{d}{d\tau} \int_0^1 dx\, x^N q(x, \tau) = \alpha_s(\tau) \int_0^1 dy\, y^N q(y, \tau) \int_0^1 dz\, z^N P(z)$$

(10.61)

and hence we have a differential equation for the moments M_N of the quark distribution functions

$$\frac{d}{d\tau} M_N(\tau) = \alpha_s(\tau) M_N(\tau) A_N$$

(10.62)

where

$$M_N(\tau) \equiv \int_0^1 dx\, x^N q(x, \tau) \equiv \int_0^1 dx\, x^{N-1} F_2(x, \tau)$$

(10.63)

$$A_N \equiv \int_0^1 dz\, z^N P(z)$$

(10.64)

The $\tau(\equiv \log Q^2)$ dependence of the moments of the structure function F_2 (equation 10.63) is governed by the differential equation (10.62) whose solution depends in particular upon the τ dependence of $\alpha(\tau)$.

If $\alpha(\tau) \xrightarrow{\tau \to \infty} \neq 0$ then we have a fixed point field theory. From equation (10.62) with α independent of τ we see that the moments have exponential τ behaviour and hence are power laws in Q^2.

If strong interactions are described by a non-Abelian gauge theory such as quantum chromodynamics then for $\alpha(\tau)$ small (compare equations 10.54 and 10.59)

$$\frac{\alpha}{\alpha(t)} = 1 + b\alpha\tau$$

(10.65)

with $\alpha = \alpha(0)$ and b a positive constant (equation 10.57). The solution of equation (10.62) then shows that the moments are logarithmically dependent upon Q^2:

$$M_N(\tau) = M_N(0)\left(\frac{\alpha}{\alpha(\tau)}\right)^{A_N/b}$$

(10.66)

(Georgi and Politzer, 1974; Gross and Wilczek, 1974).

Hence if quarks partake in an interacting field theory, the structure function $F_2(x, Q^2)$ will not scale. Instead its moments will have a well-defined pattern of scaling violation (linear in Q^2, $\log Q^2 \ldots$) determined by the dependence of $\alpha (\tau = \log Q^2)$ for that field theory. In QCD the coefficient A_N has been calculated by Altarelli and Parisi (1977) using the definition (equation 10.64) and the known $P(z)$ for the quark–gluon coupling.

The above discussion has been somewhat over-simplified since there is also a gluon distribution function in the target, $G(y, \tau)$, and a gluon may produce a quark–antiquark pair hence feeding the $q(x, \tau)$. Hence equation (10.60) generalises to

$$\frac{dq^i}{d\tau}(x, \tau) = \alpha (\tau) \int_x^1 \frac{dy}{y} \left[\sum_j q^i(y, \tau) P_{q_i q_j}(x/y) + G(y, \tau) P_{q_i G}(x/y) \right]$$

(10.67)

where the indices i and j run over quarks and antiquarks of all flavours. Similarly the gluon distributions satisfy

$$\frac{dG}{d\tau}(x, \tau) = \alpha (\tau) \int_x^1 \frac{dy}{y} \left[\sum_j q^i(y, \tau) P_{G q_i}(x/y) + G(y, \tau) P_{GG}(x/y) \right]$$

(10.68)

(the final term arising since a three-gluon vertex exists in order g in QCD). The calculation of the complete set of A_N that govern the scaling violation involves solving these integro-differential equations and is described in Altarelli and Parisi (1977) where a complete development of these ideas can be found. Notice that conservation of total momentum constrains

$$\frac{d}{d\tau} \int_0^1 dx \, x \left[\sum_i q_i(x, \tau) + G(x, \tau) \right] = 0$$

(10.69)

10.7 σ_S/σ_T in the naive parton model

In deriving the result (equations 9.72 and 9.73) for σ_S/σ_T we have explicitly neglected momentum of the parton transverse to the target. To see the effect of this on the prediction for σ_S/σ_T we will consider a $\gamma - P$ collision along the z-axis where

$$q \equiv (0; 0, 0, -2xP)$$

$$p \equiv (P; 0, 0, P)$$

and hence

$$q^2 = -4x^2P^2; \; -q^2/2P \cdot q = x$$

If the current is absorbed elastically by a parton with four momentum

$$k_\mu = (k_0; k_x, k_y, xP)$$

then the final parton four momentum will be

$$k'_\mu = (k_0; k_x, k_y, -xP)$$

We shall allow $k_{x,y}$ to be arbitrary and so shall not neglect them relative to k_0 and k_z. The photoabsorbtion cross-sections $\sigma_{S,T}$ can be calculated from equation (9.42) where $W_{\mu\nu}$ will be the tensor appropriate to electron–quark scattering, i.e. (equation 9.7)

$$L_{\mu\nu} = \tfrac{1}{2}(k_\mu k'_\nu + k'_\mu k_\nu + \tfrac{1}{2}q^2 g_{\mu\nu})$$

Hence

$$\frac{\sigma_S}{\sigma_T} = \frac{L_{00}}{\tfrac{1}{2}(L_{xx} + L_{yy})} \equiv \frac{k_0^2 + \tfrac{1}{4}q^2}{\tfrac{1}{2}(k_x^2 + k_y^2 - \tfrac{1}{2}q^2)}$$

Now, the parton mass is

$$m^2 \equiv k_0^2 - (k_x^2 + k_y^2) - x^2P^2$$

and so

$$m^2 + k_T^2 \equiv k_0^2 + \tfrac{1}{4}q^2$$

yielding finally

$$\frac{\sigma_S}{\sigma_T} = \frac{4(m^2 + k_T^2)}{Q^2 + 2k_T^2} \xrightarrow{Q^2 \to \infty} \frac{4(m^2 + k_T^2)}{Q^2}$$

Compare this with equation (28.3) in Feynman (1972). If $m^2 + k_T^2$ is of the order of $0 \cdot 25 \text{ GeV}^2$ then $\sigma_S/\sigma_T \simeq 1/Q^2 \text{ (GeV}^2)$ which is of the order of the small value seen in the data (Fig. 9.8). The possibility that k_T is x dependent has been raised by several authors (Landshoff, 1976; Gunion, 1976; Hwa *et al.*, 1977). This would imply that σ_S/σ_T is also x dependent.

11　Quark–Parton Phenomenology

11.1　Electron scattering from nuclei

The naive parton ideas described in section 9.4 are nicely illustrated in nuclear physics data. In Fig. 11.1 are exhibited the cross-sections for electrons of about 190 MeV energy scattering on carbon. Elastic scattering requires $Q^2 = 2M\nu$ hence

$$E'\{(1 - \cos\theta)E + M\} = ME \qquad (11.1)$$

For an incident energy of 190 MeV this implies that elastic scattering of the electron through $80°$ will yield electrons with energy $E' \simeq$ 186 MeV and this is clearly visible in the recoil electron spectrum. As more energy is given up by the electron, carbon nuclear resonance states are excited ($E' \simeq 180$, 177 MeV). Quasi-elastic scattering from the constituent protons of the nucleus occurs when $E' \simeq 160$ MeV (substitute $M = 940$ MeV in equation 11.1). This is not shown in Fig. 11.1(a) but the integrated quasi-elastic scattering cross-section is about $\frac{1}{2}$ of the integrated elastic nuclear cross-section here. Note the typical order of magnitude of Q^2 in this experiment: approximating $\sin^2\theta/2 \simeq \frac{1}{2}$ then

$$Q^2 \simeq 2EE' \rightarrow 0\cdot06 \text{ GeV}^2$$

for the elastic electron nuclear scattering.

At this value of Q^2 the nucleon's internal structure is not resolved.

$$G_N^E(Q^2) = (1 + Q^2/0\cdot71 \text{ GeV}^2)^{-2} \simeq 0\cdot85 \qquad (11.2)$$

whereas the exponential nuclear form factor has already suppressed the elastic nuclear cross-section significantly (the area under the quasi-elastic peak is already about $\frac{1}{2}$ of that under the elastic peak).

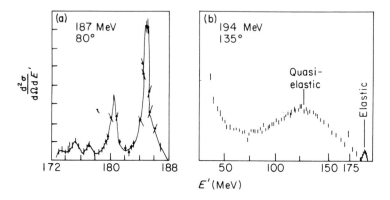

FIG. 11.1. Electron–carbon scattering: (a) through 80°; (b) through 135°.

If we now study Fig. 11.1(b) where the electron scatters through 135°, then the typical values of Q^2 are about 0·1 GeV². The elastic e-C scattering has almost completely disappeared. From equation (11.1) we expect that quasi-elastic scattering from the protons will occur when $E' \simeq 140$ MeV (modulo 10 per cent binding effects). In the data the quasi-elastic scattering peak is indeed clearly visible, the area beneath it now being about 300 times larger than that of the elastic e-C peak. The scattering is now dominated by this quasi-elastic scattering from the nuclear constituents. To see these constituents has required

i. $\sqrt{Q^2} \gg C_{12}$ size,

ii. $\nu > C_{12}$ nuclear level spacing, $\qquad\qquad\qquad\qquad\qquad$ (11.3)

iii. $\dfrac{Q^2}{2\nu} \simeq m_{\text{constituent}}$ (modulo 10–20 per cent binding effects);

since $\sqrt{Q^2} \ll$ nucleon size the nucleon structure is not yet probed and so a scaling should be seen. Actually the scaling behaviour in the quasi-elastic region is violated by about 30 per cent in the range $0·06 \leqslant Q^2 \leqslant 0·1$ GeV² since $G_E^N(Q^2 = 0·1 \text{ GeV}^2) \simeq 0·7$ (compare 0·85 when $Q^2 = 0·06$ GeV²). The Q^2 dependences of the e-C elastic and quasi-elastic peaks are compared in Fig. 11.2.

From the properties of the quasi-elastic peak we can learn about the nuclear constituents. We have already seen that the position of the peak is consistent with the average momentum (mass) of the constituents being about $\frac{1}{12}$ of that of the target. The area under the peak is

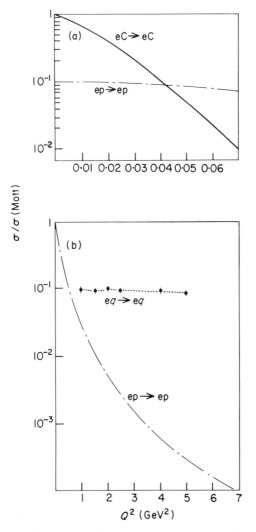

FIG. 11.2. (a) Q^2 dependence of elastic electron–carbon and deep inelastic electron–carbon scatterings. The latter is quasi-elastic electron–proton scattering. (b) Q^2 dependence of elastic electron–proton and deep inelastic (quasi-elastic electron–quark) interactions. Note the different Q^2 scales in (a) and (b).

proportional to the sum of the squared charges of the constituents weighted by any form factors that they themselves may have [$G_E^N(q^2)$] and suggests that $\sum_i Z_i^2 = 6$. The magnitude of the elastic e-C scattering suggests that $\sum_i Z_i = 6$. All of these data fit with our knowledge that C_{12} contains 6 protons and 6 neutrons as the constituents at the layer of matter immediately beneath the *nuclear* level.

By performing analogous experiments on the proton we will learn about the layer of matter immediately beneath the *nucleon* layer. Our expectations of course are that this new layer of matter will involve quarks.

From our experience with the nuclear example we may expect that to see the constituents of the proton will require

i. $\sqrt{Q^2} \gg$ proton size,

ii. $\nu >$ proton to N^* level spacing,   (11.4)

iii. $\dfrac{Q^2}{2m_p\nu} \simeq \dfrac{1}{N}$ if there are N constituents.

In the range of Q^2 between 1 and 8 $(\mathrm{GeV/c})^2$ the elastic ep $\to$ ep cross-section dies by two orders of magnitude (Fig. 9.3). The data at $x = \frac{1}{4}$ in the deep inelastic continuum are consistent with being Q^2 independent over this range of Q^2 (Fig. 9.6); hence scale invariant. This is illustrated and compared with the nuclear case in Fig. 11.2(b) (note the different scales of Q^2!). Hence over the range of $1 \leqslant Q^2 \leqslant 8\,(\mathrm{GeV/c})^2$, at least, the proton's constituents appear to be pointlike. We shall now study the evidence that they are quarks.

11.2   Quark partons and lepton scattering

11.2.1   ELECTROMAGNETIC STRUCTURE FUNCTIONS

We have seen in equation (9.72) that in the spin $\frac{1}{2}$ parton model the electromagnetic structure functions are given by

$$2xF_1(x) \equiv F_2(x) = \sum_i e_i^2 x f_i(x)$$

where the sum is over the partons whose charges are $e_i, \ldots$ We will use the notation $f_u(x) \equiv u(x)$ etc. and hence, if the partons have the quantum numbers of quarks (Table 3.1), the structure functions become

$$\frac{1}{x} F_2^{eP}(x) = \frac{4}{9}[u^P(x) + \bar{u}^P(x)] + \frac{1}{9}[d^P(x) + \bar{d}^P(x)] + \frac{1}{9}[s^P(x) + \bar{s}^P(x)] \ldots$$

$$(11.5)$$

$$\frac{1}{x} F_2^{eN}(x) = \frac{4}{9}[u^N(x) + \bar{u}^N(x)] + \frac{1}{9}[d^N(x) + \bar{d}^N(x)] + \frac{1}{9}[s^N(x) + \bar{s}^N(x)] \ldots$$

$$(11.6)$$

where the superscripts refer to the neutron (N) and proton (P) target.
Since the u, d quarks and P, N both form isospin doublets then

$$u^P \equiv d^N \text{ (call it simply } u)$$
$$d^P \equiv u^N \text{ (call it } d) \qquad\qquad (11.7)$$
$$s^P \equiv s^N \text{ (call it } s)$$

with analogous constraints for the antiquarks. Consequently

$$\frac{1}{x} F_2^{eP} = \frac{4}{9}(u + \bar{u}) + \frac{1}{9}(d + \bar{d} + s + \bar{s}) \ldots \qquad (11.8)$$

$$\frac{1}{x} F_2^{eN} = \frac{4}{9}(d + \bar{d}) + \frac{1}{9}(u + \bar{u} + s + \bar{s}) \ldots \qquad (11.9)$$

and so (Nachtmann, 1972)

$$\frac{1}{4} \leqslant \frac{F_2^{eN}}{F_2^{eP}}(x) \leqslant 4 \qquad\qquad (11.10)$$

These bounds are consistent with the data (Bloom, 1973; Bodek *et al.*,
1974) (Fig. 11.3).

We can proceed further by imposing ideas rooted in duality. Separate
the quarks (partons) into three "valence" quarks and a sea of quarks and
antiquarks along the following lines (Harari, 1971):

$$\text{3 valence quarks} + \text{Sea of } q\bar{q}$$
$$\updownarrow \qquad\qquad \updownarrow$$
$$\text{Resonances} \quad + \text{Background}$$
$$\updownarrow \qquad\qquad \updownarrow$$
$$\text{Nondiffractive} \; + \text{Diffractive}$$

Then (following, e.g. Kuti and Weisskopf, 1971, or Landshoff and Polkinghorne, 1971) write

$$q(x) \equiv q_V(x) + q_S(x) \tag{11.11}$$

The original guess was

$$u_V(x) = 2d_V(x) \tag{11.12}$$

$$s_V(x) = \bar{u}_V(x) = \bar{d}_V(x) = \bar{s}_V(x) = 0 \tag{11.13}$$

$$u_S(x) = \bar{u}_S(x) = d_S = \bar{d}_S = s_S = \bar{s}_S \equiv K \tag{11.14}$$

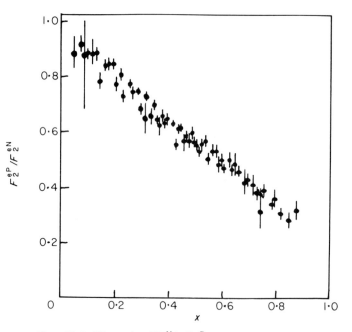

FIG. 11.3. The ratio $\nu W_2^{eN}/\nu W_2^{eP}$ as a function of x.

We will impose equations (11.13) and (11.14) but allow $u_V(x)/d_V(x)$ to be free. This gives

$$\frac{1}{x} F_2^{eN} = \frac{1}{9}(u_V + 4d_V) + \frac{12}{9}K \tag{11.15}$$

$$\frac{1}{x} F_2^{eP} = \frac{1}{9}(d_V + 4u_V) + \frac{12}{9}K \tag{11.16}$$

so that if $K(x)$ dominates

$$\frac{F_2^{\text{eN}}}{F_2^{\text{eP}}}(x) \to 1; \qquad (x \to 0?) \qquad\qquad (11.17)$$

whereas for dominance of the valence quarks (and if $u_V \geq 2d_V$)

$$\frac{F_2^{\text{eN}}}{F_2^{\text{eP}}}(x) \to \frac{2}{3} \text{ to } \frac{1}{4}; \qquad (x \geq 0 \cdot 2?) \qquad\qquad (11.18)$$

Hence we begin to have the first hints that perhaps the valence quarks are dominantly at large x while the sea is near $x = 0$. This will be reinforced in our subsequent data analyses, but first let us give an intuitive picture of why this picture is not unreasonable.

In QED the bare electron becomes dressed by diagrams such as Fig. 11.4(a). The analogue for the partons will be that vector (?) gluons (something has to hold the target together) will play the role of the photons in QED. Then a three valence quark system will be dressed in Fig. 11.4(b) where the wiggly lines denote gluons and the solid lines are quark–partons. The bremsstrahlung probability for momentum k in the gluon behaves as dk/k and hence like dx/x. This means that the gluon emission, and hence the $q\bar{q}$ structure or sea, tends to like small x.

(a)   (b)

FIG. 11.4. (a) Electron dressing in QED. (b) Analogous dressing of three quarks in QCD.

Hence in Fig. 11.5(a) we can visualise how one might start with a "primieval" model of three free quarks (for which $F_2(x)$ is essentially $\delta(x - \frac{1}{3})$) and then switch on the gluons, or place the quarks in a potential, which smears $F_2(x)$ (Fig. 11.5(b)). This far is like the nuclear analogy. Internal conversion and bremsstrahlung then produce the $q\bar{q}$ sea at small x yielding finally Fig. 11.5(c) which qualitatively resembles the data (Novikov *et al.*, 1976; Altarelli and Parisi, 1977; Altarelli *et al.*, 1974).[1]

From equations (11.15) and (11.16) we find that the difference of proton and neutron structure is given by

$$F_2^{eP}(x) - F_2^{eN}(x) = \tfrac{1}{3}x[u_V(x) - d_V(x)] \qquad (11.19)$$

The sea of $q\bar{q}$ pairs does not contribute here; only the three valence quarks play a role. The data show a peak at $x \simeq \tfrac{1}{3}$ (Fig. 11.6). A naive interpretation of this result (compare the nuclear physics case and equation 11.1) would be that the constituent masses are about 300 MeV

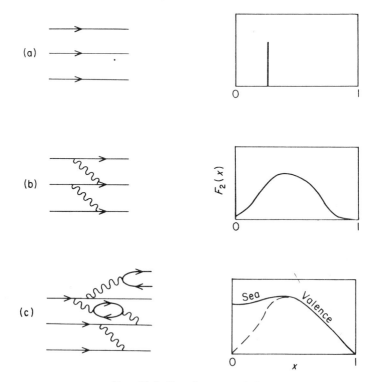

FIG. 11.5. Development of $F_2(x)$.

[1] Carried to its logical conclusion this picture gives rise to scaling violations as hinted already in sections 9.4 and 10.6. See Hinchliffe and Llewellyn Smith (1977), Buras and Gaemers (1977) and also section 11.8.

(compare also equation 15.32, section 17.2, and Chapter 18) and that a genuine quasi-elastic peak is being seen. We shall study these data again when discussing sum rules (section 11.2.4).

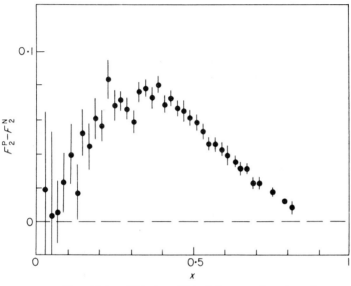

FIG. 11.6. $\nu W_2(ep) - \nu W_2(en)$ data as a function of x.

11.2.2 COMPARISON OF ELECTROMAGNETIC AND NEUTRINO INTERACTIONS

If we perform experiments at energies where charm production is absent or negligible then the charged weak current couples to the isospin of the partons and in the limit of zero Cabibbo angle the reaction is triggered by

$$\nu \binom{d}{\bar{u}} \to \mu^- \binom{u}{\bar{d}} \tag{11.20}$$

Hence

$$\frac{1}{x} F_2^{\nu P}(x) = 2[d(x) + \bar{u}(x)] \tag{11.21}$$

$$\frac{1}{x} F_2^{\nu N}(x) = 2[u(x) + \bar{d}(x)] \tag{11.22}$$

where in equation (11.22) we have used $d^N = u^P = u$ etc. The factor of 2 arises from the presence of axial as well as vector currents coupling, and in the parton model the weak current is taken to be $V - A$ as for leptons; hence the axial coupling magnitude is the same as that of the vector.

Comparing equations (11.21) and (11.22) with equations (11.8) and (11.9) yields

$$\left[\frac{F_2^{eN} + F_2^{eP}}{F_2^{\nu N} + F_2^{\nu P}}\right](x) = \frac{\frac{5}{9}(u + \bar{u} + d + \bar{d}) + \frac{2}{9}(s + \bar{s})}{2(u + \bar{u} + d + \bar{d})} \geq \frac{5}{18} \quad (11.23)$$

(the rather mysterious $\frac{5}{18}$ is of course just the average squared charge of the u, d quarks).

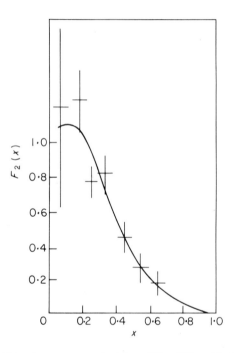

FIG. 11.7. Comparison of $(18/5)\,(F_2^{eP} + F_2^{eN})$ with $(F_2^{\nu P} + F_2^{\nu N})$.

In Fig. 11.7 we see the data from CERN–Gargamelle (Perkins, 1975) where $F_2^{\nu N + \nu P}(x)$ is compared with $\frac{18}{5}F_2^{eP + eN}$ from SLAC. The agreement supports the quark quantum numbers and the saturation of the inequality at large x suggest that $s, \bar{s}(x \geq 0\cdot 2) \approx 0$ (which is in line with our picture that strange quarks are in the $q\bar{q}$ sea which in turn is confined to small x values).

With Han–Nambu quarks (Chapter 8) the average squared charge of u and d quarks is $\frac{1}{2}$ which is considerably larger than the data allow (Fig. 11.7). However, if only colour singlet states are being produced then all transitions are from colour singlet to colour singlet and so the operative charges will be the $\widetilde{SU(3)}_c$ average charges. These are identical with the Gell-Mann and Zweig charges (equation 8.14) and so the Han–Nambu scheme is consistent with the data if the data are below threshold for producing nonsinglets of colour.

11.2.3 GLUON MOMENTUM

Since the $u(x)$, $d(x)$, $s(x)$ etc. are the probability distributions of the parton flavours in x, then $xu(x)$ etc. will be the fractional distributions of the target's momentum among the parton flavours. Since $F_2(x) \sim xf(x)$ then this structure function is a direct measure of the fractional momentum distribution of the quarks.

Below charm threshold, momentum conservation yields

$$\int_0^1 dx\, x(u + \bar{u} + d + \bar{d} + s + \bar{s}) = 1 - \varepsilon \tag{11.24}$$

where ε is the fraction of momentum carried by electrically neutral constituents, e.g. the gluons.

From equations (11.8) and (11.9) and equations (11.21) and (11.22) we see that equation (11.24) can be rewritten as

$$\int_0^1 dx\, \left(\tfrac{9}{2}F_2^{eP+eN} - \tfrac{3}{4}F_2^{\nu P+\nu N}\right) = 1 - \varepsilon \tag{11.25}$$

Inserting the data on the left-hand side we find $\varepsilon \simeq \frac{1}{2}$, i.e. about half the momentum is carried by the gluons (Llewellyn Smith, 1974).

Instead of using electron and neutrino data we could use electron data alone and, defining the ratio of momentum carried by strange quarks to nonstrange by δ,

$$\delta \equiv \frac{\int_0^1 dx\, x(s + \bar{s})}{\int_0^1 dx\, x(u + \bar{u} + d + \bar{d})} \tag{11.26}$$

then from equations (11.8), (11.9), (11.21) and (11.22) we have

$$\int_0^1 dx\, x(u + \bar{u} + d + \bar{d} + s + \bar{s}) = \frac{9(1+\delta)}{5+2\delta} \int_0^1 dx\, (F_2^{eP} + F_2^{eN})$$

(11.27)

If the sea is an SU(3) singlet then extraction of quark distribution functions from data (section 11.4) suggest $\delta \leqslant 0{\cdot}06$. If δ ranges from 0 to 0.06 then equation (11.27) with data yields $(0.54 \text{ to } 0.56) \pm 0.04$ for the total q and $\bar{q}$ momentum. Hence again we conclude that nearly half of the momentum is carried by neutrals (i.e. partons which do not take part in electromagnetic and weak interactions).

11.2.4 SUM RULES

Since a nucleon has no strangeness then

$$0 = \int_0^1 dx\, [s(x) - \bar{s}(x)]$$

(11.28)

The charges of proton and neutron give

$$1 = \int_0^1 dx [\tfrac{2}{3}(u - \bar{u}) - \tfrac{1}{3}(d - \bar{d})]$$

$$0 = \int_0^1 dx\, [\tfrac{2}{3}(d - \bar{d}) - \tfrac{1}{3}(u - \bar{u})]$$

and so

$$2 = \int_0^1 dx\, [u(x) - \bar{u}(x)]$$

(11.29)

$$1 = \int_0^1 dx\, [d(x) - \bar{d}(x)]$$

(11.30)

These state that the net excess of s, d, u quarks over $\bar{s}, \bar{d}, \bar{u}$ in the proton are 0, 1, 2, respectively.

These sum rules for the quark distributions can now be combined with the relations (11.8), (11.9), (11.21) and (11.22) to yield sum rules for the targets.

One interesting consequence is that we can check if the data indeed agree that there is a net excess of three quarks over antiquarks:

$$N(q) - N(\bar{q}) = \int_0^1 [(u + d + s) - (\bar{u} + \bar{d} + \bar{s})] \, dx \qquad (11.31)$$

Then since the nucleon has zero strangeness we utilise equation (11.28) and obtain

$$N(q) - N(\bar{q}) = \int_0^1 [(u + d) - (\bar{u} + \bar{d})] \, dx$$

$$= \frac{1}{2} \int_0^1 (F_3^{\nu P + \nu N}) \, dx \qquad (11.32)$$

where F_3 is the vector–axial vector interference term[1] that will be met in the study of neutrino interactions in section 11.3. The data (Cundy, 1974) on this integral show that it equals $3 \cdot 2 \pm 0 \cdot 6$, and thus is consistent with a net excess of three quarks over antiquarks. This is the Gross–Llewellyn Smith sum rule (1969).

Since $F_2^{\nu N}(x) - F_2^{\nu P}(x) \equiv 2x(u - \bar{u} - d + \bar{d})$ we find the Adler sum rule (Adler, 1966):

$$\int_0^1 \frac{dx}{x} [F_2^{\nu N}(x) - F_2^{\nu P}(x)] = 2 \qquad (11.33)$$

Another interesting quantity is

$$\frac{F_2^{eP}(x) - F_2^{eN}(x)}{x} \equiv \frac{1}{3}(u + \bar{u} - d - \bar{d}) \qquad (11.34)$$

If we impose duality (i.e. $u = u_V + u_S$ etc., equation 11.11) then the sum rules (11.29) and (11.30) become

$$1 = \int_0^1 dx \, d_V(x) \qquad (11.35)$$

$$2 = \int_0^1 dx \, u_V(x) \qquad (11.36)$$

and so equation (11.34) yields

$$\frac{F_2^{eP}(x) - F_2^{eN}(x)}{x} \equiv \frac{1}{3}(u_V(x) - d_V(x)) \qquad (11.37)$$

---

[1] We will sometimes employ $\mathscr{F}$ manifestly to discriminate weak interaction structure functions from their F electromagnetic analogues.

which gives

$$\int_0^1 \frac{dx}{x}(F_2^{eP}(x) - F_2^{eN}(x)) = \frac{1}{3} \equiv \sum_i [(e_i^2)_P - (e_i^2)_N] \qquad (11.38)$$

The data are consistent with this and yield (Bloom, 1973) $0 \cdot 28 \pm$? The ? is the contribution from large ω (small x). If we believe that $F^{eP} - F^{eN} \sim x^{1/2}$ when $x \leqslant 0 \cdot 1$ (Regge-like) then the data are consistent with the predicted value of $\frac{1}{3}$.

11.3 Neutrino interactions

Defining $x \equiv Q^2/2M\nu$ and $y = \nu/E$ then it is a straightforward exercise to rewrite equation (9.71) in the form

$$\frac{d^2\sigma}{dx\, dy} = \frac{4\pi\alpha^2 s}{t^2}[F_2(x)(1-y) + F_1(x)xy^2] \qquad (11.39)$$

For the process $\nu(\bar{\nu})N \rightarrow \mu^{\pm}X$ one has a similar formula (Llewellyn Smith, 1974)

$$\frac{d^2\sigma_\nu^{\bar{\nu}}}{dx\, dy} = \frac{G^2 s}{2\pi}\left[\mathscr{F}_2(x)(1-y) + \mathscr{F}_1(x)xy^2 \mp y\left(1 - \frac{y}{2}\right)x\mathscr{F}_3(x)\right]$$

$$(11.40)$$

and if $2x\mathscr{F}_1(x) = \mathscr{F}_2(x)$ (as suggested by spin $\frac{1}{2}$ partons) this becomes

$$\frac{d^2\sigma_\nu^{\bar{\nu}}}{dx\, dy} = \frac{G^2 s}{2\pi}\mathscr{F}_2(x)\left[\frac{1+(1-y)^2}{2} \mp \frac{1-(1-y)^2}{2}\frac{x\mathscr{F}_3(x)}{\mathscr{F}_2(x)}\right] \qquad (11.41)$$

In comparison with the electromagnetic case (equation 11.39) we see the absence of t^{-2} due to the assumed pointlike (no photon exchanged) nature of the neutrino interaction. Also there is the new structure function $\mathscr{F}_3$ which is due to the violation of parity in the weak interactions. Its role will be transparent when we discuss the quark–parton model for this process (section 11.3.2). First we shall discuss the above formulae and neutrino kinematics. This can be bypassed if desired, proceeding directly to the parton model.

11.3.1   NEUTRINO KINEMATICS

We shall endeavour to make comparisons with the electromagnetic case examined in section 9.1.3. The electron interaction took place by exchange of a single photon and we shall assume that the weak interaction involves the exchange of a W boson of mass m_W. The cross-section may be written by analogy to the electromagnetic case (9.27):

$$\frac{d^2\sigma}{d\Omega\,dE'} = \frac{G^2}{(2\pi)^2}\left(\frac{m_W^2}{m_W^2+Q^2}\right)^2 \frac{E'}{E} L_{\mu\nu}^{(\nu)} W^{\mu\nu} \qquad (11.42)$$

The lepton tensor is now (compare equation 9.7)

$$L_{\mu\nu} = k_\mu k'_\nu + k'_\mu k_\nu - g_{\mu\nu} k \cdot k' \pm i\varepsilon_{\mu\nu\lambda\sigma} k^\lambda k'^\sigma \qquad (11.43)$$

with upper and lower signs for the (left-handed) neutrino or (right-handed) antineutrino beam. In the unpolarised electron beam the $\varepsilon_{\mu\nu\lambda\sigma}$ cancelled in the mixture of right- and left-handed leptons.

For an unpolarised initial nucleon the hadronic tensor may be written (compare with equation 9.30)

$$W^{\mu\nu} = W_1 g^{\mu\nu} + \frac{W_2}{M^2} p^\mu p^\nu - i\varepsilon^{\mu\nu\alpha\beta} p_\alpha q_\beta \frac{W_3}{2M^2} + \frac{q^\mu q^\nu W_4}{M^2}$$

$$+ W_5\left(\frac{p^\mu q^\nu + q^\mu p^\nu}{M^2}\right) + W_6 i\frac{(p^\mu q^\nu - p^\nu q^\mu)}{2M^2} \qquad (11.44)$$

The W_4 and W_5 terms were constrained by gauge invariance in the electromagnetic case and W_6 was not present since it multiplies a tensor antisymmetric in $\mu\nu$; here they contribute in principle but we shall neglect them as they yield contributions to the cross-section only to the order of the lepton mass, and, in the absence of heavy leptons, can be dropped henceforth. The term W_3 was absent in the electromagnetic case since its contribution to the cross-section violates parity (as can be seen later—equation 11.47).

Contracting $L_{\mu\nu} W^{\mu\nu}$ we obtain

$$2EE'\left\{ 2\sin^2\frac{\theta}{2} W_1 + \cos^2\frac{\theta}{2} W_2 \mp W_3 \frac{(E+E')}{M}\sin^2\frac{\theta}{2}\right\} \qquad (11.45)$$

(compare equation 9.35 in the electron case), the upper (lower) sign coming from $\nu(\bar{\nu})$ interactions. Hence

$$\frac{\mathrm{d}^2\sigma^{(\nu,\bar{\nu})}}{\mathrm{d}\Omega\,\mathrm{d}E'} = \frac{G^2 E'^2}{2\pi^2}\left(\frac{m_{\mathrm{W}}^2}{m_{\mathrm{W}}^2 + Q^2}\right)^2$$

$$\times\left\{2\mathscr{W}_1\sin^2\frac{\theta}{2} + \mathscr{W}_2\cos^2\frac{\theta}{2} \mp \mathscr{W}_3\frac{(E+E')}{M}\sin^2\frac{\theta}{2}\right\}$$

(11.46)

and from this point on we shall send $m_{\mathrm{W}} \to \infty$ which has the effect of removing the term in the first parentheses above. Obtaining the x, y form at equation (11.40) is straightforward. The x, y terms multiplying $\mathscr{F}_{1,2}$ follow immediately by analogy with the electromagnetic case (equation 11.39). The only new feature is the $\mathscr{W}_3$ term. To manipulate this into the desired form we note first that

$$E + E' \equiv 2E(1 - y/2).$$

Then comparing with the $\mathscr{W}_1$ term that also multiplies $\sin^2(\theta/2)$ we multiply and divide by ν. This yields $\nu\mathscr{W}_3$ (which is $\mathscr{F}_3(x)$) and $E/\nu \equiv 1/y$, hence the $xy(1-y/2)$ factor in place of xy^2 in the $\mathscr{F}_1$ contribution.

The similarity and difference with the electromagnetic case (equation 9.35) is obvious. Further, in analogy with the electromagnetic case, we can define "W-absorption cross-sections" for right-, left-handed or scalar W as σ_{R}, σ_{L}, σ_{S}. The calculation of the relations between these cross-sections and the structure functions proceeds as in the electromagnetic case and we obtain

$$\mathscr{W}_1 = \frac{K}{\pi G\sqrt{2}}(\sigma_{\mathrm{R}} + \sigma_{\mathrm{L}})$$

$$\mathscr{W}_2 = \frac{K}{\pi G\sqrt{2}}\frac{Q^2}{Q^2 + \nu^2}(\sigma_{\mathrm{R}} + \sigma_{\mathrm{L}} + 2\sigma_{\mathrm{S}})$$

(11.47)

$$\mathscr{W}_3 = \frac{K}{\pi G\sqrt{2}}\frac{2M}{\sqrt{(\nu^2 + Q^2)}}(\sigma_{\mathrm{R}} - \sigma_{\mathrm{L}})$$

In the electromagnetic interaction parity invariance forces $\sigma_{\mathrm{R}} = \sigma_{\mathrm{L}}$ and hence $W_3^{\mathrm{e.m.}}$ was zero, and $\sigma_{\mathrm{T}} = \frac{1}{2}(\sigma_{\mathrm{R}} + \sigma_{\mathrm{L}})$. The other important difference with the electromagnetic case is that $\mathscr{W}_{1,2}$ contain now both vector–vector (VV) and axial–axial (AA) pieces in contrast to the electromagnetic $W_{1,2}$ which are purely VV. The vector–axial vector interference is given by $\mathcal{W}_3$ and hence the sign change in equation (11.46) upon replacing a neutrino beam $(V-A)$ by an antineutrino beam $(V+A)$.

11.3.2 PARTON MODEL AND NEUTRINO INTERACTIONS

In the electromagnetic case we were able to understand the data by postulating that the electromagnetic current (photon) interacted with partons whose electromagnetic currents were $\bar{u}\gamma_\mu u$, i.e. like the muon's electromagnetic current, hence "pointlike". The one new assumption that is invoked in the weak interaction is that the parton's weak current has the familiar form of the neutrino $\bar{u}\gamma_\mu(1-\gamma_5)u$, i.e. $V-A$, while the antiparton is $V+A$. Hence it is almost obvious by inspection that the model predicts that $\mathcal{W}_1$ and $\nu\mathcal{W}_2$ will scale (compare the electron case) and that $|\mathcal{W}_{1,2}^{VV}| = |\mathcal{W}_{1,2}^{AA}|$. As an exercise the reader should calculate this explicitly using the electron example as a guide (see also Llewellyn Smith, 1974). In the process you will find that the prediction is that the $\mathcal{W}_i$ scale as $\mathcal{W}_1(\nu, Q^2), \nu\mathcal{W}_2(\nu, Q^2), \nu\mathcal{W}_3(\nu, Q^2) \to \mathcal{F}_{1,2,3}(\omega)$ and that $\mathcal{F}_3$ is maximal and negative (i.e. the scattering is predicted to be all in the left-handed mode (compare equation 11.47). This is a consequence of the parton current being $\bar{u}\gamma_\mu(1-\gamma_5)u$; for antipartons with $\bar{u}\gamma_\mu(1+\gamma_5)u$ then the scattering would be right-handed. Hence one has a potential test of the relative importance of partons and antipartons in the data description. Stated another way, the parity violation is maximal if *only* partons or *only* antipartons are contributing, whereas if the partons and antipartons are equally important then the parity violation is minimal.

11.3.2.1 *y distributions*

The physical significance of the x, y dependences in the cross-section equation (11.41), namely

$$\frac{d^2\sigma_\nu^{\bar{\nu}}}{dx\,dy} = \frac{G^2 s}{2\pi} \mathcal{F}_2(x)\left[\frac{1+(1-y)^2}{2} \mp \frac{1-(1-y)^2}{2}\frac{x\mathcal{F}_3(x)}{\mathcal{F}_2(x)}\right] \quad (11.41\ bis)$$

is of interest when we discuss the quark–parton model for this process.

In the quark–parton model the basic interaction is a weak coupling of the lepton with the quark weak current. If the latter is $V - A$ (like $\nu \to \mu^-$) then the y dependence of neutrino–quark scattering is as follows (Fig. 11.8):

$$\frac{\mathrm{d}\sigma}{\mathrm{d}y}[\nu\bar{q}\,;\,\bar{\nu}q] \sim (1-y)^2 \tag{11.48}$$

$$\frac{\mathrm{d}\sigma}{\mathrm{d}y}[\nu q\,;\,\bar{\nu}\bar{q}] \sim 1 \text{ (isotropic)} \tag{11.49}$$

This can be derived explicitly from the form of the lepton tensor equation (11.43) (which is also $W_{\mu\nu}$ in the parton model). Perform this as an exercise.

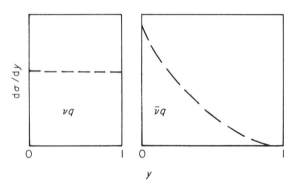

FIG. 11.8. Isotropic and $(1-y)^2$ distributions for μ^- produced in νq and $\bar{\nu}q$ interactions.

Heuristically this result can be understood as follows (Fig. 11.9). An interaction at a point is S-wave; all the angular momentum information of the νq interaction will therefore be contained in the spin structure. A νq interaction will have $J_z = 0$ in the c.m. system since both have helicity $-\frac{1}{2}$ (if $m_q = 0$). Pointlike interaction will therefore carry no memory of direction and hence an isotropic distribution can ensue. For a $\nu\bar{q}$ interaction on the other hand, $J_z = -1$ since the $\bar{q}$ has helicity $+\frac{1}{2}$ and so the total $J_z = -1$. The emerging $\bar{q}$ and μ^- are right-handed and left-handed respectively and so $J_{z'} = -1$ along the z'-axis (oriented at θ with respect to the initial z-axis). This angular momentum picture leads to $a|d^1_{11}(\theta)|^2$ distribution. Since $y = \nu/E$, then $y - 1 = -2ME'/2ME \equiv$

u/s, and so $d^1_{11}(\theta) \equiv (1+\cos\theta)/2 = 1-y$, giving the $(1-y)^2$ distribution, in which the 180° scattering is clearly suppressed, as against isotropic behaviour in the νq case.

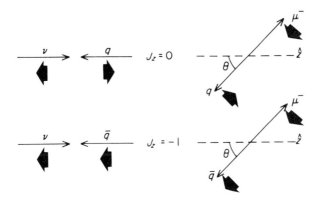

FIG. 11.9. Helicity structure of νq and $\nu \bar{q}$ interactions.

For the case of an isoscalar target, writing $q(x)$ and $\bar{q}(x)$ for the probabilities to find quarks or antiquarks at given x, then

$$\frac{d^2\sigma^\nu}{dx\,dy} \sim [q(x) + (1-y)^2\bar{q}(x)] \tag{11.50}$$

$$\frac{d^2\sigma^{\bar{\nu}}}{dx\,dy} \sim [\bar{q}(x) + (1-y)^2 q(x)] \tag{11.51}$$

where we have used equations (11.48) and (11.49). Comparing with equation (11.41) we have

$$B \equiv \frac{x\mathscr{F}_3(x)}{\mathscr{F}_2(x)} = \frac{q(x) - \bar{q}(x)}{q(x) + \bar{q}(x)} \tag{11.52}$$

and so the x distributions of quarks and antiquarks can be compared by studying the x dependence of this ratio of structure functions (more correctly, the distributions of $V \pm A$ elementary currents are revealed). The equation (11.52) also helps us to appreciate why the extra structure function $\mathscr{F}_3$ appears in the weak interaction as compared to the electromagnetic case. The parity violation causes the left- and right-handed couplings to be independent (hence $\mathscr{F}_3$) in the weak interaction, and hence the difference in q and $\bar{q}$ couplings.

The data on $\mathcal{F}_2$ and $x\mathcal{F}_3$ from Gargamelle (Perkins, 1975) ($Q^2 >$ 1 GeV^2, $W^2 > 4$ GeV^2) are shown in Fig. 11.10. We see that for $x \gtrsim$ 0·4, $\mathcal{F}_2(x) \simeq x\mathcal{F}_3(x)$ and so from equation (11.52) we have

$$\bar{q}(x)/q(x) \xrightarrow{\ x \geqslant 0\cdot4\ } 0 \qquad\qquad (11.53)$$

whereas for $x \to 0$, $x\mathcal{F}_3(x) \to 0$ and hence

$$\bar{q}(x)/q(x) \xrightarrow{\ x \to 0\ } 1 \qquad\qquad (11.54)$$

This fits in with our previous guess from the electromagnetic case, namely that (valence) quarks dominate as $x \gtrsim 0\cdot4$ while antiquarks are all in the sea with $x \to 0$ (equation 11.18 *et seq.*).

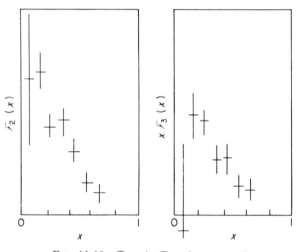

FIG. 11.10. $\mathcal{F}_2$ and $x\mathcal{F}_3$ as functions of x.

We can investigate this further by studying the y distributions for various regions of x. The data for $E_{\bar{\nu}} \lesssim 30$ GeV from Gargamelle and Fermilab are all consistent with $(1-y)^2$ distributions for $\bar{\nu}$ induced reactions and isotropy for ν interactions at large x (Perkins, 1975).

A best fit to the Gargamelle data on the y distributions yields (Perkins, 1975)

$$B \equiv \frac{\langle x\mathcal{F}_3 \rangle}{\langle \mathcal{F}_2 \rangle} \sim 0\cdot80 \qquad\qquad (11.55)$$

and hence

$$\int dx\, x[q(x) - \bar{q}(x)] = 0\cdot 8 \int dx\, x[q(x) + \bar{q}(x)] \qquad (11.56)$$

or

$$\int dx\, xq(x) \simeq 9 \int dx\, x\bar{q}(x)$$

This implies that antiquarks carry only about 5 per cent of the target momentum (45 per cent quarks and 50 per cent gluons, section 11.2.3).

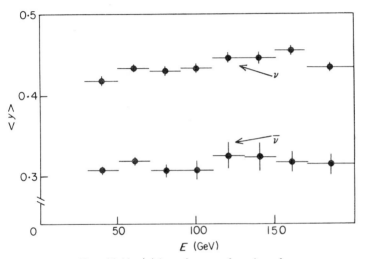

FIG. 11.11. $\langle y \rangle$ in ν data as a function of energy.

Equivalently one often sees data presented on $\langle y \rangle$ (Fig. 11.11). If quarks dominate the deep inelastic data then

$$\langle y \rangle_\nu = \langle y \rangle_{\nu q} = \tfrac{1}{2}$$
$$\langle y \rangle_{\bar{\nu}} = \langle y \rangle_{\bar{\nu}q} = \tfrac{1}{4} \qquad (11.57)$$

and the data are consistent with these.

11.3.2.2 Total cross-sections

So far we have just assumed that $\nu(\bar{\nu})$ data scale analogously to their electromagnetic cousin. This we should really check. If we integrate equation (11.40) over dx and dy, then, assuming scaling (i.e.

$\mathcal{F}_2(x, Q^2) \to \mathcal{F}_2(x))$ we have for the total cross-section

$$\sigma_\nu^{\bar\nu}(s) = \frac{G^2 s}{2\pi} \int dx\, \mathcal{F}_2^{\bar\nu,\nu}(x) \left[\frac{2}{3} \mp \frac{x\mathcal{F}_3(x)}{\mathcal{F}_2(x)} \right] \tag{11.58}$$

and hence a linear rise with energy ($s = 2ME$). This is consistent with the Gargamelle data (Perkins, 1975). Furthermore, from equations (11.58) and (11.52) we have (where $\mathcal{F}$ is written F^ν or $F^{\bar\nu}$ hereon)

$$\frac{\sigma^{\bar\nu}}{\sigma^\nu} = \frac{\int dx\, F_2^{\bar\nu}(x)(\frac{1}{3} + \frac{2}{3}\bar q(x)/[q(x) + \bar q(x)])}{\int dx\, F_2^\nu(x)(1 - \frac{2}{3}\bar q(x)/[q(x) + \bar q(x)])} \tag{11.59}$$

and hence is bounded to lie between $\frac{1}{3}$ and 3 for targets with an equal number of protons and neutrons for which $F_2^{\bar\nu} \equiv F_2^\nu$. The Gargamelle data (all Q^2, W) have this ratio $\simeq 0\cdot37$ which again fits with the dominance of quarks over antiquarks (or, rather, of left-handed parton currents).

If only data is included with $Q^2 > 1$ GeV, $W^2 > 4$ GeV^2 then from fitting the x, y distributions one has $B \sim 0\cdot80$ (equation 11.55) and hence

$$\frac{\sigma^{\bar\nu}}{\sigma^\nu} = \frac{\int dx\, x[\frac{1}{3}q(x) + \bar q(x)]}{\int dx\, x[q(x) + \frac{1}{3}\bar q(x)]} \simeq 0\cdot43 \tag{11.60}$$

which is slightly larger than when all Q^2, W were included. These data for E up to 200 GeV are shown in Fig. 11.12.

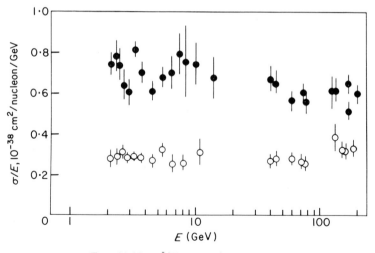

FIG. 11.12. $\sigma^{\bar\nu}/E$ and σ^ν/E for $E \leqslant 200$ GeV.

11.3.2.3 *Hydrogen versus isoscalar targets*

The previous discussion concentrated on isoscalar targets since the early experiments used heavy nuclear targets with almost the same numbers of neutrons as of protons. More recently experiments have been performed on hydrogen (i.e. proton targets).

$$F_2^{\nu P} \equiv F_2^{\bar{\nu} N} = 2x[d(x) + \bar{u}(x)] \qquad (11.21 \text{ bis})$$

$$F_2^{\nu N} \equiv F_2^{\bar{\nu} P} = 2x[u(x) + \bar{d}(x)] \qquad (11.22 \text{ bis})$$

(since only F_3 distinguishes between ν and $\bar{\nu}$).[1] Then from equation (11.41) at $y = 0$ we have

$$\left.\frac{d\sigma}{dx}\right|_{y=0}^{\bar{\nu}(\nu)} = \frac{G^2}{2\pi} s F_2^{\bar{\nu}(\nu)}(x) \qquad (11.61)$$

hence

$$\left.\sigma^{\bar{\nu}(\nu)}\right|_{y=0} = \frac{G^2}{2\pi} s \int_0^1 F_2^{\bar{\nu}(\nu)}(x) \, dx \qquad (11.62)$$

For targets with equal numbers of protons and neutrons it therefore follows from equations (11.21), (11.22) and (11.62) that

$$\left.\frac{\sigma^{\bar{\nu}P+\bar{\nu}N}}{\sigma^{\nu P+\nu N}}\right|_{y=0} = \frac{\int_0^1 dx \, F_2^{\bar{\nu}P+\bar{\nu}N}(x)}{\int_0^1 dx \, F_2^{\nu P+\nu N}(x)} = 1 \qquad (11.63)$$

This can be exploited experimentally to relatively normalise the $\bar{\nu}$ and ν cross-section data.

For targets with differing numbers of protons and neutrons one cannot obtain such a relation. For example

$$\left.\frac{\sigma^{\bar{\nu}P}}{\sigma^{\nu P}}\right|_{y=0} = \frac{\int_0^1 dx \, [u(x) + \bar{d}(x)]x}{\int_0^1 dx \, [d(x) + \bar{u}(x)]x} \qquad (11.64)$$

From $d\sigma/dx|_{y=0}$ we can immediately obtain information on the flavour distributions without needing to separate $F_3(x)$. In particular for $x \geqslant 0\cdot2$, where antiquarks can be neglected,

$$\left.\frac{d\sigma}{dx}\right|_{y=0}^{\bar{\nu}P} \bigg/ \left.\frac{d\sigma}{dx}\right|_{y=0}^{\nu P} = u(x)/d(x) \qquad (11.65)$$

[1] We are continuing to work below charm production threshold and ignore the Cabibbo angle.

From the electroproduction data it appeared that the u flavour dominates as $x \to 1$ ($F_2^{\gamma N}/F_2^{\gamma P} \to \frac{1}{4}$ in Fig. 11.3 and compare this with equations 11.8 and 11.9). This in turn suggests that

$$\frac{\dfrac{d\sigma}{dx}\Big|_{y=0}^{\nu P}}{\dfrac{d\sigma}{dx}\Big|_{y=0}^{\bar{\nu} P}}$$

should tend to zero at large x. The data do show that this ratio is falling with increasing x, consistent with the $u(x)/d(x)$ that one extracts from the large x electroproduction data (Derrick, 1977).

11.4 Quark distribution functions

The comparison of electromagnetic and neutrino data has suggested that for $x \geqslant 0.2$ antiquarks and strange quarks play a negligible role. In this domain one can invert equations (11.8) and (11.9) and immediately obtain the u, d quark distribution functions

$$u(x) \simeq \frac{1}{x} \frac{9}{15} [4F_2^{eP}(x) - F_2^{eN}(x)]$$

$$\tag{11.66}$$

$$d(x) \simeq \frac{1}{x} \frac{9}{15} [4F_2^{eN}(x) - F_2^{eP}(x)]$$

If we assume that the sea is an SU(3) singlet, then defining $K(x)$ by (11.14) yields in place of equation (11.66)

$$F_2^{eN+eP}(x) = \frac{x}{9} [5(u_V + d_V)(x) + 24K(x)] \tag{11.67}$$

$$F_2^{eP}(x) - F_2^{eN}(x) = \frac{x}{3}(u_V - d_V)(x) \tag{11.68}$$

From (11.52) and from (11.21) and (11.22) we have

$$\frac{F_2^{\nu P + \nu N}(x)}{xF_3^{\nu P + \nu N}(x)} = 1 + \frac{4K(x)}{u_V(x) + d_V(x)} \tag{11.69}$$

and from Fig. 11.10 we see that $K(x)$ is negligible for $x \geqslant 0.1$ but dominates over u_V, d_V as $x \to 0$. Also one finds (equation 11.23)

$$\frac{F_2^{eP+eN}(x)}{F_2^{\nu P+\nu N}(x)} = \frac{5}{18} + \frac{1}{18}\left[\frac{4K(x)}{(u_V+d_V)(x)+4K(x)}\right] \tag{11.70}$$

We see that equation (11.70) is not a very sensitive indicator of the importance of the sea since if the sea is absent then the ratio of electromagnetic to weak structure functions is 5/18 and the ratio rises only to 6/18 if the sea dominates. The weak structure functions in equation (11.69) provide a good way of resolving the sea contribution due to the V $\pm$ A structures of $\bar{q}$ and q being probed. The quality of electromagnetic data implies that equations (11.67) and (11.68) are perhaps the most direct way of attacking the extraction of the quark distribution functions. In addition there is the useful ratio

$$\frac{F^{eN}(x)}{F^{eP}(x)} = \frac{(u_V+4d_V)(x)+12K(x)}{(4u_V+d_V)(x)+12K(x)} \tag{11.71}$$

(in principle the neutrino data in equation (11.65) provide the most direct test of $u_V/d_V(x)$).

In Fig. 11.13 we exhibit (without error bars) the parton distribution functions as extracted by Barger and Phillips (1974). The qualitative features are obvious and their origins clear: the sea is negligible for

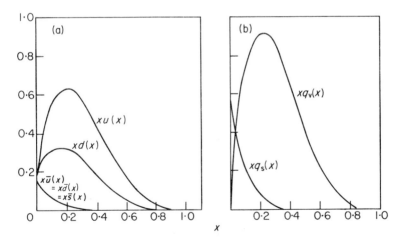

FIG. 11.13. x Dependence of parton distribution functions: (a) all the u and d quarks in the proton and for $\bar{u}$, $\bar{d}$, $\bar{s}$ in the sea; (b) valence quarks $q_v(x)$ and sea $q_s(x)$. The sea is assumed to be an SU(N) singlet.

$x \geqslant 0 \cdot 3$ (equation 11.69 and Fig. 11.10), the u quark dominates significantly at large x (equation 11.71 and Fig. 11.3), the $u - d$ difference being given directly by equation (11.68), and is positive at all $x > 0 \cdot 11$ (Fig. 11.6).

The discovery of charmed states has uncovered new ways of studying the sea (sections 11.6).

11.4.1 COUNTING RULES AND PARTON X DISTRIBUTIONS

If $\nu W_2(x) \to$ constant as $x \to 0$ then the parton probability $f(x) \sim x^{-1}$ which corresponds to Pomeron $\alpha = 1$ Regge behaviour in the Compton amplitude (equation 10.53). The nondiffractive structure function contributions (e.g. $\nu W_2^{eP} - \nu W_2^{eN}$) are controlled by Regge exchanges with $\alpha \simeq \frac{1}{2}(f, A_2)$ and hence $f^{ND}(x) \sim x^{-1/2}$. If one associates valence quarks with nondiffractive and $q\bar{q}$ sea with diffractive (Pomeron) components then as $x \to 0$

$$q_{\text{sea}}(x) \sim x^{-1}; \qquad q_{\text{valence}}(x) \sim x^{-1/2} \qquad (11.72)$$

The x dependence of $q(x)$ as $x \to 1$ is also of interest. From the Drell–Yan–West relation in equation (10.42) we see that the $q^2 \to \infty$ elastic form factor behaviour is correlated with $\nu W_2(x \to 1)$. For the proton we anticipated $\nu W_2(x \to 1) \sim (1 - x)^3$ (equation 10.36 et seq.). Hence

$$q(x) \sim \frac{\nu W_2(x)}{x} \sim (1 - x)^3 \qquad (11.73)$$

as $x \to 1$ in the proton. This need not imply that all quark flavours behave this way, only that the leading behaviour is $(1 - x)^3$. Indeed if

$$\frac{F_2^{eN}}{F_2^{eP}}(x) \xrightarrow{ x \to 1 } \frac{1}{4}$$

then it is possible (Gunion, 1974) that

$$u(x) \sim (1 - x)^3; \, d(x) \sim (1 - x)^4 \qquad (11.74)$$

in this limit.

It has been argued that $(1 - x)^3$ is the natural expectation for the leading behaviour as $x \to 1$ if the proton is viewed as three valence quarks. This result follows from counting rules for the energy dependence of large-scale scattering of composite systems (Matveev et al., 1973; Brodsky and Farrar, 1973) and is discussed in more detail in section 14.3.4. The essential idea is that an inclusive cross-section $AB \to CX$ is given by the sum of the cross-sections for contributing subprocesses $ab \to cd$ at large p_T and weighted by the probabilities $f_{a/A}, f_{b/B}, \tilde{f}_{C/c}$ for the systems A, B to fragment into constituents a, b and for c to produce C. These fragmentation probabilities are scale invariant and so the scaling behaviour of $E d\sigma/d^3 p (AB \to CX)$ is intimately related to the scaling behaviour of the elementary subprocess $ab \to cd$ (see also section 14.1). The electron–proton elastic scattering cross-section is then the sum of the electron–quark cross-sections. The greater the number of quarks in the proton then the smaller the probability for elastic scattering (since elastic scattering is the probability that in a hard collision all of the constituents change direction so that the system does not break up). The counting rules give

$$F(t) \sim t^{1-n} \tag{11.75}$$

(this result is derived in equation 14.28) and so $n = 3$ yields $F(t) \sim t^{-2}$ and by the relation equation 10.42 yields $(1-x)^3$ for $q(x)$.

Blankenbecler and Brodsky (1974) have shown that for inclusive reactions the counting rules can be formulated as follows. The elementary fields taking part in the elementary subprocess will be called "active", the remainder "passive". Then

$$\frac{E d\sigma}{d^3 p}(AB \to CX)\Big|_{s,t,u \to \infty} \sim p_T^{-2N} f(\theta_{\text{c.m.}}, \varepsilon) \underset{\varepsilon \to 0}{\sim} p_T^{-2N} \varepsilon^F f(\theta_{\text{c.m.}}) \tag{11.76}$$

for $\varepsilon = m_X^2/s$ fixed. Here

$$N \equiv n_{\text{active}} - 2 \tag{11.77}$$

$$F \equiv 2n_{\text{passive}} - 1 \tag{11.78}$$

Physically the suppression for large N is as described above while the "forbiddenness", F, increases as n_{passive} increases since these spectators use up the available phase space.

For $ep \to eX$ we have $n_{\text{active}} = 4$ for $eq \to eq$ and $n_{\text{passive}}^{\text{hadronic}} = 2$ which yields $\nu W_2(x) \sim (1-x)^3$ as $x \to 1$.

From this relation one may consider $e\bar{q} \to e\bar{q}$ to determine the x dependence of $\bar{q}(x)$ as $x \to 1$. Here $n_{\text{passive}} = 4$ (since $qqqq\bar{q}$ is the minimal configuration) and hence $\nu W_2(x)_{\bar{q}} \sim (1-x)^7$ which has been used (Gunion, 1974; Farrar, 1974) to suggest that

$$q_{\text{val}}(x) \sim (1-x)^3; \qquad q_{\text{sea}}(x) \sim (1-x)^7 \qquad (11.79)$$

as $x \to 1$. However one can in principle consider $e\bar{q} \to eq$ as the basic subprocess ($e + \bar{q}qqqq \to e + qqq$ where $\bar{q}q$ annihilate into a gluon which is absorbed by a valence quark, Fig. 11.14). This will yield $q_{\text{sea}}(x) \sim \nu W_2(x)_{\bar{q}} \sim (1-x)^5$.

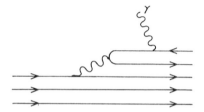

FIG. 11.14. $(1-x)^5$ sea contribution arising from $qqqq\bar{q} \leftrightarrow qqq$.

A derivation of the counting rules is given in the aforementioned papers and in section 14.3.4. A detailed discussion of the related phenomenology can be found in Sivers *et al.* (1977).

11.5 Scale invariance in electron–positron annihilation

We shall be primarily concerned here with the annihilation of an electron and positron (each carrying energy E in their centre of mass frame) into a single photon. For this process we have

$$s \equiv (p_{e^-} + p_{e^+})^2 = 4E_{\text{c.m.}}^2 \equiv Q^2 \qquad (11.80)$$

where Q^2 is the squared mass of the photon produced in the annihilation.

Since the photon has $J^{PC} = 1^{--}$, then electron–positron annihilation is a useful tool for studying the vector meson spectroscopy ($J^{PC} = 1^{--}$).

At small s we see *three* vector mesons (ρ, ω, ϕ), which is already a hint that there are *three* flavours of light quarks, since the three vector mesons reflect these three degrees of freedom:

$$\left.\begin{array}{c}\rho\\\omega\end{array}\right\}(u\bar{u} \mp d\bar{d})$$

$$\phi \quad (s\bar{s})$$

The observation of a fourth vector meson (ψ) at $E_{\text{c.m.}} = 3.1$ GeV is therefore suggestive of a fourth flavour

$$\psi(c\bar{c})$$

(The vector meson at 3·7 GeV is more naturally interpreted as a radially excited $c\bar{c}$ state rather than a state made from a fifth flavour—see section 16.)

Not only does the existence of these vector mesons ρ, ω, ϕ, ψ suggest four flavours of quarks, but their leptonic widths are related to the squared charges of the quarks contained within them and empirically

$$\Gamma^{e^+e^-}_{\rho+\omega\,:\phi\,:\psi} \sim 10:2:8 \sim \langle e^2 \rangle_{\text{u+d:s:c}}$$

(though why this quantity as against, say, $m\,\Gamma^{e^+e^-}$ should be related to e^2 is not known).

Apart from the study of vector meson properties, the e^+e^- annihilation into hadrons through a single photon is a useful tool for investigating the $J^{PC} = 1^{--}$ hadronic continuum. Three questions arise here:

a. What is the Q^2 dependence of $\sigma(e^+e^- \rightarrow \text{hadrons})$?
b. How big is it?
c. What is it made up of?

We will discuss questions (a) and (b) now, and (c) will be examined in later sections.

In order to lead in to questions (a) and (b) it is instructive to first study a simple example that nicely illustrates the relation between scaling and pointlike behaviour.

11.5.1 $e^+e^- \to \mu^+\mu^-$

The cross-section for this process can be calculated exactly from QED in the one photon annihilation. One has

$$\sigma_{\text{QED}}^{(e^+e^- \to \mu^+\mu^-)}(Q^2) = \frac{4\pi\alpha^2}{3Q^4}\left(1 - \frac{4m^2}{Q^2}\right)^{1/2}(2m^2 + Q^2) \qquad (11.81)$$

where m is the muon mass, and there is a threshold behaviour $(1 - 4m^2/Q^2)^{1/2}$ appropriate to production of $\mu^+\mu^-$ in a relative S-wave $(J^{PC} = 1^{--}$ can produce fermion–antifermion in S-wave because fermion and antifermion have opposite intrinsic parity).

If $Q^2 \gg m^2$, then

$$\sigma_{\text{QED}}^{(e^+e^- \to \mu^+\mu^-)}(Q^2) \approx \frac{4\pi\alpha^2}{3Q^2} \qquad (11.82)$$

(this is derived explicitly in equation 14.9). Notice that this expression contains no scale of length or mass associated with the muon. The quantity $Q^2\sigma$ is dimensionless and independent of Q^2 (scales).

Do all cross-sections scale? The answer is easily seen to be "no" by considering the very similar production of the fermion–antifermion pair, proton–antiproton.

11.5.2 $e^+e^- \to p\bar{p}$

The cross-section for this process is very similar to that of $e^+e^- \to \mu^+\mu^-$ and reads

$$\sigma^{(e^+e^- \to p\bar{p})}(Q^2) = \frac{4\pi\alpha^2}{3Q^4}\left(1 - \frac{4M^2}{Q^2}\right)^{1/2}[2M^2 G_E^2(Q^2) + Q^2 G_M^2(Q^2)] \qquad (11.83)$$

where now M is the proton mass and $G_{E,M}$ are the electric and magnetic form factors of the proton. If $G_E \equiv G_M \equiv 1$ then the equation looks like that for $e^+e^- \to \mu^+\mu^-$.

When $Q^2 \gg M^2$, then (assuming G_E and G_M are not too different in their Q^2 dependence)

$$\sigma^{(e^+e^- \to p\bar{p})}(Q^2) \approx \frac{4\pi\alpha^2}{3Q^2} G_M^2(Q^2)$$

$$\overset{?}{\approx} \frac{4\pi\alpha^2}{3Q^2}\left(1 + \frac{Q^2}{M^2}\right)^{-4} G_M^2(0) \qquad (11.84)$$

This differs from the previous case $(e^+e^- \to \mu^+\mu^-)$ by the presence of the form factor, in which the "size" (mass) of the proton manifestly appears. The dimensionless quantity $Q^2\sigma(Q^2)$ now depends manifestly on the scale—the proton mass or size—and hence is not scale invariant. As Q^2 increases, so $Q^2\sigma(e^+e^- \to p\bar{p})$ dies away in magnitude.

Similarly the cross-section for $e^+e^- \to \pi^+\pi^-$ has an explicit dependence on the pion form factor

$$F_\pi(Q^2) \simeq \left(1 + \frac{Q^2}{m_\rho^2}\right)^{-1}$$

$$\sigma^{(e^+e^- \to \pi^+\pi^-)}(Q^2) = \frac{\pi\alpha^2}{3Q^2}|F_\pi(Q^2)|^2$$

(11.85)

and so $Q^2(e^+e^- \to \pi^+\pi^-)$ dies out with Q^2.

It is expected that any (quasi) two-body channel behaves in a similar fashion, i.e. $Q^2\sigma$ decreases with increasing (large) Q^2 due to the finite size of the coherently produced final state particles. If this is so, then what is the behaviour of $Q^2\sigma(e^+e^- \to \text{all hadrons})$?

11.5.3 $\sigma_t(e^+e^- \to \text{hadrons})$

In the parton model we expect that $e^+e^- \to \text{hadrons}$ takes place by $e^+e^- \to \text{parton} + \text{antiparton}$, and the partons then fragment into the observed hadrons by some unknown mechanism. Then at large Q^2

$$\sigma(e^+e^- \to \text{hadrons}) = \sum_{i=\text{udsc}\dots} \sigma(e^+e^- \to q_i\bar{q}_i) \quad (11.86)$$

$$= \sum_i e_i^2 \sigma(e^+e^- \to \mu^+\mu^-) \quad (11.87)$$

and hence

$$R \equiv \frac{\sigma(e^+e^- \to \text{hadrons})}{\sigma(e^+e^- \to \mu^+\mu^-)} = \sum_i e_i^2 \quad (11.88)$$

so we expect to find this quantity *constant* in Q^2 (Feynman, 1972; Cabibbo et al., 1970) and its magnitude measures *directly* the sum of the squared charges of the fundamental fermion fields. Hence below charm threshold, the u, d, s degrees of freedom are operative, and as they come in three colours (e.g. section 15.2) we have

$$R_{uds} = 3(\tfrac{4}{9} + \tfrac{1}{9} + \tfrac{1}{9}) = 2 \tag{11.89}$$

At higher Q^2 we will cross the threshold for production of charmed mesons. The first feature in the data will be the appearance of narrow vector mesons in the e^+e^- channel (identified with the ψ at $3 \cdot 1$ GeV and $3 \cdot 7$ GeV), followed by the charm production threshold where R will rise and show complicated structure (around 4 GeV). At higher Q^2 one anticipates that R will again show scaling (become constant) with value

$$R = 2 + 3e^2_{charm} \tag{11.90}$$

If

$$e_c = \tfrac{2}{3} \quad \text{then } R \to 3\tfrac{1}{3} \tag{11.91}$$

The data do indeed show scaling behaviour (Fig. 11.15). Frascati data at $\sqrt{Q^2} < 3$ GeV is unclear but not inconsistent with constant ~ 2 to 3 in magnitude. Better data from SPEAR below $3 \cdot 5$ GeV suggest $R \sim 2 \cdot 5$ to 3 with *no obvious* structures. After the 4 GeV structures R appears to have settled down again to a value around $5\tfrac{1}{2}$. One unit of this is believed to be due to pair production of a new heavy lepton (Perl, 1977). Is the remaining $4\tfrac{1}{2}$ consistent with uds and c or are more quarks needed?

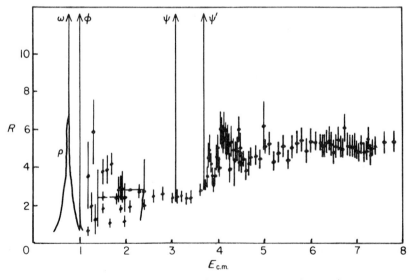

FIG. 11.15. Data on $\sigma(e^+e^- \to \text{hadrons})/\sigma(e^+e^- \to \mu^+\mu^-)$.

In non-Abelian gauge theories with asymptotic freedom one expects the asymptotic value of R to be approached slowly from above. Explicitly (Appelquist and Georgi, 1973; Zee, 1973)

$$R = \sum_i e_i^2 [1 + b/\log(Q^2/Q_0^2) + \cdots] \tag{11.92}$$

where for three colours

$$b = \tfrac{4}{9} \tag{11.93}$$

and Q_0^2 is *a priori* unknown. Hence the e^+e^- annihilation data appears to be a remarkable manifestation of the scaling idea and perhaps even of the simple quark–parton model.

If the partons have spin $\tfrac{1}{2}$ we expect $\sigma_T \gg \sigma_L$ at large Q^2 and the partons to be produced with a $(1 + \cos^2\theta)$ angular distribution relative to the e^+e^- axis. The hadron fragments from the partons will, at high energies, be produced in a cone along the direction of motion of the parent parton. Hence we expect to see jets of hadrons with a $(1 + \cos^2\theta)$ distribution. Remarkably this appears to be manifested by the high-energy data. We will show this in section 12.4.

11.6   Charm production in νN and eN

The narrow vector meson J/ψ and the rise in the e^+e^- annihilation cross-section around 4 GeV provide the first evidence for a new massive hadronic degree of freedom. It now appears clear that these phenomena are due to the existence of a fourth flavour of quark (charmed quark, denoted c) (Glashow *et al.*, 1970; Gaillard *et al.*, 1975) whose weak interaction is

$$\bar{\nu}c \to \mu^+(-d\sin\theta_c + s\cos\theta_c) \tag{11.94}$$

(Gell-Mann, 1964; Hara, 1964; Glashow *et al.*, 1970). The theoretical background to the charm hypothesis and the phenomenology of the emerging spectroscopy are described in Chapters 15 and 16. Here we will be concentrating on the phenomenological consequences of charm in deep inelastic scattering.

If the lowest mass charmed particles have typical branching ratios B to decay weakly into a final state including a lepton, e.g. $D \to K\mu\nu$, then upon crossing the charm production threshold in ν or $\bar{\nu}$ induced

processes a "dilepton" signal will be seen. With the GIM current of
equation (11.94) the rates for seeing these dileptons will be

$$\nu + d \rightarrow c + \mu^- + \cdots$$
$$\qquad \qquad \rightarrow \mu^+ + \cdots \qquad \qquad \propto \sin^2 \theta_c \times B \qquad (11.95)$$

$$\nu + s \rightarrow c + \mu^- + \cdots$$
$$\qquad \qquad \rightarrow \mu^+ + \cdots \qquad \qquad \propto \varepsilon_s \cos^2 \theta_c \times B \qquad (11.96)$$

where ε_s is the momentum carried by the strange quarks (in the sea)
relative to the (valence) d quarks and is given by

$$\varepsilon_s = \frac{\int xs(x)\, dx}{\int xd(x)\, dx} \simeq 5 \text{ per cent} \qquad (11.97)$$

Hence the dilepton signal for charm production arises roughly equally
from the valence d and sea s quarks ($\tan^2 \theta_c \sim 1/25$) when a neutrino
beam is employed.

With an incident antineutrino beam the dominant mechanism will be

$$\bar{\nu} + \bar{s} \rightarrow \bar{c} + \mu^+ + \cdots$$
$$\qquad \qquad \rightarrow \mu^- + \cdots \qquad \qquad \propto \varepsilon_s \cos^2 \theta_c \times B \qquad (11.98)$$

The W^- transferred necessarily lowers the charge at the hadronic
vertex. This forces the sea to be involved; there is no way that charm can
appear from valence quarks interacting with $\bar{\nu}$ (other than by associated
production ($\bar{\nu}u \rightarrow \mu^+d + c\bar{c} + \cdots$) which we can safely ignore). The x
distributions of the dimuons are shown in Fig. 11.16 (Steinberger,
1977) and are consistent with coming from the sea (Fig. 11.17(a) and
equation 11.98) and sea with suppressed valence (Fig. 11.17(b) and
equations 11.95 and 11.96).

The rate of producing opposite sign dimuons via a neutrino beam is
about 1 per cent of single muon production. Since $\sin^2 \theta_c + \varepsilon_s \cos^2 \theta_c \sim$
$\frac{1}{10}$ (in equations 11.95 and 11.96) then $B \approx 10$ per cent. More quantitative estimates need detailed consideration of experimental cuts and
detection criteria. The order of 10 per cent branching ratio is consistent
with the results found in e^+e^- production of charmed mesons (Brandelik et al., 1977b) and hence a measure of self-consistency emerges in
the data.

The charm production will be more dramatic in $\bar{\nu}$ induced reactions
than ν reactions because $\bar{q}$ are selected (equation 11.98). The $\bar{\nu}\bar{q}$

angular distribution is isotropic in contrast to the $(1-y)^2$ $\bar{\nu}q$ which dominated the low-energy data. In ν induced reactions the data are dominantly isotropic both below and above charm threshold and so the threshold effect will be less pronounced.

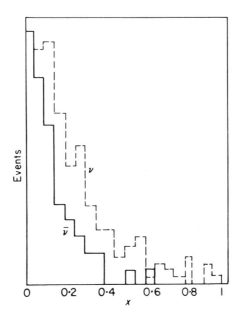

FIG. 11.16. x Distributions of dimuons produced in $\bar{\nu}N - \mu^+\mu^-$ .... If these arise from weak decays of charmed hadrons then these distributions are also the x distribution of antistrange quarks. $\nu N \to \mu^+\mu^-$ is shown (– – –) for comparison.

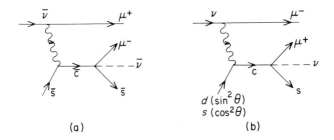

FIG. 11.17. Dimuon production: (a) $\bar{\nu}$ interacting with the sea; (b) ν interacts with sea + $(\sin^2\theta_c) \times$ valence quarks.

Notice that if the u quark had a right-handed coupling to some quark, b, with charge $-\frac{1}{3}$, e.g.

$$\bar{\nu}u \xrightarrow{\text{R.H.}} \mu^{+}b \qquad (11.99)$$

then a dramatic threshold phenomenon would be expected because now a valence quark is involved as well as the isotropic y distribution. Hence $\bar{\nu}$ interactions can be rather sensitive tests of new quarks and especially of postulated right-handed currents (Fritzsch *et al.*, 1975).

In deep inelastic electroproduction (or muon production) the charm production threshold should be seen at small x where

$$\frac{\Delta\sigma}{\sigma} = \frac{\frac{4}{9}[c(x)+\bar{c}(x)]}{\frac{4}{9}[u(x)+\bar{u}(x)]+\frac{1}{9}[d(x)+\bar{d}(x)]+\frac{1}{9}[s(x)+\bar{s}(x)]} \qquad (11.100)$$

In the case of an SU(4) symmetric sea one therefore would expect that $\Delta\sigma/\sigma = \frac{2}{3}$. While this may be true as $Q^2 \to \infty$ (where all mass scales are probably irrelevant), presumably at finite Q^2 the charm quark will be less important (being associated typically with heavier mass scales).

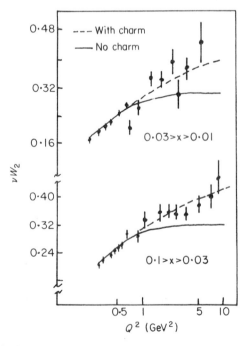

FIG. 11.18. Scaling violation at small x compared with charm production predictions.

Hence

$$\frac{\Delta\sigma}{\sigma}(x \to 0; Q^2) = \frac{2}{3}\varepsilon(Q^2); \qquad (11.101)$$

where ε is the probability that a charmed quark is present in the sea,

$$\varepsilon \equiv \frac{c(x \to 0; Q^2)}{u, d, s(x \to 0; Q^2)} < 1 \qquad (11.102)$$

A guess for the relative importance of charm to uncharmed quarks as a function of Q^2 might be something like

$$\varepsilon = \frac{M_\rho^2 + Q^2}{M_\psi^2 + Q^2} \qquad (11.103)$$

which is about 5 per cent at $Q^2 = 0$ (like estimates from VMD) rising through 50 per cent by $Q^2 = 10 \text{ GeV}^2$. The hypothesis of equation (11.103) enables one to compute the scaling violation effects that might be expected as $x \to 0$ above charm production threshold. Scaling violations similar to these have been seen in the data (Fig. 11.18) (Mo, 1975; Anderson *et al.*, 1976; Bharadwaj, 1977). It is not yet clear whether these scaling violations are evidence for charm production or are of the type expected if the quarks interact or have substructure (sections 9.4 and 10.6 and Fig. 9.13).

11.7   Neutral currents

In addition to the charge changing weak currents discussed in section 11.3 *et seq.*, neutral current effects have been observed in $ve \to ve$ (Blietschau *et al.*, 1976; Faissner *et al.*, 1976; Reines *et al.*, 1976) and also in exclusive processes on nuclei (Cline *et al.*, 1976; Lee *et al.*, 1977).

We shall analyse inclusive neutral current processes using the quark–parton model and supposing that the current is vector and axial vector with arbitrary amounts of right- and left-handed parts. This is in contrast with the charged current which appears to be purely left-handed. (*A priori* the neutral current could have scalar, tensor etc. components.)

Hence the neutral quark current may be written

$$j_\mu^{\text{N.C.}} = \sum_{i=udsc} (C_V^i \bar{q}_i \gamma_\mu q_i + C_A^i \bar{q}_i \gamma_\mu \gamma_5 q_i)$$

$$\equiv \sum_i \bar{q}_i \gamma_\mu \left(\frac{C_V^i + C_A^i}{2}(1+\gamma_5) + \frac{C_V^i - C_A^i}{2}(1-\gamma_5) \right) q_i$$

$$\equiv \sum_i \bar{q}_i (C_R^i \gamma_\mu (1+\gamma_5) + C_L^i \gamma_\mu (1-\gamma_5)) q_i$$

(11.104)

From our discussion of neutrino kinematics in section 11.3 (in particular equations 11.48 and 11.49) we know that for a left-handed neutrino interacting with a left-handed quark the distribution in y is isotropic whereas it is $(1-y)^2$ when the interaction is with a right-handed quark current. Hence if we consider a system of u and d quarks (neglecting s, c and antiquarks) then for an average nucleon the charged and neutral current cross-sections will be

$$\frac{d^2\sigma}{dx\,dy}(\nu \to \mu^-) = \frac{G^2 s}{2\pi} xq(x)$$

(11.105)

$$\frac{d^2\sigma}{dx\,dy}(\bar{\nu} \to \mu^+) = \frac{G^2 s}{2\pi} xq(x)(1-y)^2$$

(11.106)

$$\frac{d^2\sigma}{dx\,dy}(\nu \to \nu) = \frac{G^2 s}{2\pi} xq(x) \sum_{i=u,d} \left\{ \left(\frac{C_V^i + C_A^i}{2} \right)^2 + \left(\frac{C_V^i - C_A^i}{2} \right)^2 (1-y)^2 \right\}$$

(11.107)

$$\frac{d^2\sigma}{dx\,dy}(\bar{\nu} \to \bar{\nu}) = \frac{G^2 s}{2\pi} xq(x) \sum_{i=u,d} \left\{ \left(\frac{C_V^i + C_A^i}{2} \right)^2 (1-y)^2 + \left(\frac{C_V^i - C_A^i}{2} \right)^2 \right\}$$

(11.108)

We notice immediately that in our present approximation the x and y dependences have factorised and so the following total cross-section ratios immediately obtain

$$R^\nu \equiv \frac{\sigma(\nu \to \nu)}{\sigma(\nu \to \mu^-)} = \tfrac{1}{4}[(C_V + C_A)^2 + \tfrac{1}{3}(C_V - C_A)^2]$$

$$\equiv C_L^2 + \tfrac{1}{3}C_R^2$$

(11.109)

$$R^{\bar{\nu}} \equiv \frac{\sigma(\bar{\nu} \to \bar{\nu})}{\sigma(\bar{\nu} \to \mu^{+})} = \tfrac{3}{4}[\tfrac{1}{3}(C_{V} + C_{A})^{2} + (C_{V} - C_{A})^{2}]$$

$$\equiv C_{R}^{2} + \tfrac{1}{3}C_{L}^{2} \tag{11.110}$$

where a sum over $i = u, d$ is implicit in $C_{V,A}$ (compare equations 11.107 and 11.108).

These neutral current phenomena are of particular significance in the light of the recent interest in unified models of weak and electromagnetic interactions (see section 15.1). Consider first the ν_{e} and e^{-} leptons. The left-handed component of the electron forms a doublet of "weak isospin" with the neutrino

$$\chi_{L} = \begin{pmatrix} \nu_{e} \\ e_{L}^{-} \end{pmatrix} \tag{11.111}$$

The charged weak current may be written

$$J_{\mu}^{+} = \bar{\nu}_{e}\gamma_{\mu}(1 - \gamma_{5})e \equiv \bar{\chi}_{L}\begin{pmatrix} 0 & 1 \\ 0 & 0 \end{pmatrix}\gamma_{\mu}\chi_{L} \equiv \bar{\chi}_{L}\tau^{+}\gamma_{\mu}\chi_{L} \tag{11.112}$$

where τ^{+} is the raising operator for weak isospin (equation 2.16).

The electromagnetic current will be

$$J_{\mu}^{\text{e.m.}} = -\bar{e}\gamma_{\mu}e \tag{11.113}$$

$$\equiv -\bar{e}_{L}\gamma_{\mu}e_{L} + \tfrac{1}{2}\bar{\nu}_{e}\gamma_{\mu}\nu_{e} - \tfrac{1}{2}\bar{\nu}_{e}\gamma_{\mu}\nu_{e} - \bar{e}_{R}\gamma_{\mu}e_{R} \tag{11.114}$$

$$\equiv \bar{\chi}_{L}\left(\frac{\tau_{3}}{2} - \frac{1}{2}\right)\gamma_{\mu}\chi_{L} - \bar{e}_{R}\gamma_{\mu}e_{R} \tag{11.115}$$

where the right-handed piece of the electron also takes part in contrast to the weak interaction in equation (11.112).

The u quark forms a weak isodoublet with the Cabibbo rotated d_{θ} ($= d\cos\theta + s\sin\theta$). Their left-handed charged weak current has the same form as equation (11.112) where now

$$\chi_{L} = \begin{pmatrix} u \\ d_{\theta} \end{pmatrix}_{L} \tag{11.116}$$

in contrast to (11.111). Their electromagnetic current will be

$$J_{\mu}^{\text{e.m.}} = \left[\bar{\chi}_{L}\left(\frac{\tau_{3}}{2} + \frac{1}{6}\right)\gamma_{\mu}\chi_{L} + \frac{2}{3}\bar{u}_{R}\gamma_{\mu}u_{R} - \frac{1}{3}\bar{d}_{R}\gamma_{\mu}d_{R}\right]e \tag{11.117}$$

If we introduce "weak-hypercharge", $Y_{wk} \equiv Q - I_3$ for left-handed fermions and $Y_{wk} \equiv Q$ for right-handed fermions, then

$$J_\mu^{e.m.} = e\left[\bar{\chi}_L(I_3 + Y_{wk})\gamma_\mu\chi_L + Y_{wk}\bar{\chi}_R\gamma_\mu\chi_R\right] \qquad (11.118)$$

Clearly there are two primieval neutral currents corresponding to the weak isospin

$$\mathscr{W}_3 \sim (u\bar{u} - d\bar{d}) \qquad (11.119)$$

and weak hypercharge

$$B^0 \sim (u\bar{u} + d\bar{d}) \qquad (11.120)$$

with arbitrary couplings g and g' respectively. The electromagnetic current is one combination of these and its strength is e. The orthogonal combination is a new neutral current Z^0 which couples (see equation 15.11)

$$J_\mu^{N.C.} = \frac{2e}{\sin 2\theta}\left[\bar{\chi}(I_3 - Q\sin^2\theta)\gamma_\mu\chi\right] \qquad (11.121)$$

where Q is the charge and θ is an arbitrary parameter commonly known as the Weinberg–Salam angle ($\tan\theta = g'/g$).

Given equation (11.121) we can compute $C_{V,A}$ in this model. For u, d quarks the left- and right-handed couplings will be

$$
\begin{aligned}
u_L &\sim \tfrac{1}{2} - \tfrac{2}{3}\sin^2\theta \\
d_L &\sim -\tfrac{1}{2} + \tfrac{1}{3}\sin^2\theta \\
u_R &\sim -\tfrac{2}{3}\sin^2\theta \\
d_R &\sim +\tfrac{1}{3}\sin^2\theta
\end{aligned}
\qquad (11.122)
$$

Hence in this model, equations (11.109) and (11.110) yield

$$
\begin{aligned}
R^\nu &= \tfrac{1}{2} - \sin^2\theta + \tfrac{20}{27}\sin^4\theta \\
R^{\bar{\nu}} &= \tfrac{1}{2} - \sin^2\theta + \tfrac{20}{9}\sin^4\theta
\end{aligned}
\qquad (11.123)
$$

Hence R^ν and $R^{\bar{\nu}}$ are given as a function of one parameter, θ, and so must lie on a curve in $R^\nu, R^{\bar{\nu}}$ space (Fig. 11.19). Data from various experiments are consistent with $0\cdot2 \leqslant \sin^2\theta \leqslant 0\cdot4$ (Blietschau et al., 1977; Barish et al., 1977; Benvenuti et al., 1977; Steinberger, 1977). The solid curve is that corresponding to our illustrative analysis where antiquarks have been ignored. If some 10–15 per cent of antiquarks are included (compare equation 11.56) then the curve should be shifted to the dotted curve.

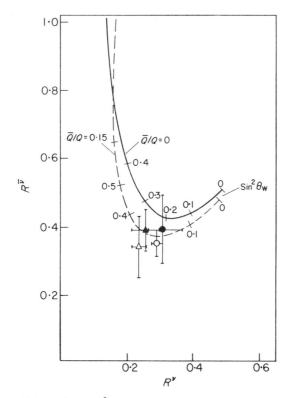

FIG. 11.19. R^ν and $R^{\bar\nu}$ in Weinberg–Salam model as a function of θ_W.

11.8   Scaling violations

We have presented arguments that the structure function $F_2(x, Q^2)$ should shift its $\langle x \rangle$ from large to small values as Q^2 increases if quarks have substructure being revealed at large Q^2 or if they take part in interactions with gluons (sections 9.4, 10.6 and Fig. 9.13). The qualitative effect is that at small x, $F_2(x, Q^2)$ will rise as Q^2 increases while at large x it will decrease. There are indications that the data indeed show such a behaviour (Taylor, 1975; Mo, 1975; Anderson *et al.*, 1977). A quantitative comparison with the scaling violations expected in field theories has been made by Tung (1976). Perkins *et al.* (1977) have shown that the neutrino data and electromagnetic data show similar

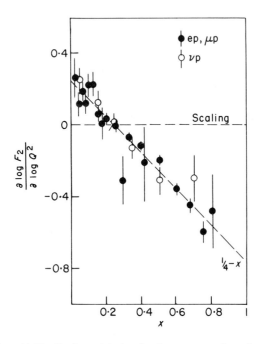

FIG. 11.20. Scaling violation in electromagnetic and neutrino data.

patterns of scaling violation and that the scaling violation may have power law rather than logarithmic behaviour. Their parametrisation is

$$F_2(x, Q^2) = F_2(x, Q_0^2)\left(\frac{Q^2}{Q_0^2}\right)^{(1/4-x)} \qquad (11.124)$$

hence rising with increasing Q^2 when $x < \frac{1}{4}$ and falling when $x > \frac{1}{4}$. To compare this parametrisation with data note that it can be rewritten

$$\frac{\partial[\log F_2(x, Q^2)]}{\partial[\log Q^2]} = \frac{1}{4} - x \qquad (11.125)$$

This is indeed consistent with the scaling violation seen in both the electromagnetic and weak interactions (Fig. 11.20).

12 Inclusive Production of Hadrons and the Quark–Parton Model

In Chapter 11 we saw how the quark–parton model made testable predictions for the total cross-sections in lepton–hadron scattering and e^+e^- annihilation. In all of this, no discussion was made of the nature of the final hadronic state.

Interesting tests of the quark–parton model arise from inclusive hadron production experiments like $e^+e^- \to h + $ anything, and $e(\nu)N \to e(\mu^-)h + $ anything. In particular the production of detected hadrons h in the current fragmentation region (defined below) of the $e(\nu)$ scattering experiments is intimately related with the production in e^+e^- annihilation, and there is some support that this correlation is in fact realised in the data.

12.1 $lN \to l'h \ldots (l \equiv e, \mu, \nu)$

This process is illustrated in Fig. 12.1. The hadron and nucleon momenta are p_h, p_N respectively and $x \equiv Q^2/2M\nu$ as usual. There is great similarity with the lN inclusive process discussed earlier, but now we have an extra kinematical degree of freedom associated with p_h, the momentum of the hadron h detected. We will work in the centre of mass system of the current (electromagnetic or weak) and nucleon, and will define the positive z-axis to be the direction of the current. Then we choose variables to characterise the problem:

$$Q^2, \qquad x(\equiv Q^2/2M\nu), \qquad p_T^h, z(\equiv p_N \cdot p_h/p_N \cdot q) \qquad (12.1)$$

There are two rather different regions, $z \lessgtr 0$. In both of these $q \cdot p_h \sim O(Q^2)$ while $p_N \cdot p_h$ is finite for $z < 0$ but grows as $O(Q^2)$ for

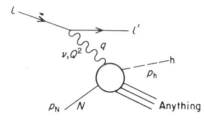

FIG. 12.1. Inclusive hadron production in lepton scattering.

$z > 0$. The former is intuitively the target fragmentation region and can be represented by Fig. 12.2(a). This is intimately related to the diagram met in the total cross-section at large Q^2 (Figs 9.9 and 12.2(b)).

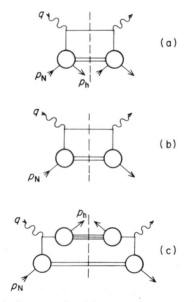

FIG. 12.2. Parton model diagrams for (a) target fragmentation, and (c) current fragmentation regions. The total cross-section is in (b) for comparison.

Therefore one expects scaling in this region (technically, one can argue that the light-cone dominates here).

The natural picture for $z > 0$ is shown in Fig. 12.2(c) with p_h emerging along the direction of q and $p_h \cdot q \sim 0(Q^2)$. In the light-cone formalism one can say very little about this region since it is not light-cone dominated (the fragmentation takes place between the two currents in the Figure). Hence the parton model has extra power here if we define functions $D_i^h(z, Q^2, p_T)$ to represent the fragmentation probabilities for (quark)-parton of flavour i to produce hadron h.

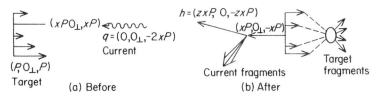

FIG. 12.3. Inclusive hadron production in the current and quark Breit frame. The four momenta are denoted (E, p_T, p_z).

We will concentrate on the current fragmentation region $(z > 0)$. The parton model analysis of this process is illustrated in the Breit frame of the current and the parton with which it interacts (Fig. 12.3). The nucleon carries a large longitudinal momentum P and is treated as a collection of independent pointlike constituents (partons). The current, with momentum

$$q = (0; 0, 0, -2xP) \qquad (12.2)$$

interacts incoherently with a parton whose momentum vector is

$$p = (xP; 0, 0, xP) \qquad (12.3)$$

and so its momentum is reversed. This is analogous to the total cross-section description of Chapter 9, and this part of the process is described by the quark–parton distribution functions $u(x) \, dx$ etc. (the average number of u quarks in an interval dx of x).

In Fig. 12.3(b) we exhibit the fragmentation of the quark–parton into hadrons, one of which, h, is observed. The struck parton is separated by a large momentum from the nucleon fragments and so we shall assume that the fragmentation is independent of the earlier current interaction. Hence we shall assume it to be independent of x and only dependent upon z, which, in a frame in which the parton moves fast (c.m., Breit, etc.), is the fraction of the parton's longitudinal momentum carried off by the observed hadron (Fig. 12.3). Hence

$$p_h = (zxP; 0, 0, -zxP) \qquad (12.4)$$

and so we introduce a set of "parton fragmentation functions" $D_i^h(z) \, dz$ which represent the probability that parton type i produces a hadron in an interval dz about z (Feynman, 1972).

12.2   The quark fragmentation functions

In terms of the known quark distribution functions $u(x)$ ... and the unknown $D_i^h(z)$ fragmentation functions, we can discuss hadron inclusive production in a variety of current-induced processes, e.g. $e^+e^- \to h \dots$, $ep(n) \to eh \dots$, $\nu p(n) \to \mu^- h \dots$ etc. We can obtain relations among these various processes due to the $\sum_q (x)D_q^h(z)$ structure and constrain the relative production rates of various hadrons by limiting the number of independent $D_i^h(z)$ using isospin and charge-conjugation invariance. This yields, for π production,

$$D_u^{\pi^+} = D_d^{\pi^-} = D_{\bar{d}}^{\pi^+} = D_{\bar{u}}^{\pi^-} \qquad (12.5)$$

$$D_d^{\pi^+} = D_u^{\pi^-} = D_{\bar{u}}^{\pi^+} = D_{\bar{d}}^{\pi^-} \qquad (12.6)$$

$$D_s^{\pi^+} = D_s^{\pi^-} = D_{\bar{s}}^{\pi^+} = D_{\bar{s}}^{\pi^-} \qquad (12.7)$$

(Here, for simplicity, we have ignored any contributions from new heavy quarks. These will in general be necessary when discussing very high energy data, but for our present introduction we will restrict our attention to data that are believed to be below threshold for production of heavy hadronic degrees of freedom such as charm.)

The way these fragmentation functions enter in comparison with data depends upon the process under study. We list these below; their derivation is obvious.

$e^+e^- \to h \dots$

$$\frac{1}{\sigma_{\text{TOT}}} \frac{d\sigma \, (e^+e^- \to h \dots)}{dz} = \frac{\sum_i e_i^2 [D_i^h(z) + D_{\bar{i}}^h(z)]}{\sum_i e_i^2} \qquad (12.8)$$

$$(i = \text{quark flavours})$$

(Note here that the photon produces a parton–antiparton pair, either of which could have produced the observed hadron, hence the D_i and $D_{\bar{i}}$ appear, in contrast to the next examples.)

$ep \to eh \dots$

$$\frac{1}{\sigma_{\mathrm{TOT}}} \frac{d\sigma}{dz}(ep \to eh \dots) = \frac{\sum_i e_i^2 f_i(x) D_i^{\mathrm{h}}(z)}{\sum_i e_i^2 f_i(x)} \qquad (12.9)$$

(i = quark *and* antiquark flavours)

(where $f_i(x)$ are the quark–parton distribution functions of Chapter 11).

$\nu p \to \mu^- h \dots$

$$\frac{1}{\sigma_{\mathrm{TOT}}} \frac{d\sigma}{dz}(\nu p \to \mu^- h \dots) = \frac{d(x) D_u^{\mathrm{h}}(z) + \frac{1}{3}\bar{u}(x) D_{\bar{d}}^{\mathrm{h}}(z)}{d(x) + \frac{1}{3}\bar{u}(x)} \qquad (12.10)$$

(where we have approximated $\theta_c = 0$ and ignored charm).

Note that the d quark turns into a u quark before fragmenting. The $\frac{1}{3}$ is due to the left-handed current coupling to antiquarks (integration over dy having been performed as in Chapter 11).

12.2.1 $\nu p \to \mu^- h \dots$

From the nature of these expressions we see that the neutrino data are a *direct* measurement of the fragmentation functions for pions since from equations (12.5) and (12.6)

$$D_u^{\pi^\pm} \equiv D_{\bar{d}}^{\pi^\pm} \qquad (12.11)$$

and so the $d(x) + \frac{1}{3}\bar{u}(x)$ cancels in numerator and denominator of equation (12.10) yielding

$$\frac{1}{\sigma_{\mathrm{TOT}}} \frac{d\sigma}{dz}(\nu p \to \mu^- \pi^\pm \dots) = D_u^{\pi^\pm}(z) \qquad (12.12)$$

Also

$$\frac{1}{\sigma_{\mathrm{TOT}}} \frac{d\sigma}{dz}(\bar{\nu} p \to \mu^+ \pi^\pm \dots) = D_{\bar{d}}^{\pi^\pm}(z) = D_u^\pi(z) \qquad (12.13)$$

Data from Gargamelle (Cundy, 1974) on the ratio of π^+/π^- production with ν beams (and equivalently π^-/π^+ with $\bar{\nu}$) are shown in

Fig. 12.4. These directly yield

$$\eta(z) \equiv \frac{D_u^{\pi^+}(z)}{D_u^{\pi^-}(z)} \qquad (12.14)$$

and we see this is of order 3 for $0\cdot3 \leqslant z \leqslant 0\cdot7$, arising for $z > 0\cdot7$. That this ratio is greater than 1 is intuitively reasonable since π^+ is $u\bar{d}$ in the

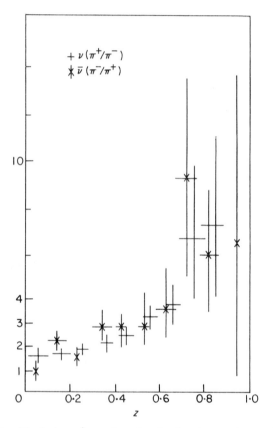

FIG. 12.4. Ratio of inclusive π^+ to π^- production in ν interactions and π^- to π^+ in $\bar{\nu}$ interactions.

simplest configuration. It has been widely argued that, as $z \to 1$, $\eta(z) \to \infty$ due to the presence of the u valence quark in π^+ whereas u in π^- is in the sea. Whether or not these data support this is unclear, since at any finite energy $\eta(z \to 1) \to \infty$ due to the fact that $\nu p \to \mu^- + (\text{charge } 2)$ in the quasi-exclusive limit. What is of immediate interest is that

$$\eta(z)_{0.3 \leqslant z \leqslant 0.7} \simeq 2 \text{ to } 3 \qquad (12.15)$$

is consistent with the inelastic electroproduction data (discussed below), and also that the data support the implication of equation (12.12), viz.

$$\frac{(d\sigma/dz)(\nu p \to \mu^- \pi^+ \ldots)}{(d\sigma/dz)(\nu p \to \mu^- \pi^- \ldots)} = \text{independent of } x \qquad (12.16)$$

The data with $0.3 < z < 0.7$ are shown as a function of x in Fig. 12.5 and are indeed consistent with this prediction (Cundy, 1974).

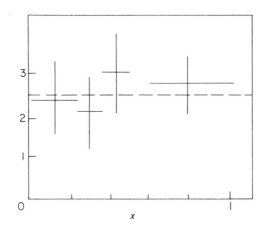

FIG. 12.5. Average ratio for π^+/π^- for ν and π^-/π^+ for $\bar{\nu}$ interactions. The data have $0.3 \leqslant z \leqslant 0.7$ and are consistent with being x independent.

Data from the 15 ft Hydrogen bubble chamber at Fermilab yield information on the production of positives and negatives separately (Fig. 12.6) (Berge, 1975). The ratio of $+/-$ production is qualitatively in agreement with the lower energy Gargamelle data, namely $+/- > 1$ and rising as z increases, though the difference between positive and negative production appears to be rather larger at Fermilab than the Gargamelle data at a comparable z. One reason may be due to the Fermilab experiment being *all* positive (negative) charges whereas Gargamelle is explicitly $\pi^\pm$; also there may be some contamination from quasi-exclusive channels that have a $Q^2(E_\nu)$ dependence that has to be taken into account before a proper comparison can be made.

If the hadrons are dominantly π and K then

$$\frac{\langle n^+\rangle}{\langle n^-\rangle} = \frac{D_u^{\pi^+}(z) + [d(x)D_u^{K^+}(z) + \frac{1}{3}\bar{u}(x)D_{\bar{d}}^{K^+}(z)]/[d(x) + \frac{1}{3}\bar{u}(x)]}{D_u^{\pi^-}(z) + [d(x)D_u^{K^-}(z) + \frac{1}{3}\bar{u}(x)D_{\bar{d}}^{K^-}(z)]/[d(x) + \frac{1}{3}\bar{u}(x)]} \quad (12.17)$$

The contribution from antiquarks is believed to be very small (Chapter

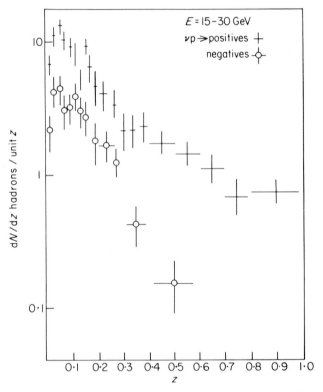

FIG. 12.6. Inclusive production of positive and negative charge hadrons in νp interactions in the 15 to 30 GeV range of laboratory energy.

11); so, neglecting them for simplicity, we have

$$\frac{\langle n^+\rangle}{\langle n^-\rangle} = \frac{D_u^{\pi^+}(z) + D_u^{K^+}(z)}{D_u^{\pi^-}(z) + D_u^{K^-}(z)} \quad (12.18)$$

Hence

$$\frac{\langle n^+ \rangle}{\langle n^- \rangle} > \frac{\langle n^{\pi^+} \rangle}{\langle n^{\pi^-} \rangle} \text{ if } \frac{D_u^{K^+}}{D_u^{K^-}} > \frac{D_u^{\pi^+}}{D_u^{\pi^-}}$$ (12.19)

12.2.2 $ep(n) \rightarrow eh \ldots$

The analysis of inelastic electron scattering is slightly more involved than for neutrinos due to the contributions from all the charged quarks:

$$\frac{1}{\sigma_{TOT}} \frac{d\sigma}{dz}(eN \rightarrow eh \ldots) = \frac{\sum_i e_i^2 f_i(x) D_i^h(z)}{\sum_i e_i^2 f_i(x)}$$ (12.20)

For ease of notation we normalise to the total cross-section (W_1) and write

$$N^h(x, z) = \sum_i e_i^2 f_i(x) D_i^h(z)$$ (12.21)

In their original analysis of the data of Bebek *et al.* (1973) ($Q^2 = 2 \text{ GeV}^2$, $\omega = 4$) Cleymans and Rodenberg (1974) ignored the contribution from all but the valence quarks, which is reasonable for $\omega \simeq 4$. Hence in $ep \rightarrow e\pi^{\pm} \ldots$ they have (writing $u(x) \equiv f_u(x)$ etc.)

$$\frac{N^{\pi^+}(x, z)}{N^{\pi^-}(x, z)} = \frac{\frac{4}{9}u(x)D_u^{\pi^+}(z) + \frac{1}{9}d(x)D_d^{\pi^+}(z)}{\frac{4}{9}u(x)D_u^{\pi^-}(z) + \frac{1}{9}d(x)D_d^{\pi^-}(z)}$$ (12.22)

$$= \frac{4u(x)\eta(z) + d(x)}{4u(x) + d(x)\eta(z)}$$ (12.23)

where we have used equations (12.5) to (12.7) and (12.14).

The Bebek data (1973) are consistent with scaling in the range $0 \cdot 2 \leqslant z \leqslant 0 \cdot 7$ and so the analysis was limited to this region for which $\langle n^{\pi^+} \rangle / \langle n^{\pi^-} \rangle \simeq 2$ independent of z. For $\omega \simeq 4$, $u(x) \simeq 2d(x)$ and so

$$2 \simeq \frac{8\eta(z) + 1}{8 + \eta(z)}$$ (12.24)

which yields $\eta(z) \simeq 2 \cdot 5$, $0 \cdot 2 \leqslant z \leqslant 0 \cdot 7$. This is in perfect agreement with the Gargamelle data on $\pi^{\pm}$ production by neutrino beams (equation 12.15) and so we have strong support here for the quark–parton picture of the semi-inclusive hadron production.

Dakin and Feldman (1973) refined and extended the above analysis by incorporating later data in the range $0.5 < Q^2 < 2.5$ GeV^2 and $3 \leqslant \omega \leqslant 60$ and allowing for the contribution of valence and sea quarks. They parametrised the longitudinal momentum distributions of the quarks as follows

$$u(x) = u_V(x) + K(x)$$

$$d(x) = d_V(x) + K(x)$$

$$s(x) = \bar{s}(x) = \bar{u}(x) = \bar{d}(x) = K(x) \tag{12.25}$$

where $u_V(x), d_V(x)$ represent the distribution functions for valence quarks, and the sea was hypothesised to be SU(3) symmetric (it turns out that this is not a very crucial assumption for their analysis).

These functions $u(x)$, $K(x)$ etc. were taken from the McElhaney and Tuan (1973) fits to the total cross-section data (this is essentially the Kuti–Weisskopf model modified to take account of the fact that $\nu W_2^{eN}/\nu W_2^{eP} < \frac{2}{3}$ as $x \to 1$). (Kuti and Weisskopf, 1971; Landshoff and Polkinghorne, 1971).

Then one has, in place of equation (12.23),

$$\frac{N^{\pi^+}(x, z)}{N^{\pi^-}(x, z)} = \frac{4u_V(x)\eta(z) + d_V(x) + [5\eta(z) + 7]K(x)}{4u_V(x) + \eta(z)d_V(x) + [5\eta(z) + 7]K(x)} \tag{12.26}$$

The Cleymans–Rodenberg formula, equation (12.23), is obtained when $K(x) \to 0$ (and hence $u_V \equiv u$ etc.). The effect is to raise $\eta(z)$ slightly as compared to $K(x) = 0$:

$$\eta(z) \simeq 3.0 \pm 0.6 \tag{12.27}$$

(compare $\eta(z) \simeq 2.5$ when $K(x) = 0$ as in Cleymans–Rodenberg). Qualitatively it is obvious that this should be so since the sea populates π^+ and π^- equally, and hence tends to dilute the ratio. To have the same ratio as in the data, $\eta(z)$ must be larger than in the analysis where the sea was ignored.

Having determined $\eta(z)$ and knowing the $f_i(x)$ from the McElhaney–Tuan parametrisation of the total cross-section data then one can predict the $x(\omega)$ dependence of the π^+/π^- production ratio using equation (12.26). This quantity is compared with the data in Fig. 12.7.

Due to the dominance of $u(x)$ as $x \to 1$, more positive charge is predicted to be forward produced.

The production from neutron targets is immediately obtained by interchanging u_V and d_V in equation (12.26), while $K(x)$ is the same as before (the sea has $I = 0$). Hence from a neutron target,

$$\frac{N^{\pi^+}(x, z)}{N^{\pi^-}(x, z)} = \frac{4d_V(x)\eta(z) + u_V(x) + [5\eta(z) + 7]K(x)}{4d_V(x) + \eta(z)u_V(x) + [5\eta(z) + 7]K(x)} \quad (12.28)$$

so that with $\eta(z) \approx 3$ we immediately predict the curve of Fig. 12.7 which is compared with the data.

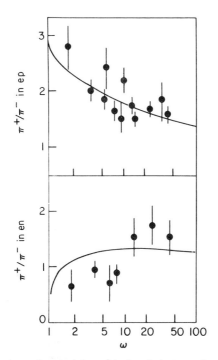

FIG. 12.7. Omega dependence of the π^+/π^- ratio in ep and en interactions compared with quark–parton model expectations.

Note the general feature that as $\omega \to \infty (x \to 0)$ the π^+/π^- ratio tends to unity (sea dominance). Coming to smaller ω the ratio rises and then as $\omega \to 1$ falls below 1 due to dominance of the u_V quark. In general with u quark dominance

$$\frac{\pi^+}{\pi^-}(x \to 1) \to \frac{1}{\eta} \text{ in en}$$
$$\to \eta \text{ in ep} \quad (12.29)$$

A cautionary note is provided by Hanson at the Stanford Symposium (Hanson, 1975). Plotting the π^+/π^- ratios against x and also W for various Q^2 (W is the mass of hadronic system) one cannot yet tell if (i) π^+/π^- is a function only of ω, i.e. scales in $\omega(x)$, or (ii) a function only of W. The variation of π^+/π^- with W or x over the measured range of parameters is too small to see a significant difference between the two.

Further orientation on the significance of these production ratios is obtained by noting that in the photon fragmentation region at $Q^2 = 0$, $\pi^+/\pi^- = 0\cdot8$. This is quite different from the values $1\cdot2$ to $1\cdot3$ predicted at moderate ω in the present model for $Q^2 \neq 0$.

i. *Sum Rules in $eN \rightarrow eh$*. Normalising to $F_1^{eN}(x)$, the number of hadrons h in the current fragmentation region, with momentum z in an experiment done at fixed x reads

$$N^h(z, x) = \tfrac{4}{9}[u(x)D_u^h(z) + \bar{u}(x)D_{\bar{u}}^h(z)]$$
$$+ \tfrac{1}{9}[d(x)D_d^h(z) + \bar{d}(x)D_{\bar{d}}^h(z)] + \tfrac{1}{9}[s(x)D_s^h(z) + \bar{s}(x)D_{\bar{s}}^h(z)]$$

(12.31)

plus further possible contributions from charmed quarks etc. We can simplify this messy expression by studying, for example, the excess of π^+ over π^-

$$N_{eP}^{\pi^+}(z, x) - N_{eP}^{\pi^-}(z, x) = [D_u^{\pi^+}(z) - D_u^{\pi^-}(z)][\tfrac{4}{9}(u - \bar{u}) - \tfrac{1}{9}(d - \bar{d})](x)$$

(12.32)

where we used relations like $D_s^{\pi^+} = D_s^{\pi^-}$ etc. (equation 12.7). (This expression is true in general, since further quarks with $I = 0$ will not contribute to the $\pi^+\pi^-$ difference.)

Since we know from the proton and neutron charge sum rules (equations 11.29 and 11.30) that

$$\int_0^1 [u(x) - \bar{u}(x)]\,dx = 2\int_0^1 [d(x) - \bar{d}(x)]\,dx = 2 \qquad (12.33)$$

then

$$\int_0^1 dx[N_{eP}^{\pi^+}(z, x) - N_{eP}^{\pi^-}(z, x)] = \tfrac{7}{9}[D_u^{\pi^+}(z) - D_u^{\pi^-}(z)] \qquad (12.34)$$

Similarly on neutron targets one derives (interchanging u, d in equation 12.32)

$$N_{eN}^{\pi^+}(z, x) - N_{eN}^{\pi^-}(z, x) = [D_u^{\pi^+}(z) - D_u^{\pi^-}(z)][\tfrac{4}{9}(d - \bar{d}) - \tfrac{1}{9}(u - \bar{u})](x)$$

(12.35)

and so

$$\int_0^1 dx \, [N_{eN}^{\pi^+}(z, x) - N_{eN}^{\pi^-}(z, x)] = \tfrac{2}{9}[D_u^{\pi^+}(z) - D_u^{\pi^-}(z)] \quad (12.36)$$

Consequently, independent of z,

$$\frac{\int dx \, [N^{\pi^+}(z, x) - N^{\pi^-}(z, x)]_{eN}}{\int dx \, [N^{\pi^+}(z, x) - N^{\pi^-}(z, x)]_{eP}} = \frac{2}{7} \quad (12.37)$$

Experimentally it is more useful to integrate over all z, and since

$$\frac{\int dz \, N_{eN}^{\pi^+}(z)}{\int dz \, N_{eP}^{\pi^+}(z)} \equiv \frac{\int dz \, (d\sigma/dz)_{eN}^{\pi^+}}{\int dz \, (d\sigma/dz)_{eP}^{\pi^+}} \quad (12.38)$$

then

$$\frac{\int_0^1 dx \, F_1^{eN}(x)[\langle n^{\pi^+} \rangle_{eN} - \langle n^{\pi^-} \rangle_{eN}]}{\int_0^1 dx \, F_1^{eP}(x)[\langle n^{\pi^+} \rangle_{eP} - \langle n^{\pi^-} \rangle_{eP}]} = \frac{2}{7} \quad (12.39)$$

where $\langle n^h \rangle$ is the average multiplicity of particle h as a function of x. This sum rule was derived by Gronau et al. (1973) but is not yet well tested by data.

12.2.3 $e^+ e^- \to h \ldots$

In equation (12.8) we have

$$\frac{1}{\sigma_{TOT}} \frac{d\sigma}{dz}(e^+ e^- \to h \ldots) = \frac{\sum_i e_i^2 [D_i^h(z) + D_i^h(z)]}{\sum_i e_i^2} \quad (12.40)$$

Allowing for 3 colours of quarks

$$\sigma_{TOT}/\sigma_{\mu\mu} \equiv R = 3 \sum_i e_i^2$$

so

$$\frac{1}{\sigma_{\mu\mu}} \frac{d\sigma}{dz}(e^+ e^- \to h \ldots) = 3 \sum_i e_i^2 [D_i^h(z) + D_i^h(z)] \quad (12.41)$$

If for small z, $D(z) \sim 1/z$ (e.g. by analogy with the $f(x) \sim 1/x$ for the probability of finding given quarks in the hadron) then a logarithmic rise in multiplicity is predicted, since on integrating over z

$$R\langle n^{\mathrm{h}}\rangle = \int_{z_{\min}}^{1} \frac{1}{\sigma_{\mu\mu}} \frac{\mathrm{d}\sigma}{\mathrm{d}z}\,\mathrm{d}z = 3\sum_i e_i^2 \int_{2m/\sqrt{Q^2}}^{1} [D_i^{\mathrm{h}} + D_{\bar{i}}^{\mathrm{h}}]\,\mathrm{d}z \quad (12.42)$$

and the lower limit on the z integral generates the logarithmic growth in Q^2. We have already seen (Fig. 12.6) some hint that $D(z) \sim 1/z$ as $z \to 0$

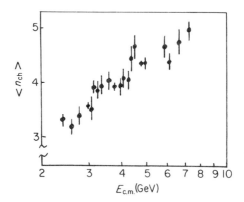

FIG. 12.8. Charged particle multiplicity as a function of $E_{\mathrm{c.m.}}$ for e^+e^- annihilation.

and so it is interesting to find that there may be a logarithmic growth of the multiplicity in $e^+e^- \to \mathrm{h}\ldots$ (Fig. 12.8).

Since $\sigma_{\mu\mu} = 4\pi\alpha^2/3s$ (derived in equation 14.9) we can rewrite equation (12.41) to read

$$s\frac{\mathrm{d}\sigma}{\mathrm{d}z}(e^+e^- \to \mathrm{h}\ldots) = \frac{4\pi\alpha^2}{3}\cdot 3\sum_i e_i^2 [D_i^{\mathrm{h}}(z) + D_{\bar{i}}^{\mathrm{h}}(z)]$$

$$\simeq 88 \times 3\sum_i e_i^2 [D_i^{\mathrm{h}}(z) + D_{\bar{i}}^{\mathrm{h}}(z)]\mathrm{nb}\cdot\mathrm{GeV}^2$$

$$(12.43)$$

The distributions in $s\,(\mathrm{d}\sigma/\mathrm{d}z)$ are shown (Schwitters, 1975) in Fig. 12.9 and do show the possibility of scaling for $z > 0.5$. We do not expect scaling for all z here because R is rising as one passes through this complicated region. It does, however, appear that the data scale for all s

when $z \gtrsim 0.5$. This, and the z, s dependence of the scaling violation, are nicely seen in Fig. 12.10 which plots $s(d\sigma/dz)$ versus $E_{c.m.}$ for various z intervals. Scaling would imply that $s(d\sigma/dz)$ should be independent of $E_{c.m.}$ for any fixed z.

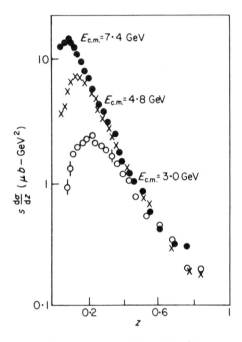

FIG. 12.9. $s(d\sigma/dz)$ at 3, 4·8, and 7·4 GeV $e^{+}e^{-}$ centre of mass energies.

If the entire rise in R is due to pair production of new particles $e^{+}e^{-} \rightarrow U^{+}U^{-}$, which decay immediately into the observed hadrons, then the final decay products at *threshold* should be limited to $z < 0.5$ since each new U is carrying half the energy. If each of these then decays, clearly half the momentum of any single decay product cannot exceed $\frac{1}{4}$ of the total energy and hence $z < 0.5$. For U production slightly above threshold a *few* decay products can have $z > 0.5$ but their effect will be negligible so the argument holds true.

Bearing this in mind, look again at Fig. 12.10. For $z > 0.5$, we see scaling (independence of $s(d\sigma/dz)$) for the full range of $3 \leqslant E_{c.m.} \leqslant 8$ GeV. For $z < 0.5$ the data have rescaled above 4 GeV except at the smallest values of z. Here the finite energy means that we are still seeing threshold effects and so we don't expect scaling to set in until PEP/PETRA energies. Hence, semi-quantitatively we can understand the observed behaviour as a combination of threshold and scaling phenomena.

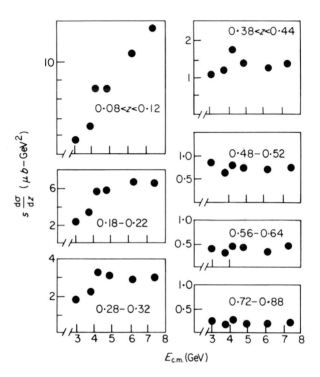

FIG. 12.10. $s\,(\mathrm{d}\sigma/\mathrm{d}z)$ versus $E_{\mathrm{c.m.}}$ for various z intervals.

Consequently we may suppose that the $s(\mathrm{d}\sigma/\mathrm{d}z)$ distribution is a superposition of "old" and "new".

If this is indeed true, then the data at 3 GeV is due entirely to "old" physics and moveover is exhibiting (for $z \geqslant 0\cdot2$) the scaling behaviour of the uds quark degrees of freedom. Hence we might analyse this data in terms of the relation

$$\frac{1}{\sigma_{\mu\mu}}\frac{\mathrm{d}\sigma}{\mathrm{d}z}(\mathrm{e}^{+}\mathrm{e}^{-}\to\mathrm{h}\ldots)=3\sum_{i=\mathrm{uds}}e_{i}^{2}\cdot[D_{i}^{\mathrm{h}}(z)+D_{i}^{\mathrm{h}}(z)]\qquad(12.41)$$

and compare with the analogous data on inclusive hadron production in lepto-induced reactions as discussed previously.

As orientation and to simplify matters let us just make the approximation that only the u quark is important (Gilman, 1975) (it is the most probable quark in the proton and also has the biggest squared charge by a factor of four). Then for $ep \to eh \ldots$

$$\frac{1}{\sigma_T}\frac{d\sigma}{dz} = \frac{\sum_i e_i^2 f_i(x) D_i^h(z)}{\sum_i e_i^2 f_i(x)} \to D_u^h(z) \tag{12.44}$$

and hence

$$\frac{1}{\sigma_{TOT}}\left[\frac{d\sigma}{dz}(ep \to eh^+ \ldots) + \frac{d\sigma}{dz}(ep \to eh^- \ldots)\right] = D_u^{h^+}(z) + D_u^{h^-}(z) \tag{12.45}$$

For e^+e^- annihilation at $\sqrt{Q^2} < 3\cdot 5$ GeV (where by hypothesis only the uds degrees of freedom contribute), taking $u\bar{u}$ as the largest contributor to R, then

$$\frac{1}{2\sigma_T}\left[\frac{d\sigma}{dz}(e^+e^- \to h^+ \ldots) + \frac{d\sigma}{dz}(e^+e^- \to h^- \ldots)\right]$$

$$= \tfrac{1}{2}[D_u^{h^+}(z) + D_u^{h^-}(z) + D_{\bar{u}}^{h^+}(z) + D_{\bar{u}}^{h^-}(z)] = D_u^{h^+}(z) + D_u^{h^-}(z) \tag{12.46}$$

and so finally one has the immediate comparison

$$\frac{1}{\sigma_T}\frac{d\sigma}{dz}(ep \to eh^\pm \ldots) = \frac{1}{2\sigma_T}\left(\frac{d\sigma}{dz}\right)(e^+e^- \to h^\pm \ldots) \tag{12.47}$$

This comparison is shown in Fig. 12.11 and agreement is excellent at large z where different choices for the comparison variable are less important (Gilman, 1975).

12.3 Angular distributions of hadrons produced in e^+e^- annihilation

In order to appreciate some features of the discovery of heavy (charmed) mesons in e^+e^- annihilation when $\sqrt{s} \gtrsim 3\cdot 9$ GeV and the phenomena of scaling and of jets in $e^+e^- \to$ hadron + anything for

$\sqrt{s} \gtrsim 5$ GeV, it will be useful to study the kinematics for $e^+e^- \to AB$ elaborating on our treatment of Chapter 11. Let us study this process in two pieces: (i) the formation process $e^+e^- \to \gamma$ along the z-axis, and (ii) the decay $\gamma \to AB$ along axis z' oriented by angle θ with respect to the z-axis.

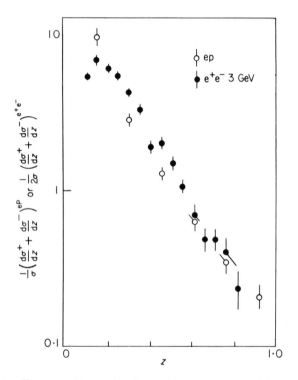

FIG. 12.11. Test of scaling and u flavour dominance in ep interactions and in e^+e^- annihilation at $E_{\text{c.m.}} = 3$ GeV.

If E (the beam energy) $\gg m_e$ (which is certainly true in practice), then the produced γ is polarized transverse to the z-axis, i.e. $J_z = \pm 1$ (this is a property of the γ_μ coupling to the e^+e^-). This is the essential feature from which all the results flow and so the reader should verify this fact for herself/himself (e.g. write $\bar{\psi}\gamma_\mu\psi$ with your favourite representation of the γ matrices and expand out for $\mu = 3$, $1 \pm i2$. The $\mu = 0$ component does not contribute since we are in the e^+e^- centre of mass and so $\mathbf{q}_\gamma = 0$, hence $q \cdot j = 0 \Rightarrow q_0 j_0 = 0$).

The photon produced in e^+e^- annihilation has therefore $J_z = \pm 1$, but along the decay axis (z') will be a mixture of $J_{z'} = 0, \pm 1$. The precise mixture will depend upon θ (e.g. at $\theta = 0$, $z' = z$ and so only $J_{z'} = \pm 1$ results). Explicitly it is given in terms of the $d^J_{mm'}(\theta)$ angular rotation functions (where in this case $J \equiv J_\gamma = 1$ and m, m' are the photon's spin projected along the z, z'-axes).

If $A^{AB}_{m'}$ is the amplitude for $\gamma \to AB$, with total projection of spin m' along z', then we define

$$\sigma_{m'}(e^+e^- \to AB) \sim |A^{AB}_{m'}|^2 \qquad (12.48)$$

Now

$$A_{m'}(\theta) = d^1_{mm'}(\theta) A_m(\theta = 0) \qquad (12.49)$$

and at $\theta = 0$ we have only $m = \pm 1$ (the photon had $J_z = \pm 1$ only). For simplicity concentrate first on the $J_z = +1$ state. Then

$$\frac{d\sigma}{d\theta}(e^+e^- \to AB) \sim \sum_{m'=\pm 1,0} |d^1_{1m'}(\theta)|^2 \sigma_{m'}(e^+e^- \to AB) \qquad (12.50)$$

where $d^1_{1\pm 1} = (1 \pm \cos\theta)/2$, $d^1_{10} = \sin\theta/\sqrt{2}$. Substituting into equation (12.50), we find (parity kills terms linear in $\cos\theta$)

$$\frac{d\sigma}{d\theta}(e^+e^- \to AB) = \left(\frac{1+\cos^2\theta}{2}\right)\frac{\sigma_1 + \sigma_{-1}}{2} + \left(\frac{\sin^2\theta}{2}\right)\sigma_0 \qquad (12.51)$$

The case $J_z = -1$ yields an identical form. If we define, conventionally, the transverse (T) and longitudinal (L) cross-sections as

$$\sigma_T \equiv (\sigma_1 + \sigma_{-1})/2; \qquad \sigma_L \equiv \sigma_0 \qquad (12.52)$$

and sum over initial polarizations, then

$$\frac{d\sigma}{d\theta}(e^+e^- \to AB) = (1+\cos^2\theta)\sigma_T + (1-\cos^2\theta)\sigma_L \qquad (12.53)$$

AB could be a genuine two-body final state, or A could be a detected particle while B represents the recoiling system. In general the observed angular distribution depends upon the relative magnitudes of σ_T and σ_L. Clearly the angular distribution can be written $(1 + \alpha \cos^2\theta)$ where $-1 < \alpha < +1$; this is a result of the $J = 1$ state that produced the system AB. Terms odd in $\cos\theta$ would violate parity. Terms of higher order in $\cos\theta$ would require $J > 1$ (they could arise from $e^+e^- \to 2\gamma \to$

hadrons for instance and are not discussed further here where we are concerned solely with the e^+e^- annihilation into a *single* photon). Integrating equation (12.53) over all θ we find

$$\sigma = \tfrac{4}{3}(2\sigma_T + \sigma_L) \equiv \tfrac{4}{3}(\sigma_1 + \sigma_{-1} + \sigma_0) \tag{12.54}$$

i.e. an equally weighted sum over each of the possible helicity states of the system AB. Note that this equal kinematic weighting of the various helicity states arose *only after integrating over all θ*. At any fixed θ, or after integrating over a fraction of 2π, this equal population will *not* occur.

12.3.1 $e^+e^- \to p\bar{p}$

The transverse and longitudinal production cross-sections are given in terms of two form factors $G_{E,M}(Q^2)$ commonly referred to as the electric and magnetic form factors (compare $e^-p \to e^-p$, Chapter 9).

$$\sigma_T = \left(\frac{4\pi\alpha}{Q^2}\right)^2 \cdot \frac{1}{8\pi} \left(1 - \frac{4m^2}{Q^2}\right)^{1/2} \cdot Q^2 G_M^2(Q^2) \tag{12.55}$$

$$\sigma_L = \left(\frac{4\pi\alpha}{Q^2}\right)^2 \cdot \frac{1}{8\pi} \left(1 - \frac{4m^2}{Q^2}\right)^{1/2} 4m^2 G_E^2(Q^2) \tag{12.56}$$

Notice the threshold factor $[1 - (4m^2/Q^2)]^{1/2}$ appropriate to production of equal mass m objects in a relative S-wave. This is because the $J^{PC} = 1^{--}$ photon can produce the fermion–antifermion system in a relative S-wave since

$$P(p\bar{p})_L = (-1)^{L+1}$$

$$C(p\bar{p})_{L,S} = (-1)^{L+S}$$

allows $S = 1$ and $L = 0$.

The ratio of longitudinal to transverse cross-sections for $p\bar{p}$ production is

$$\frac{\sigma_L}{\sigma_T} = \frac{4m^2}{Q^2} \frac{G_E^2(Q^2)}{G_M^2(Q^2)} \tag{12.57}$$

Notice that if $G_E(Q^2) = G_M(Q^2) = 1$ (which would be the case for $e^+e^- \to \mu^+\mu^-$ or any pair of elementary spin $\frac{1}{2}$ particles, e.g. quark–partons) then

$$\left(\frac{\sigma_L}{\sigma_T}\right)^{e^+e^- \to q\bar{q}} = \frac{4m^2}{Q^2} \to 0 \quad \text{as} \quad Q^2 \to \infty \tag{12.58}$$

Near to threshold one has the constraint

$$G_E(Q^2 \simeq 4m^2) \simeq G_M(Q^2 \simeq 4m^2)$$

This can be seen if one writes $G_{E,M}$ in terms of the Pauli and Dirac form factors $F_{1,2}$ (which at $Q^2 = 0$ are normalised to the charge and anomalous magnetic moment of the spin $\frac{1}{2}$ object). This relation may be written (equation 9.23)

$$G_E(Q^2) = F_1 + \frac{Q^2}{4m^2}\kappa F_2 \tag{12.59}$$

$$G_M(Q^2) = F_1 + \kappa F_2 \tag{12.60}$$

and so for $Q^2 \to 4m^2$ we see $G_E \to G_M$. Consequently for spin $\frac{1}{2}$ pair production near to threshold one has

$$\frac{\sigma_L}{\sigma_T} \to \frac{4m^2}{Q^2} \to 1$$

12.3.2 $e^+e^- \to \pi^+\pi^-$ OR $D\bar{D}$

Production of two spinless objects implies that along their axis of motion (z')

$$J_{z'} = 0 \tag{12.61}$$

Hence one has in this process

$$\sigma_T(e^+e^- \to \pi\pi, D\bar{D}) = 0$$

and the angular distribution is that of longitudinal production

$$\frac{d\sigma}{d\theta}(e^+e^- \to \pi\pi, D\bar{D}) \sim \sin^2\theta \tag{12.62}$$

The cross-section is written

$$\sigma_L(e^+e^- \to D\bar{D}) = \frac{\pi\alpha^2}{3Q^2}\left(1 - \frac{4m^2}{Q^2}\right)^{3/2} |\mathcal{G}_E(Q^2)|^2 \qquad (12.63)$$

Notice here the threshold factor $[1-(4m^2/Q^2)]^{3/2}$ appropriate to P-wave production of equal mass objects (the $J^P = 1^-$ produces a pair of pseudoscalars in P-wave). The form factor $\mathcal{G}_E(Q^2)$ is for production of a $D\bar{D}$ pair and we have included the subscript E to remind us that this production is entirely in the longitudinal mode (cf. equation 12.56).

So we find similar results for e^+e^- annihilation as for spacelike lepton scattering. Namely, that for spin-zero partons $\sigma_T = 0$ (equation 12.63) while for spin $\frac{1}{2}(\sigma_L/\sigma_T) \xrightarrow{Q^2 \to \infty} 0$ (equation 12.58).

12.4   Angular distributions of hadrons in $e^+e^- \to h + \text{anything}$ and polarisation of the beams

The stored $e^\pm$ beams circulate in a magnetic field whose direction $(\hat{y})$ is perpendicular to the plane $(\hat{x}\hat{z})$ of the storage ring. After a period of time the positrons (electrons) tend to populate the state where their spins are parallel (antiparallel) to the guide magnetic field, this state having lower energy than the opposite spin orientation. Consequently the storage ring beams are polarised in the $\hat{y}$ direction. If the polarisation is 100 per cent then the photon created by the e^+e^- annihilation has zero helicity along the $\hat{y}$ direction,

$$J_y^\gamma = 0 \qquad (12.64)$$

We will calculate the angular distribution of a hadron h produced by such a polarised photon in

$$e^+e^- \to \gamma \to h + \text{anything}$$

If the hadron emerges at angle θ relative to the $\hat{z}$-axis (the e^+ direction) and ϕ relative to the $\hat{x}\hat{z}$ plane of the ring then the direction of its momentum vector is (Fig. 12.12)

$$\hat{p}_h = (\sin\theta\cos\phi, \sin\theta\sin\phi, \cos\theta) \qquad (12.65)$$

so that the angle β between the hadron momentum vector and the $\hat{y}$-axis is given by

$$\cos \beta = \sin \theta \sin \phi \qquad (12.66)$$

The expression for the angle β enables us to immediately calculate $(d\sigma/d\Omega)(e^+e^- \to h \ldots)$. If λ is the spin projection of the photon along $\hat{p}_h$

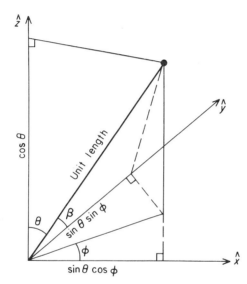

FIG. 12.12. Hadron momentum vector and components.

then, since the photon spin projection $= 0$ along $\hat{y}$, we have

$$\frac{d\sigma}{d\Omega} \sim \sum_{\lambda = \pm 1, 0} |A_\lambda|^2 |d^1_{\lambda_0}(\beta)|^2 \qquad (12.67)$$

$$\text{Parity forces } |A_1| = |A_{-1}| \qquad (12.68)$$

and so

$$\frac{d\sigma}{d\Omega} \sim |A_1|^2 \sin^2 \beta + |A_0|^2 \cos^2 \beta \qquad (12.69)$$

We define the transverse and longitudinal cross-sections proportional to $|A_1|^2$ and $|A_0|^2$ respectively, and these are functions of Q^2 and p_h or z.

These functional dependences are implicit in all the following equations. We can therefore write

$$\left(\frac{d\sigma}{d\Omega}\right)_{pol} = \sigma_T \sin^2 \beta + \sigma_L \cos^2 \beta = \sigma_T + (\sigma_L - \sigma_T) \sin^2 \theta \sin^2 \phi$$

$$(12.70)$$

The subscript "pol" is to remind us that this is the cross-section arising from a completely polarised photon (e^+e^- beams).

In practice the beams are not 100 per cent polarised. The $e^\pm$ not only couple their intrinsic spin magnetic moments with the magnetic field but also couple to the field because they are rotating (orbital angular momentum). This leads to a depolarising effect. The detailed treatment of this phenomenon is complicated (Jackson, 1976a) and there is complete depolarisation at energies satisfying

$$\frac{g-2}{2} \cdot \frac{E}{m} = N \qquad (12.71)$$

where N is an integer, E, m are the energy and mass of the e^- (or the e^+) and $g-2$ is the anomalous magnetic moment. $N = 7$ yields

$$E_{c.m.} = 2E = 6 \cdot 16 \text{ GeV}$$

and depolarisation is indeed seen at SPEAR at this energy (Schwitters et al., 1975).

Let us now return to our kinematics, equation (12.70), and consider instead completely unpolarised beams. Then since $\langle \sin^2 \phi \rangle = \frac{1}{2}$ we have

$$\left(\frac{d\sigma}{d\Omega}\right)_{unpol} = \sigma_T + \frac{1}{2}(\sigma_L - \sigma_T) \sin^2 \theta \qquad (12.72)$$

This is, of course, identical in structure to equation (12.53) that we derived previously when we assumed unpolarised $e^\pm$ beams.

We can write down finally the cross-section for beams with arbitrary degree of polarisation P. We have then

$$\frac{d\sigma}{d\Omega} = (1 - P^2)\left(\frac{d\sigma}{d\Omega}\right)_{unpol} + P^2\left(\frac{d\sigma}{d\Omega}\right)_{pol} \qquad (12.73)$$

and so

$$\frac{d\sigma}{d\Omega} = \frac{1}{2}(\sigma_T + \sigma_L)[1 + \alpha(\cos^2 \theta + P^2 \sin^2 \theta \cos 2\phi)] \qquad (12.74)$$

with

$$\alpha \equiv \frac{\sigma_T - \sigma_L}{\sigma_T + \sigma_L} \tag{12.75}$$

Equation (12.74) is the most general angular distribution for the inclusive hadron production in electron–positron annihilation through one photon. From the observed angular distributions of h one can determine $\sigma_L/\sigma_T(z, Q^2)$ which contains the interesting dynamical information. (In the model where the hadrons are the fragmentation products of spin $\frac{1}{2}$ partons then $\sigma_L/\sigma_T \sim 1/Q^2$.)

In *principle* one can determine σ_L/σ_T, or equivalently α, from the θ distribution alone and so the polarisation P gives no additional information. In *practice* since the SPEAR detector has rather limited acceptance in θ, $|\cos \theta| \leqslant 0\cdot6$ (due to the open ends of the cylindrical detector which allows the beams to enter and depart) while there is complete acceptance in ϕ, then it is easier to separate σ_L/σ_T from the ϕ dependence, i.e. exploiting the polarised beams. This is illustrated clearly in the data (Schwitters, 1975). Integrating over ϕ one has

$$\frac{d\sigma}{d\Omega} = \tfrac{1}{2}(\sigma_T + \sigma_L)(1 + \alpha \cos^2 \theta) \tag{12.76}$$

and the θ distributions very poorly determine α.

The inclusive azimuthal distributions for particles with $z > 0\cdot3$ and $|\cos \theta| \leqslant 0\cdot6$ are exhibited in Fig. 12.13 at $7\cdot4$ GeV in the c.m. There is a very clear $\cos 2\phi$ dependence. (At $6\cdot2$ GeV there is a depolarising resonance in the SPEAR ring as discussed in equation (12.71). Hence at $6\cdot2$ GeV the beams are "accidentally" unpolarised and an isotropic ϕ distribution emerges—Schwitters *et al.*, 1975.)

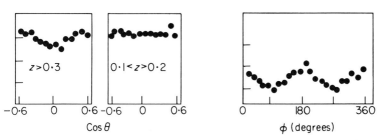

FIG. 12.13. Azimuthal distributions of hadrons produced in $7\cdot4$ GeV e^+e^- annihilation.

Using the $E_{c.m.} = 7.4$ GeV data with its clear $\cos 2\phi$ dependence we can determine α by making a best fit to the form of $d\sigma/d\Omega$ once we have obtained the magnitude of P^2. This quantity is found by fitting the distributions for $e^+e^- \to \mu^+\mu^-$ data which are collected at the same time as the hadronic production data and as $(\sigma_L/\sigma_T)_{\mu^+\mu^-} \simeq 0$ then

$$\left(\frac{d\sigma}{d\Omega}\right)_{e^+e^- \to \mu^+\mu^-} = \tfrac{1}{2}\sigma_T(1 + \cos^2\theta + P^2\sin^2\theta\cos 2\phi) \quad (12.77)$$

Hence P^2 is determined and found to be 0.46 ± 0.05 at this energy. One now uses this information in fitting the hadronic sample and α (or σ_L/σ_T) is obtained for $e^+e^- \to h \dots$.

The results for σ_L/σ_T (and α) as functions of z at $E_{c.m.} = 7.4$ GeV are shown in Fig. 12.14. At low z where the hadron h is recoiling against a

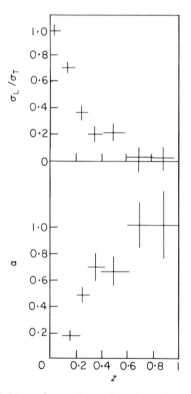

FIG. 12.14. σ_L/σ_T and α as functions of z at 7.4 GeV.

high mass system near to threshold (it is produced nearly at rest) σ_L and σ_T are almost equal. At $z \geqslant 0\cdot2$, where Bjorken scaling was observed (section 12.2.3), σ_T dominates, characteristic of production of pairs of spin $\frac{1}{2}$ particles (cf. $\mu^+\mu^-$).

Hence the data are consistent with the model where the observed hadrons are emitted by spin $\frac{1}{2}$ partons.

Further support for the idea that the hadrons are parton fragments comes from a study of the multiprong hadronic events, where it is found that these have a "jet" structure (limited momentum transverse to some axis). This phenomenon is familiar in hadron physics and is a natural consequence of the parton model. The picture is that at high $E_{c.m.}$ the spin $\frac{1}{2}$ partons are produced with angular distribution typical of $\alpha = 1$ and that the final state observed hadrons will limit momenta transverse to the direction θ in which the partons were produced. Hence two jets of particles will be expected, the jet axis being the memory of the original parton direction.

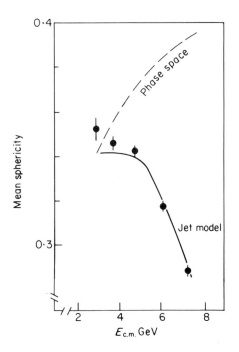

FIG. 12.15. Mean sphericity as a function of energy compared with jet and phase space models.

In those events with $\geqslant 3$ hadrons a search was made for an axis which minimised the sum of the squares of the momenta perpendicular to it. For any event, having found this axis, then a quantity S called the "sphericity" is defined (Hanson *et al.*, 1975)

$$S \equiv 3 \sum_i (p_T^i)^2 / 2 \sum_i (p^i)^2; \qquad 0 \leqslant S \leqslant 1$$

where p^i, p_T^i are the i-th particle's momentum and its momentum transverse to the jet axis. Events with $S \to 0$ are jetlike while $S \to 1$ are spherical. Mean sphericity as a function of energy is shown in Fig. 12.15. As the energy increases one can see the increasingly jetlike character of the events.

While much work still remains to be done here, these effects do appear to be more than just correlations arising from energy–momentum conservation and are genuine *multiparticle* effects. A Monte Carlo jet model with $\langle p_T \rangle = 350$ MeV/c is an excellent description, whereas the Monte Carlo phase space is a poor realisation of the data. The inclusive angular distributions are also well fitted by the jet model and for $E_{c.m.} = 7 \cdot 4$ GeV the comparison is shown in Fig. 12.16 with $\alpha_{jet} =$

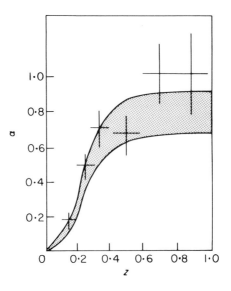

FIG. 12.16. Comparison of inclusive hadron z distributions at $7 \cdot 4$ GeV with jet model expectations.

0.78 ± 0.12. The momenta of the hadrons relative to the jet produce the range of α values in the shaded area of the Figure.

While jets are observed, apparently in nice agreement with the quark–parton model, one should perhaps worry here that things are working too well. For instance, taken literally the model appears to predict quark quantum numbers in the final state and this is not the case empirically. If the quark–parton model is the correct description then the jets' observation tells us that the produced quark dresses itself with $q\bar{q}$ pairs in such a way that (i) "conventional" quantum numbers are detected in the final state, (ii) the dressing does not destroy the memory of the original axis of quark–antiquark production.

What one would like to see is a consistent picture of quarks incorporating confinement and with the jets appearing naturally. One possible model is the massive quark model of Preparata (1975).

13   Polarised Electroproduction

13.1   Kinematic preliminaries

Inelastic lepton scattering gave information on the quantum numbers of the partons and suggested that they could be spin $\frac{1}{2}$ quarks. By polarising the lepton beam and the target one can gain information on the way the spin of the polarised proton is distributed among the flavours of quark and the role that the gluon spins might play in building up the proton spin. To describe these phenomena it is necessary to make a small kinematic introduction.

The electron scattering cross-section has been expressed in terms of the lepton and hadron tensors $L_{\mu\nu}$, $W_{\mu\nu}$ in equation (9.28). The unpolarised $L_{\mu\nu}$ is symmetric in k and k', the initial and final lepton momenta. For initial helicity $\pm\frac{1}{2}$ leptons

$$L_{\mu\nu}^{\pm} \equiv \tfrac{1}{2}(L_{\mu\nu}^{(S)} + L_{\mu\nu}^{(\pm)(A)}) \tag{13.1}$$

where

$$L_{\mu\nu}^{(\pm)(A)} = \mp i\varepsilon_{\mu\nu\lambda\sigma}k^{\lambda}k'^{\sigma} \tag{13.2}$$

Compare equation (11.43) for neutrino beams.

Consider now a polarised nucleon (with four momentum p and covariant-spin vector s^{σ} such that $s \cdot p = 0$, $s \cdot s = -1$. The hadron tensor is

$$W_{\mu\nu} = W_{\mu\nu}^{(S)} + W_{\mu\nu}^{(A)} \tag{13.3}$$

where $W_{\mu\nu}^{(S)}$ is given in equation (9.33) and

$$W_{\mu\nu}^{(A)} = i\varepsilon_{\mu\nu\lambda\sigma}q^{\lambda}\left\{s^{\sigma}[MG_1(\nu, q^2) + \frac{p \cdot q}{M}G_2(\nu, q^2)] - s \cdot qp^{\sigma}\frac{G_2(\nu, q^2)}{M}\right\}$$

(13.4)

where $G_{1,2}$ are two new structure functions (Bjorken, 1966).

Notice that the contributions of $G_{1,2}$ change sign when the nucleon polarisation is reversed whereas $W_{1,2}$ are polarisation independent. Hence $W_{1,2}$ alone are probed in unpolarised experiments whereas to extract $G_{1,2}$ requires both polarised beam and target. This follows because $G_{1,2}$ are in the antisymmetric piece of $W_{\mu\nu}$ and so contract into $L_{\mu\nu}^{(A)}$, i.e. the piece dependent upon the lepton polarisation.

The double differential cross-sections $\uparrow\uparrow, \uparrow\downarrow$ where the beam and target spins are polarised parallel (anti-parallel) to each other along the beam direction are

$$\frac{d^2\sigma}{d\Omega\, dE'}(\uparrow\uparrow + \uparrow\downarrow) = \frac{8\alpha^2 E'^2}{Q^4}\left\{2\sin^2\frac{\theta}{2}W_1 + \cos^2\frac{\theta}{2}W_2\right\}$$

(13.5)

$$\frac{d^2\sigma}{d\Omega\, dE'}(\uparrow\uparrow - \uparrow\downarrow) = \frac{4\alpha^2 E'}{EQ^2}\{(E + E'\cos\theta)MG_1 - Q^2G_2\}$$

(13.6)

Where (13.5) is (twice) the familiar spin averaged cross-section (section 9.1, equation 9.35). Polarising the target transverse to the lepton enables us to obtain a different weighting of the G_1, G_2:

$$\frac{d^2\sigma}{d\Omega\, dE'}(\uparrow\rightarrow - \uparrow\leftarrow) = \frac{4\alpha^2 E'}{EQ^2}E'\sin\theta\{MG_1 + 2EG_2\}$$

(13.7)

It is very important to be able to separate both of these structure functions in order to test various theoretical models.

To appreciate the physical significance of these structure functions let us relate the above kinematics to polarised (off-shell) photoabsorption cross-sections. This will also show why a total of four independent amplitudes are needed in the description of the experiment.

The total photoabsorption cross-sections are related by the optical theorem to the imaginary part of the forward Compton scattering amplitudes. In Table 13.1 we tabulate all possible ways the photon and nucleon helicities can combine (one could also flip each and every spin and obtain five more possibilities but by parity these are related to the

TABLE 13.1
Forward Compton helicity amplitudes

	Before			After	
	$J_z = \pm 1, 0$	$J_z = \pm\frac{1}{2}$		$J_z = \pm\frac{1}{2}$	$J_z = \pm 1, 0$
	Initial state		Intermediate state	Final state	
	γ_V	P	J_z	γ_V	P
(A)	$+1$	$+\frac{1}{2}$	$+\frac{3}{2}$	$+1$	$+\frac{1}{2}$
(B)	$+1$	$-\frac{1}{2}$	$+\frac{1}{2}$	$+1$	$-\frac{1}{2}$
(C)	$+1$	$-\frac{1}{2}$	$+\frac{1}{2}$	0	$+\frac{1}{2}$
$(\tilde{C})$	0	$+\frac{1}{2}$	$+\frac{1}{2}$	1	$-\frac{1}{2}$
(D)	0	$+\frac{1}{2}$	$+\frac{1}{2}$	0	$+\frac{1}{2}$

tabulated set). Since we are looking at forward Compton scattering with equal mass particles then the configurations C and $\tilde{C}$ are related by time reversal and so we are left with four independent helicity amplitudes

$$W_{1(1/2),1(1/2)}, \qquad W_{1(-1/2),1(-1/2)}, \qquad W_{0(1/2),0(1/2)}, \qquad W_{1(-1/2),0(1/2)}$$

where our notation is $W_{ij,i'j'}$ with $i(i')$, $j(j')$ the initial (final) spin projections of the photon and nucleon. The relations between these helicity amplitudes and the electroproduction structure functions are

$$W_1 = \tfrac{1}{2}(W_{1(1/2),1(1/2)} + W_{1(-1/2),1(-1/2)}) \sim \sigma^T_{3/2} + \sigma^T_{1/2} \qquad (13.8)$$

$$(1 + \nu^2/Q^2)W_2 - W_1(\equiv W_L) = W_{0(1/2),0(1/2)} \sim \sigma^S_{1/2} \qquad (13.9)$$

$$\nu M G_1 - Q^2 G_2 = \tfrac{1}{2}(W_{1(1/2),1(1/2)} - W_{1(-1/2),1(-1/2)}) \sim \sigma^T_{3/2} - \sigma^T_{1/2} \qquad (13.10)$$

$$\sqrt{2Q^2}(MG_1 + \nu G_2) = W_{1(-1/2),0(1/2)} \sim \sigma^{TS}_{1/2} \qquad (13.11)$$

where the σ are the photoabsorption cross-sections into states with $J_z = \frac{1}{2}, \frac{3}{2}$ and with transverse (helicity ± 1) or scalar (helicity 0) photons.

A more detailed discussion of the kinematics can be found in Dombey (1969, 1971) or Hey (1974a,b). For the purposes of our parton model discussion we will focus attention on the transverse asymmetry

$$A = \frac{\sigma^T_{1/2} - \sigma^T_{3/2}}{\sigma^T_{1/2} + \sigma^T_{3/2}} = \frac{-Q^2 G_2 + \nu M G_1}{W_1} \qquad (13.12)$$

From equations (13.5), (13.6) and (9.46) verify that (Dombey, 1969, 1971)

$$\frac{(\mathrm{d}^2\sigma/\mathrm{d}\Omega\,\mathrm{d}E')(\uparrow\uparrow-\uparrow\downarrow)}{(\mathrm{d}^2\sigma/\mathrm{d}\Omega\,\mathrm{d}E')(\uparrow\uparrow+\uparrow\downarrow)} = \sqrt{(1-\varepsilon^2)}A \tag{13.13}$$

where $\varepsilon(Q^2, \nu, \theta)$ is given in equation (9.47) and we have neglected longitudinal photon contributions. The cross-sections are for lepton and hadron spins parallel and antiparallel. From equation (13.13) we see that the experiments directly measure A. We shall concentrate on this quantity because it has a fairly immediate physical interpretation being directly related to the (virtual) photon–target interaction.

13.2 Physical interpretation of A in the parton model

Consider a proton with $J_z = +\tfrac{1}{2}$ denoted $\mathrm{P}^\uparrow$. A photon–proton collision occurs along the colinear z-axis. For a transversely polarised photon $J_z = \pm 1$ and so the absorption has total $J_z = \tfrac{3}{2}$ or $\tfrac{1}{2}$:

$$\begin{aligned} \gamma(J_z = +1): \quad & \gamma^\uparrow + \mathrm{P}^\uparrow \to \sigma_{3/2} \\ \gamma(J_z = -1): \quad & \gamma_\downarrow + \mathrm{P}^\uparrow \to \sigma_{1/2} \end{aligned} \tag{13.14}$$

Now consider a photon–quark interaction. If the quarks are moving along the z-axis (hence $k_\mathrm{T} = 0$ in the proton), then the photon–quark collision is colinear and the transverse photon will necessarily flip the S_z of the quark with which it interacts (since $L_z = 0$ for a colinear collision). Hence

$$\begin{aligned} \gamma^\uparrow + q_\downarrow &\to q^\uparrow \\ \gamma_\downarrow + q^\uparrow &\to q_\downarrow \end{aligned} \tag{13.15}$$

but $\gamma^\downarrow q^\downarrow$ and $\gamma^\uparrow q^\uparrow$ cannot interact, and so with equation (13.14) we see that

$$\sigma_{3/2} \sim \gamma^\uparrow \mathrm{P}^\uparrow \sim \sum_i e_i^2 q_\downarrow \tag{13.16}$$

$$\sigma_{1/2} \sim \gamma_\downarrow \mathrm{P}^\uparrow \sim \sum_i e_i^2 q^\uparrow$$

hence

$$A \equiv \frac{\sigma_{1/2} - \sigma_{3/2}}{\sigma_{1/2} + \sigma_{3/2}} = \frac{\sum_i e_i^2 [q_i^\uparrow - q_{i\downarrow}]}{\sum_i e_i^2 [q_i^\uparrow + q_{i\downarrow}]} \tag{13.17}$$

The probabilities $q^\uparrow$, $q_\downarrow$ are for quarks with spin parallel (antiparallel) to the target.

As a qualitative example consider a proton with spin $J_z = +\frac{1}{2}$ built from three quarks. If there is negligible L_z in the system then two quarks will have $S_z = +\frac{1}{2}$ and one $S_z = -\frac{1}{2}$. Hence the quarks spins will be dominantly aligned along the direction of the proton's spin and so we expect that $A > 0$. To quantify A we need a model for the distribution of flavours (hence e_i^2) among the $q^\uparrow$ and $q_\downarrow$ probabilities. If we take a **56**, $L_z = 0$ wavefunction for the proton (Tables 3.7 and 4.3) then the probabilities to find $u^\uparrow, u_\downarrow, d^\uparrow, d_\downarrow$ in the proton will be

$$u^\uparrow = \tfrac{5}{9}, \qquad u_\downarrow = \tfrac{1}{9}, \qquad d^\uparrow = \tfrac{1}{9}, \qquad d_\downarrow = \tfrac{2}{9} \tag{13.18}$$

For the neutron $u^\uparrow$ is replaced by $d^\uparrow$ etc. (compare equation 11.7). Hence from equation (13.17) we have (Kuti and Weisskopf, 1971)

$$A^{\gamma P} = \frac{\tfrac{4}{9}(\tfrac{5}{9} - \tfrac{1}{9}) + \tfrac{1}{9}(\tfrac{1}{9} - \tfrac{2}{9})}{\tfrac{4}{9}(\tfrac{5}{9} + \tfrac{1}{9}) + \tfrac{1}{9}(\tfrac{1}{9} + \tfrac{2}{9})} = \frac{5}{9} \tag{13.19}$$

and similarly for the neutron one finds

$$A^{\gamma N} = \frac{\tfrac{1}{9}(\tfrac{5}{9} - \tfrac{1}{9}) + \tfrac{4}{9}(\tfrac{1}{9} - \tfrac{2}{9})}{\tfrac{1}{9}(\tfrac{5}{9} + \tfrac{1}{9}) + \tfrac{4}{9}(\tfrac{1}{9} + \tfrac{2}{9})} = 0 \tag{13.20}$$

For the unpolarised ratio one finds from the denominators of equations (13.19) and (13.20) that $\sigma^{\gamma N}/\sigma^{\gamma P} = \tfrac{2}{3}$ which checks the arithmetic since the unpolarised proton and neutron have been taken as uud and ddu respectively, the sum of squared charges being in the ratio 2:3.

It is instructive to see how this naive picture satisfies the sum rule of Bjorken (1966):

$$\frac{1}{3} \frac{g_A}{g_V} = \int_0^1 \frac{dx}{x} [A^{\gamma P}(x) F_2^{\gamma P}(x) - A^{\gamma N}(x) F_2^{\gamma N}(x)] \tag{13.21}$$

From equation (13.20) we have that $A^{\gamma N} = 0$. The $\int_0^1 (dx/x) F_2^{\gamma P}(x)$ is given by $\sum_i e_i^2$ (equation 11.38) which is unity for uud. Hence the sum rule is satisfied if $g_A/g_V = \tfrac{5}{3}$. From the discussion in section 6.4 we see that this is indeed the correct magnitude for g_A/g_V given our model with an $L_z = 0$ three quark 56plet nucleon. In that discussion we saw how

g_A/g_V is reduced from $\frac{5}{3}$ by angular momentum or quark transverse momentum effects in the nucleon. In turn these effects will reduce $A^{\gamma P}$ from $\frac{5}{9}$. Since g_A/g_V is a coherent property of the nucleon then its reduction from $\frac{5}{3}$ tells us about $\langle \sigma_z \rangle$ of the quarks when integrated over all momentum fractions x. The polarisation asymmetry $A^{\gamma P}$ shows how $\langle \sigma_z(x) \rangle$ is distributed in x. This may become clearer when we derive Bjorken's sum rule using the parton model.

We will study the difference of proton and neutron structure functions and so will consider only u and d valence quarks. Recall that in unpolarised electroproduction (equation 11.37)

$$x^{-1}[F_2^{eP}(x) - F_2^{eN}(x)] = \tfrac{1}{3}[u(x) - d(x)] \qquad (13.22)$$

The coherent constraint that $\langle 2I_3(\text{proton}) \rangle = 1$ yielded a quark sum rule

$$I \equiv \langle 2I_3^P \rangle = \int_0^1 dx\, [u(x) - d(x)] \qquad (13.23)$$

and so, on combining equation (13.22) with equation (13.23), we found a sum rule for the target

$$\int_0^1 \frac{dx}{x}(F_2^{eP}(x) - F_2^{eN}(x)) = \tfrac{1}{3} \qquad (13.24)$$

From equation (13.17) we see that, for a given quark flavour,

$$A(x)(q^\uparrow(x) + q_\downarrow(x)) \equiv (q^\uparrow(x) - q_\downarrow(x)) \qquad (13.25)$$

Since $q(x) \equiv q^\uparrow(x) + q_\downarrow(x)$ in the unpolarised case, then defining $\tilde{q}(x) \equiv q^\uparrow(x) - q_\downarrow(x)$, we have by analogy with equation (13.22) that

$$x^{-1}[A^{eP}(x)F_2^{eP}(x) - A^{eN}(x)F_2^{eN}(x)] = \tfrac{1}{3}[\tilde{u}(x) - \tilde{d}(x)] \qquad (13.26)$$

In place of equation (13.23) we use the coherent constraint that for a proton with $J_z = +\frac{1}{2}$, $\langle 2I_3\sigma_z \rangle = g_A/g_V$ and hence we obtain a quark sum rule

$$g_A/g_V \equiv \langle 2I_3\sigma_z \rangle = \int_0^1 dx\, [\tilde{u}(x) - \tilde{d}(x)] \qquad (13.27)$$

in place of equation (13.23). Finally, combining equation (13.27) with (13.26) yields

$$\frac{1}{3}\frac{g_A}{g_V} = \int_0^1 \frac{dx}{x}[A^{eP}(x)F_2^{eP}(x) - A^{eN}(x)F_2^{eN}(x)] \qquad (13.28)$$

which is Bjorken's sum rule. Derived this way we see that it is the spin-dependent analogue of the Gottfried or squared charge sum rule for unpolarised scattering equations (13.24) and (11.38).

If A^N and A^P each receive contributions only from u and d flavours (i.e. if $\bar{u}$, $\bar{d}$, s, $\bar{s}$, etc. are in an unpolarised sea) then we can form sum rules for proton and neutron individually. If $\bar{u}$, $\bar{d}$, s, $\bar{s}$ are unpolarised then the J_z of the nucleon is built from the u and d flavours alone. Hence

$$\langle \sigma_z \rangle = \int_0^1 dx \, [\tilde{u}(x) + \tilde{d}(x)] \tag{13.29}$$

Consequently we can combine with equation (13.27) to derive two quark sum rules:

$$\int_0^1 dx \, \tilde{u}(x) = \frac{1}{2}\left[\langle \sigma_z \rangle + \frac{g_A}{g_V} \right]$$

$$\int_0^1 dx \, \tilde{d}(x) = \frac{1}{2}\left[\langle \sigma_z \rangle - \frac{g_A}{g_V} \right] \tag{13.30}$$

Since

$$x^{-1}A^{eP}(x)F_2^{eP}(x) = \tfrac{4}{9}\tilde{u}(x) + \tfrac{1}{9}\tilde{d}(x) \tag{13.31}$$

then from equation (13.30) we have

$$\int_0^1 \frac{dx}{x} A^{eP}(x)F_2^{eP}(x) = \frac{5}{18}\langle \sigma_z \rangle + \frac{1}{6}\frac{g_A}{g_V} \tag{13.32}$$

and similarly for the neutron

$$\int_0^1 \frac{dx}{x} A^{eN}(x)F_2^{eN}(x) = \frac{5}{18}\langle \sigma_z \rangle - \frac{1}{6}\frac{g_A}{g_V} \tag{13.33}$$

The difference of these yields Bjorken's sum rule (equation 13.28). If $g_A/g_V = \tfrac{5}{3}\langle \sigma_z \rangle$ then

$$\int_0^1 \frac{dx}{x} A^{eN}(x)F_2^{eN}(x) = 0$$

$$\int_0^1 \frac{dx}{x} A^{eP}(x)F_2^{eP}(x) = \frac{1}{3}\frac{g_A}{g_V} = \frac{5}{9}\langle \sigma_z \rangle \tag{13.34}$$

Notice that the simple model equations (13.18, 13.19 and 13.20) indeed satisfy the separate sum rules with $\langle \sigma_z \rangle = 1$. In general there will be gluons in the proton and these may carry some amount of $\langle J_z \rangle$ in the polarised target. Also the sea may be polarised. These phenomena can be taken into account by replacing $\langle \sigma_z \rangle$ in equations (13.29) to (13.34) by $\langle \sigma_z^{\text{valence}} \rangle$ (see also Sehgal, 1974; Ellis and Jaffe, 1974; and Gourdin, 1972).

In all of the foregoing discussion the quarks have been assumed to have $k_T = 0$ (equation 13.15 *et seq.*). A quark described by a two-component spinor $\chi^{\uparrow}$ at rest is described by the four-component spinor

$$q^{\uparrow} = \sqrt{\left(\frac{E+m}{2E}\right)} \left(\frac{\chi^{\uparrow}}{\dfrac{P_z\chi^{\uparrow} + (P_x + iP_y)\chi_{\downarrow}}{E+m}} \right) \qquad (13.35)$$

if it is boosted to momentum $P_{x,y,z}$. By explicitly acting on this spinor with

$$\gamma_{\pm} = \begin{pmatrix} 0 & \sigma_{\pm} \\ -\sigma_{\pm} & 0 \end{pmatrix}$$

one sees the effect of absorbing a photon with $J_z = 1$. There is now a nonvanishing amplitude for $\gamma^{\uparrow}q^{\uparrow}$ (contrast equation 13.15) due to the $\chi_{\downarrow}$ in the lower components carrying transverse momenta ($P_{x,y}$).

To exhibit the effects of the transverse momenta we can consider the quark asymmetry (13.17) for a single flavour of quark. In the original case when $k_T = 0$ then this $A^{\text{(quark)}} \equiv 1$. In the case of $k_T \neq 0$ then from equation (13.35) we see that

$$A^{\text{(quark)}} = \frac{(E+m)^2 + P_z^2 - p_T^2}{(E+m)^2 + P_z^2 + p_T^2} = 1 - \frac{p_T^2}{E(E+m)} \qquad (13.36)$$

Notice also that

$$\langle q^{\uparrow}|\sigma_z|q^{\uparrow}\rangle = 1 - \frac{p_T^2}{E(E+m)} \qquad (13.37)$$

and so the sum rule for this quark target follows. Embedding this quark in a general target p_T and E will in general be functions of x and hence $A(x)$ will replace A in (13.36). Then

$$\frac{g_A}{g_V} = \int dx \, [\langle u^{\uparrow}(x)|\sigma_z|u^{\uparrow}(x)\rangle - \langle d^{\uparrow}(x)|\sigma_z|d^{\uparrow}(x)\rangle] \qquad (13.38)$$

and so the sum rule in equation (13.34) follows where now $\langle \sigma_z \rangle$ is given by (13.37) or g_A/g_V by (6.90).

13.3 $x \to 1$ behaviour

The unpolarised data suggest that the u quark dominates in the $x \to 1$ region (Chapter 11). The x dependence of $A \equiv (\sigma_{1/2} - \sigma_{3/2})/(\sigma_{1/2} - \sigma_{3/2})$ will tell us whether it is $u^{\uparrow}$ or $u_{\downarrow}$ that is preferred as $x \to 1$. If the unpolarised data attain the bound of $\frac{1}{4}$, i.e.

$$\frac{\sigma_{1/2}^{N} + \sigma_{3/2}^{N}}{\sigma_{1/2}^{P} + \sigma_{3/2}^{P}} \to \frac{1}{4} \tag{13.39}$$

then both $\sigma_{1/2}^{N}/\sigma_{1/2}^{P}$ and $\sigma_{3/2}^{N}/\sigma_{3/2}^{P}$ must separately reach this bound since the σ are positive and only one quark is contributing. Explicitly if $u^{\uparrow}$ and $u_{\downarrow}$ dominate over $d^{\uparrow}d_{\downarrow}$ etc. then

$$\frac{(\sigma_{1/2} + \sigma_{3/2})^{N}}{(\sigma_{1/2} + \sigma_{3/2})^{P}} \to \frac{u^{\uparrow} + u_{\downarrow}}{4(u^{\uparrow} + u_{\downarrow})} \tag{13.40}$$

$$\frac{\sigma_{1/2}^{N}}{\sigma_{1/2}^{P}} = \frac{u^{\uparrow}}{4u^{\uparrow}}, \qquad \frac{\sigma_{3/2}^{N}}{\sigma_{3/2}^{P}} = \frac{u_{\downarrow}}{4u_{\downarrow}} \tag{13.41}$$

Hence

$$A^{\gamma N} = A^{\gamma P} = \frac{u^{\uparrow} - u_{\downarrow}}{u^{\uparrow} + u_{\downarrow}} \tag{13.42}$$

This equality arises as a consequence of single flavour dominance. If $u^{\uparrow}$ also dominates over $u_{\downarrow}$ then $A^{N} = A^{P} \to 1$ modulo gluons or angular momenta contributing in this region.

Notice that there is another reasonable way in which the asymmetry could attain the maximum of unity without the unpolarised ratio reaching the bound of $1:4$ (Farrar and Jackson, 1975). If there is no orbital angular momentum in the nucleon and if gluons are absent as $x \to 1$ then the $1:4$ bound implies that there is only one quark participating (the u quark) and that it carries all the target spin (i.e. the remaining quarks have net spin of zero). This is the physical reason for the maximising of the asymmetry. All that is required to maximise it however is the weaker demand that $S_z = 0$ for the spectators. In this case the interacting quark still has the same spin orientation as the proton but has a probability to be a $d^{\uparrow}$ and not just a $u^{\uparrow}$. The resulting ratio for the unpolarised cross-sections in this case becomes

$$\frac{\sigma^{N}}{\sigma^{P}} \to \frac{3}{7} \tag{13.43}$$

as follows. From the wavefunctions of Table 3.7 verify that

$$\sigma^P_{3/2} = \tfrac{2}{3}A^2, \qquad \sigma^N_{3/2} = A^2, \qquad \sigma^P_{1/2} = \tfrac{7}{3}B^2, \qquad \sigma^N_{1/2} = B^2$$

$$(13.44)$$

where A and B are the amplitudes to find the diquark with $S_z = 0$ or 1. Hence

$$\frac{\sigma^N}{\sigma^P} = \frac{A^2 + B^2}{\tfrac{2}{3}A^2 + \tfrac{7}{3}B^2} \begin{array}{l} \nearrow^{A=B} \tfrac{2}{3} \\ \searrow_{A=0} \tfrac{3}{7} \end{array} \qquad (13.45)$$

the SU(6) result arising when $A = B$ but the minimum is now $\tfrac{3}{7}$ when the $A = 0$. In a group theory sense the above is SU(6) broken down to SU(3) × SU(3).

The reason why we have dwelt on this $x \to 1$ behaviour is that there is no fully satisfactory understanding as to why the unpolarised cross-section ratio should fall below 2:3 in this limit. If the nucleon is a three-quark structure then any pair of quarks can have $I = 0$ or 1. If the $I = 0$ is the only combination present, then the third quark must carry the isospin of the target, and hence be a u in the proton and d in the neutron. This yields a ratio of 1:4 for the cross-sections but has just pushed the problem to a different place—no reason being given for the suppression of the isovector diquark component.

It is plausible that the coupling of the quarks and gluons in the nucleon may prefer the diquarks to be in a given state of spin as a function of x. If one could show that as $x \to 1$ the diquarks prefer to have $S = 0$ then from the SU(6) nature of the wavefunction in equation (4.29) one would immediately have that the diquarks have $I = 0$ and so the 1:4 ratio would arise.

In the laboratory frame a spin–spin force between pairs of quarks splits the Δ and nucleon masses (sections 15.2.2 and 17.2). The nucleon mass is pulled down relative to the mean of N and Δ due to this force. The $\mathbf{s \cdot s}$ force pulls down quark pairs in $S = 0$ more than $S = 1$ (analogous to the π being lighter than ρ). From Tables 3.7 and 4.3 we see that the u quark is in an $S = 1$ pair more than $S = 0$, in contrast to the d quark. Hence in the rest frame the u quark contributes a greater amount to the nucleon mass than a d quark. In general this will yield $\langle x \rangle_u > \langle x \rangle_d$. However it is not clear how this picture survives a boost $P_z \to \infty$ since the $S_z = 0$ components of $S = 0$ and 1 get mixed up. Farrar and Jackson (1975) argued that quark–gluon interactions cause the $S = 1$, $S_z = 1$ diquark component, rather than the full $S = 1$ diquark system, to be suppressed as $x \to 1$. A summary of the $x \to 1$ phenomena as a function of the diquark spin states contributing is given below.

	A^N	A^P	σ^N/σ^P
$S = 1$ and $S = 0$ equiprobable (SU(6))	0	$\frac{5}{9}$	$\frac{2}{3}$
$S_z = 1$ suppressed ($S = 1$, $S_z = 0$ and $S = 0$ retained)	1	1	$\frac{3}{7}$
$S = 1$ suppressed ($S = 0$ retained)	1	1	$\frac{1}{4}$

As $x \to 0$ we expect the $q\bar{q}$ sea to dominate and the polarisation to tend to zero (Kuti and Weisskopf, 1971). If the naive 56plet model works empirically for $x \simeq \frac{1}{3}$ (as in the unpolarised case) and the asymmetry tends to be large as $x \to 1$ then predictions for $A^{\gamma P}(x)$ follow like those in

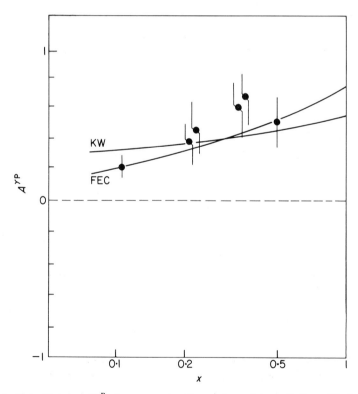

FIG. 13.1. Data on $A^{\gamma P}(x)$ compared with quark model predictions (Kuti and Weisskopf, 1971; Close, 1974a).

Fig. 13.1. The simple quark model again appears to be in good agreement with the data (Alguard *et al.*, 1976).

13.4 Inclusive hadron production in polarised electroproduction

In equation (12.23) we saw that the relative rate for $\pi^{\pm}$ production in the current fragmentation region with a proton target is

$$\frac{N^{\pi^+}(x, z)}{N^{\pi^-}(x, z)} = \frac{4u(x)\eta(z) + d(x)}{4u(x) + d(x)\eta(z)} \tag{13.46}$$

where $\eta(z) \equiv D_u^{\pi^+}(z)/D_u^{\pi^-}(z)$ is the relative probability that a u quark fragments into π^+ or π^- and empirically is of order 3 (equation 12.27). As $x \to 1$ the u quark dominance causes $N^{\pi^+} \simeq 12N^{\pi^-}$. When $x \simeq \frac{1}{3}$, $u \simeq 2d$ in the unpolarised case and so $N^{\pi^+} \simeq 2 \cdot 5 N^{\pi^-}$.

In polarised scattering we expect that when $x \simeq \frac{1}{3}$ then $u^{\uparrow} : u_{\downarrow} : d^{\uparrow} : d_{\downarrow} = 5 : 1 : 1 : 2$. Since from equation (13.16) we recall that $\sigma_{3/2(1/2)}$ probe $q_{\downarrow(\uparrow)}$ respectively then

$$\left(\frac{N^{\pi^+}}{N^{\pi^-}}\right)_{1/2} = \frac{4u^{\uparrow}(x)\eta(z) + d^{\uparrow}(x)}{4u^{\uparrow}(x) + d^{\uparrow}(x)\eta(z)} \tag{13.47}$$

while the ratio for $J_z = \frac{3}{2}$ has $u^{\uparrow}$ replaced by $u_{\downarrow}$ and $d^{\uparrow}$ by $d_{\downarrow}$, $\eta(z)$ being polarisation independent for the spinless pion. Then when $x \simeq \frac{1}{3}$ we expect

$$\left(\frac{N^{\pi^+}}{N^{\pi^-}}\right)_{1/2} \simeq 2\tfrac{2}{3}, \qquad \left(\frac{N^{\pi^+}}{N^{\pi^-}}\right)_{3/2} \simeq 1\tfrac{1}{2} \tag{13.48}$$

which is only a small polarisation dependent effect. If as $z \to 1 \eta(z) \to \infty$ (equation 13.14 *et seq.*) then a dramatic effect may be observed. In the unpolarised case for $x \simeq \frac{1}{3}$ as $z \to 1$, $N^{\pi^+}/N^{\pi^-} \simeq 8$, whereas in the polarised cases as $z \to 1$

$$\left(\frac{N^{\pi^+}}{N^{\pi^-}}\right)_{1/2} \to 20, \qquad \left(\frac{N^{\pi^+}}{N^{\pi^-}}\right)_{3/2} \to 2 \tag{13.49}$$

Hence the forward production of fast pions should depend strongly on the polarisation. Similar calculations can be performed for the neutron. One can take into account the sea contributions as $x \to 0$ and predict the π^+/π^- rates as a function of x and polarisation (Heimann, 1977).

14   Large p_T Phenomena

In deep inelastic electron scattering the basic interaction was seen to be between the electron and point structures in the target. These pointlike constituents (partons) appeared to have the same quantum numbers as quarks.

Hadron–hadron scattering is more complicated than electron–hadron because:

1. Both particles are composite whereas the electron is already a pointlike probe.
2. In electron interactions the scattering mechanism is known to be photon exchange. In hadron interactions the scattering mechanism is less clear. For example at high energies and *small* momentum transfers Regge trajectories are exchanged, the nature of the exchanged trajectory depending critically on the quantum numbers of the particles involved. For larger momentum transfers the picture is less clear and cuts play a role.

At high energies and very *large* momentum transfers the hadron interactions do appear to be more simple. The interaction appears to be due to a hard scattering of the hadron constituents and some qualitative similarities to electron–proton scattering emerge.

We shall describe some of the theoretical models for large p_T processes based on the idea that large-angle scattering comes about as a result of a hard scattering of the hadronic constituents (partons). Obviously we will be interested in the question as to whether these partons are again the quarks that manifested themselves in the lepton scattering.

14.1 Basic picture of a large p_T process

A typical example of a large p_T process is $A + B \rightarrow C + $ anything where
the detected particle C has large momentum transverse to the $A - B$
axis. Kinematically we have the situation shown in Fig. 14.1 where

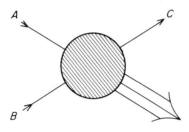

FIG. 14.1. $A + B \rightarrow C + $ anything.

$p_T \equiv |\mathbf{p}| \sin \theta$. If $s = (p_A + p_B)^2$ then it is conventional to define $x_T \equiv$
$2p_T/\sqrt{s}$. In the centre of mass frame of AB one has

$$t[\equiv (p_B - p_C)^2] = -\frac{s}{2} x_T \tan \frac{\theta}{2} \tag{14.1}$$

and

$$u[\equiv (p_A - p_C)^2] = -\frac{s}{2} x_T \cot \frac{\theta}{2} \tag{14.2}$$

so that

$$\frac{tu}{s} \equiv p_T^2 \tag{14.3}$$

Finally one defines $\varepsilon \equiv M^2/s$ where M is the missing mass of the
"anything" in $AB \rightarrow C + $ anything.

If s, t, u, M^2, p_T are all large (i.e. larger than $m_{A,B,C}^2$) then one may
hope that no intrinsic mass scales are governing the dynamics and that
the basic scattering is described as in Fig. 14.2.

The essential hypotheses are:

   i. Soft fragmentations $A \rightarrow a, B \rightarrow b, c \rightarrow C, d \rightarrow$ hadrons occur where
      the fragments carry finite fractions x of the parents' momenta. It is

supposed that these scale, i.e. $f_a^A \equiv f_a^A(x)$ etc. and that the transverse momenta are limited.

ii. All the large p_T then arises from the hard $2 \to 2$ scattering $ab \to cd$.

One might hypothesise that $abcd$ are quarks. If so we can calculate the cross-sections $AB \to C + $ anything if we know:

a. *The quark distributions in the hadrons $f_i^{A,B}(x)$*

These describe the vertices in Fig. 14.2(a), (b), and in principle can be determined from deep inelastic electron scattering since (Fig. 14.3(a)) $x^{-1}F_2^{eA} = \sum_i e_i^2 f_i^A$. If $A = $ proton or neutron this is feasible in practice. If

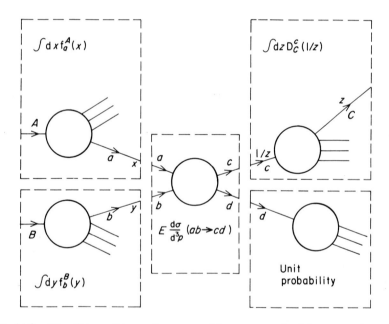

FIG. 14.2. $AB \to C + $ anything at large p_T arising as a result of fragmentations $A \to a \ldots$, $B \to b \ldots$, hard scattering of constituents $ab \to cd$, recombinations $c \to C \ldots$, $d \to$ hadrons.

$A = \pi$, K some further assumption may be required. Alternatively one might hope to learn about these distribution functions *a posteriori* from the data.

b. *The hadron distributions in a quark $D_i^h(z)$*

These control the vertex in Fig. 14.2(c), and can be determined from
$e^+e^- \to C +$ anything (Chapter 12; Fig. 14.3(b)).

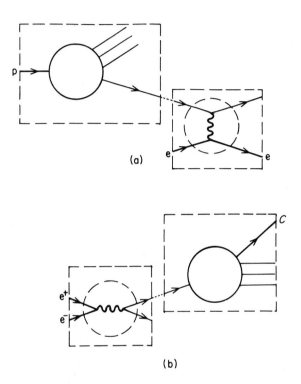

(a)

(b)

FIG. 14.3. Inelastic electron scattering and $e^+e^- \to C +$ anything viewed as large p_T processes (compare with the template in Fig. 14.2). (a) $A \to a \ldots$ followed by $ea \to ea$. (b) $e^+e^- \to c\bar{c}$ followed by $c \to C +$ anything.

c. *The hard scattering subprocess $ab \to cd$*

Given knowledge of the distribution functions (a) and (b) then a model for $ab \to cd$ will completely specify the cross-section for $AB \to C +$ anything. Conversely, data on $AB \to C +$ anything will enable us to extract the behaviour $ab \to cd$ and hence learn about basic quark dynamics and the nature of the strong interactions. We can therefore visualise the large p_T arena as one where the scattering of quark beams is being investigated.

The cross-section explicitly becomes

$$E\frac{d\sigma}{d^3p}\bigg|_{AB\to CX} = \int dx\, f_a^A(x) \int dy\, f_b^B(y)\left[E\frac{d\sigma}{d^3p}(ab\to cd)\right]\int dz\, D_C^c\left(\frac{1}{z}\right)$$
(14.4)

The exclusive process $ab \to cd$ can be rewritten using

$$E\frac{d\sigma}{d^3p}\bigg|_{ab\to cd} \equiv \frac{s^{ab}}{\pi}\delta[(a+b-c)^2 - m_d^2]\frac{d\sigma}{dt}\bigg|_{ab\to cd}$$
(14.5)

where $a \equiv p_a$ etc. and $s^{ab} \equiv (a+b)^2 \simeq 2a.b = 2xyA.B \simeq xys$. Similarly $t^{ac} \simeq xzt$ and $u^{bc} \simeq yzu$. Hence

$$E\frac{d\sigma}{d^3p}\bigg|_{AB\to CX} = \int dx\, dy\, dz\, f_a^A(x)f_b^B(y)f_C^c\left(\frac{1}{z}\right)xy\frac{d\sigma}{dt}$$

$$\times(ab\to cd)\bigg|_{\substack{s'=xys\\t'=xxt\\u'=yzu}} \quad \frac{1}{\pi}\delta\left(xy + yz\frac{t}{s} + zx\frac{u}{s}\right)$$
(14.6)

where a sum over all possibilities is implied.

14.2   Hadron production at large p_T

As a consequence of the basic assumptions (1) and (2) the following behaviours follow just from kinematic considerations and are independent of the quantum numbers of a, b, c, d:

a. The $2\to2$ process $ab \to cd$ is necessarily coplanar. Hence supplemented with the assumption that p_T is limited in $A \to a$, $B \to b$, $c \to C$ we expect that $AB \to C + anything$ *will be coplanar*.
b. There will be a *jet of particles* on the *same side* as the trigger particle C.
c. There will be a *jet on the opposite side* due to $d \to$ hadrons.
d. There will be a *low p_T background* of particles coming from "$A-a$" and "$B-b$" and moving along the A and B directions. These particles will look like those in an "ordinary" small p_T collision of A and B at energy $\sqrt{s_{\text{reduced}}} \simeq \sqrt{s} - 2p_T$.

A summary of the data bearing on these points is as follows:

a. If the z-axis is the AB collision axis and x the axis defining p_T then noncoplanarity is measured by studying distributions in p_y of those events with large p_x. It seems that

$$\frac{\mathrm{d}N}{\mathrm{d}y} \sim \exp\left(-\frac{p_y}{\langle p_y \rangle}\right)$$

where $\langle p_y \rangle \sim 300$–$500$ MeV. Since we expect a in A and b in B to have p_T limited by about 300–500 MeV then the observed $\langle p_y \rangle$ is consistent with originating there and the $ab \to cd$ indeed being coplanar.

b and c. There is some evidence to suggest that other high p_T particles follow the trigger and is an effect not entirely due to resonance production. Also there is some evidence supporting a correlation of rapidity on the opposite side to a high p_T trigger.

More detailed discussion of these points can be found in Sivers *et al.* (1977). For our present purposes it is sufficient to note that they do appear to be present in the data and hence the basic kinematics suggest the hard scattering mechanism to be realised. This means that it is now meaningful to concentrate attention on the nature of *abcd* (are they quarks?) and the dynamics of the hard scattering $ab \to cd$.

14.2.1 THE HARD SCATTERING $ab \to cd$

What do we expect for the behaviour of the $ab \to cd$ large p_T process? This will depend in particular upon the nature of $ab \to cd$; mechanisms suggested include quark–quark scattering, quark–hadron scattering, quark–antiquark fusion and quark interchange. More recently quark–gluon and gluon–gluon scatterings have also been considered (Cutler and Sivers, 1977; Combridge *et al.*, 1977).[1]

In order to gain a feeling for the behaviours expected for each of these it will be useful to return to some calculable familiar hard scattering processes like $e\mu \to e\mu$, $ep \to ep$, $ep \to eX$, $e^+e^- \to \mu^+\mu^-$. We will explicitly compute the large p_T cross-sections for the "elementary" processes and discuss their relation with the first three hadronic subprocesses suggested above (the quark interchange will be discussed separately in

[1] Gluon–gluon fusion for producing charmonium states has been suggested by Einhorn and Ellis (1975).

section 14.6). Having done this spadework we can insert the cross-sections for the elementary processes into the full diagram for $AB \rightarrow CX$ and so calculate $E(d\sigma/d^3p)|_{AB \rightarrow CX}$ from equation (14.6).

14.3 Some familiar large p_T examples

14.3.1 $e\mu \rightarrow e\mu$

This is the template for $qq \rightarrow qq$ in Fig. 14.2. We have seen in equation (9.58) that

$$\frac{d\sigma}{dt} = 2\pi\alpha^2 \frac{s^2 + u^2}{s^2 t^2} \rightarrow \frac{1}{s^2} f(t/s) \tag{14.7}$$

at large s, t, u.

The $e^+e^- \rightarrow \mu^+\mu^-$ (template for $q\bar{q}$ fusion) then follows at once and we find

$$\frac{d\sigma}{dt} = 2\pi\alpha^2 \frac{t^2 + u^2}{s^4} \rightarrow \frac{1}{s^2} f(t/s) \tag{14.8}$$

Notice the scaling behaviour, namely that the dimensionless quantity $s^2(d\sigma/dt)$ is dependent only upon angle ($f(t/s)$) and not energy at large s, t, u. Integrating equation (14.8) over $d(t/s)$ yields

$$\sigma(e^+e^- \rightarrow \mu^+\mu^-) = \frac{4\pi\alpha^2}{3s} \tag{14.9}$$

14.3.2 $e\pi \rightarrow e\pi$

This differs from $e\mu \rightarrow e\mu$ in its angular distribution (compare F_2 in equation 9.71) and the presence of the elastic pion from factor $F_\pi(t)$. Hence

$$\frac{d\sigma}{dt} = -2\pi\alpha^2 \frac{us}{s^2 t^2} F_\pi^2(t) \tag{14.10}$$

At large t we suppose that

$$F_\pi(t) \sim t^{-N}$$

and so

$$\frac{d\sigma}{dt} \to \frac{1}{s^2} \frac{1}{s^{2N}} f(t/s) \qquad (14.11)$$

The dimensionless $s^2(d\sigma/dt)$ now has an energy dependence s^{-2N} at large s, t, u for fixed angle. The scaling behaviour of $e\mu \to e\mu$ is now broken by the pion form factor (which in turn is a measure of the nonpointlike nature of the pion—i.e. it is a $q\bar{q}$ composite state). It is probable that $N_\pi \sim 1$.

The spin averaged elastic electron–proton scattering cross-section has a similar form but now $N_P \sim 2$. The proton's form factor is related to its composite nature just as in the pion example. The higher value of N, hence more dramatic fall off in t or s, is probably related to the fact that the proton is a composite of (at least) three quarks whereas the pion is only two ($q\bar{q}$). Indeed it has been suggested that the spin averaged electromagnetic form factor of a system with a total of N quarks and/or antiquarks should behave at large t as (Matveev et al., 1973; Brodsky and Farrar, 1973, 1975)

$$F(t) \sim t^{1-N} \qquad (14.12)$$

Hence mesons with $N = 2$ have a t^{-1} behaviour while baryons with $N = 3$ have dipole t^{-2} dependence. The pion form factor is consistent with a t^{-1} behaviour at large $|t|$ both for $t > 0$ and $t < 0$ (Fig. IV.C.2 of Sivers et al., 1977). The quantity $t^2 G_M(t)$ for the proton is shown in Fig. 9.3.

14.3.3   ELASTIC MM→MM MESON SCATTERING

The $e\mu$ and eM elastic scattering had different energy dependence at fixed angle due to the presence of the meson form factor. In turn, the photon exchange contribution to MM → MM will be related to eM → eM. Hence

$$s^2 \frac{d\sigma}{dt}(\text{MM} \to \text{MM}) \sim |F_M(t)|^4 f(t/s)$$

$$\sim s^{-4N} g(t/s) \qquad (14.13)$$

If M is a pion then $N \sim 1$ yields

$$\frac{d\sigma}{dt}(\pi\pi \to \pi\pi) \sim s^{-6}g(t/s) \tag{14.14}$$

whereas for a proton $N \sim 2$ yields

$$\frac{d\sigma}{dt}(pp \to pp) \sim s^{-10}h(t/s) \tag{14.15}$$

At large angles there does appear to be evidence supporting this energy dependence (Fig. IV.B.1 in Sivers *et al.*, 1977) but the magnitude of the observed cross-section is too big for photon exchange. Perhaps a vector gluon is being exchanged? In fact various models give s^{-10} behaviour (see, e.g. the quark counting rules in section 14.3.4).

14.3.4 QUARK COUNTING

In the scattering of two elementary fermions, e.g. $e\mu \to e\mu$, the dimensionless quantity $s^2(d\sigma/dt)$ was seen to be a function of angle and not explicitly of energy at large s, t, u. For every additional elementary fermion in the amplitude we saw that the amplitude dies by one additional power of s. The quark counting ideas applied to large-angle scattering of composite systems $AB \to CD$ state that the energy dependence at fixed angle is governed by the total number of quarks or leptons (Matveev *et al.*, 1973; Brodsky and Farrar, 1973, 1975). Explicitly

$$\frac{d\sigma}{dt}(AB \to CD) \sim s^{2-(N_A+N_B+N_C+N_D)}f(t/s)$$

$$\equiv s^{-N}f(t/s) \tag{14.16}$$

and so we have

$$
\left.
\begin{aligned}
e\mu \to e\mu : &\quad N = 2 \\
eM \to eM : &\quad N = 4 \quad \text{(any meson)} \\
eB \to eB : &\quad N = 6 \quad \text{(any baryon)} \\
MM \to MM : &\quad N = 6 \\
MB \to MB : &\quad N = 8 \\
BB \to BB : &\quad N = 10
\end{aligned}
\right\} \tag{14.17}
$$

as the slowest energy dependences for each process since $N_e^{min} = 1$, $M_M^{min} = 2$, $N_B^{min} = 3$. Where available, data do seem to support these counting rules and we shall utilise them when making hypotheses for the s-dependence of subprocesses inserted into Fig. 14.2 for $(d\sigma/dt)$ $(ab \rightarrow cd)$.

For elastic scattering the target must not break up. If one constituent is given a large kick then the system will break up unless all the other constituents recoil along with the struck one. The larger the number of constituents the smaller will be the chance that the whole system recoils along with the struck constituent. Hence the elastic form factor dies faster with q^2 the larger the value of N (Fig. 14.4).

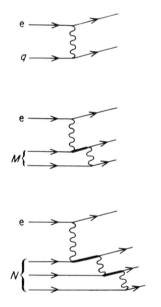

FIG. 14.4. Large t scattering of electrons on increasingly complex systems.

The mathematical formulation of this has been given by Matveev *et al.* (1973). The dimension of a single particle state vector is m^{-1} if relativistic normalisation is used. If this state is resolved and seen to be an n-constituent system then its state vector may be written

$$|a\rangle = \sqrt{N_a}|n_a\rangle \qquad (14.18)$$

and the dimension of N will be

$$[N] = m^{2(n_a-1)} \qquad (14.19)$$

in order that the m^{-1} dimension of $|a\rangle$ is preserved.

Since the elastic differential cross-section for $ab \to ab$ is given by

$$\frac{d\sigma}{dt}(ab \to ab) = s^{-2}|(ab|T|ab)|^2 \qquad (14.20)$$

then

$$(N_a N_b)^{-2} \frac{d\sigma}{dt}(ab \to ab) = s^{-2}|(n_a n_b|T|n_a n_b)|^2 \equiv F_{ab}(s, t) \quad (14.21)$$

and the dimensions of either side are

$$m^{-4(n_a+n_b-1)} \qquad (14.22)$$

Matveev *et al.* then hypothesise ("automodelity hypothesis") for $s, t \to \infty$, t/s fixed that all dimensional constants are contained in $N_{a,b}$. Hence the right-hand side of equation (14.21) depends only on kinematical variables s and t but not on any dimensional constants. Hence under a change of momentum scale:

$$\left.\begin{array}{r} p_i \to \lambda p_i \\ s \to \lambda^2 s \\ t \to \lambda^2 t \end{array}\right\} i = a, b \qquad (14.23)$$

this quantity must transform homogeneously with the relevant dimensions. Hence

$$F_{ab}(\lambda^2 s, \lambda^2 t) = \lambda^{-4(n_a+n_b-1)} F_{ab}(s, t) \qquad (14.24)$$

and so

$$\frac{d\sigma}{dt}(ab \to ab) \to s^{-2(n_a+n_b-1)} f_{ab}(t/s) \qquad (14.25)$$

The energy dependence is a function of the number of constituents, the angular dependence is *a priori* undetermined.

If particle a is elementary, e.g. an electron, then $n_a = 1$ and so

$$\frac{d\sigma}{dt}(eb \to eb) \to s^{-2n_b} f_{ab}(t/s)$$

$$\to t^{-2} t^{2-2n_b} g_{ab}(t/s) \qquad (14.26)$$

Defining the spin averaged electromagnetic form factor of system b conventionally (cf. equations 9.17 and 9.22) as

$$\frac{\mathrm{d}\sigma}{\mathrm{d}t} \sim t^{-2}|f(t)|^2 \qquad (14.27)$$

then

$$f(t) \xrightarrow{t \to \infty} t^{1-n} \qquad (14.28)$$

hence t^{-1} for a meson and t^{-2} for the proton. Further development of these ideas can be found in Broksky and Farrar (1975), Blankenbecler and Brodsky (1974), Sivers et al. (1977).

14.3.5   ep → eX

In section 9.2.7 we showed that this cross-section can be written (assuming scaling with $2xF_1 = F_2$ etc., equation 9.71)

$$\frac{E\,\mathrm{d}\sigma}{\mathrm{d}^3p} \equiv \frac{s}{\pi}\frac{\mathrm{d}^2\sigma}{\mathrm{d}t\,\mathrm{d}u} = \frac{2\alpha^2}{t^2}\frac{s^2+u^2}{s(s+u)}F_2(\omega) \qquad (14.29)$$

and so at fixed t/s, u/s (note $\omega = (s+u)/-t$) the behaviour is

$$\frac{E\,\mathrm{d}\sigma}{\mathrm{d}^3p} \sim \frac{1}{s^2}f(t/s) \qquad (14.30)$$

and the s dependence is that of elementary $e\mu$ (or e-quark) scattering (equations 9.59 or 14.7) in contrast to the s^{-6} of *elastic* e-proton scattering. It is this scaling behaviour (viz. $s^2(E\,\mathrm{d}\sigma/\mathrm{d}^3p)$ being dimensionless and independent of energy scales) that suggested that inelastic electron–proton scattering occurs by the elementary subprocess $eq \to eq$ (section 9.2).

In the quark counting picture we can now see an intuitive picture of scaling in deep inelastic electron scattering. In the elastic eA scattering, which is coherent, the target reveals itself as a system of three (or more) quarks and the fast fall off of the cross-section with s (or t) results (equation 14.11). In the incoherent inelastic process only individual quarks reveal themselves in the target. Hence the number of operative quarks is smaller than the coherent case and so the slower sort dependence obtains.

14.4   Specific models for $ab \to cd$ and $AB \to CX$

In section 9.2.1.1 we calculated the cross-section for the elementary process $eq \to eq$ and inserted it into the full diagram $eA \to eX$ and hence "predicted" scaling. It is this same idea that we will exploit in the hadronic collisions $AB \to CX$, namely inserting the subprocesses in hadronic diagrams so as to predict the behaviour of the hadronic large p_T process.

14.4.1 $qq \to qq$

One contribution to this process will be the single photon exchange and from the $e\mu \to e\mu$ calculation we see that

$$s^2 \frac{d\sigma}{dt}(qq \to qq)_{1\gamma} = f(t/s) \tag{14.31}$$

Hence (Berman *et al.*, 1971)

$$\frac{E \, d\sigma}{d^3 p}\bigg|_{AB \to CX} = \frac{1}{s^2} f\left(\frac{t}{s}, \frac{u}{s}\right) = \frac{1}{p_T^4} g(X_T, \theta) \tag{14.32}$$

Exchange of a single vector gluon will therefore give p_T^{-M}, $M = 4$ as the only difference from the photon exchange example will be the gluon–quark coupling constant α_s replacing $\alpha = \frac{1}{137}$.

The data seem to suggest that M is larger than 4, and this has raised questions as to whether gluon exchange is operative. Clearly if a coloured gluon is exchanged then there must be further flow of colour if colour singlet hadrons are produced in the final state.

We can admit the difficulty of calculating the qq scattering and instead extract it from the data. Within the hypothesis of $qq \to qq$ one can make predictions that are independent of $d\sigma/dt$ (e.g. particle production ratios, correlations etc.). The reader interested in this is referred to Field and Feynman (1977) and Feynman *et al.* (1977).

14.4.2 $qM \to qM$

In $e\mu \to e\mu$ and $qq \to qq$, we found $(d\sigma/dt) \sim s^{-2}$ and hence a p_T^{-4} behaviour in $AB \to CX$. These results followed from photon or vector gluon exchange or from lepton–quark counting.

In $eM \to eM$ and $qM \to qM$ we will analogously expect $d\sigma/dt \sim s^{-4}$ and hence

$$E \left.\frac{d\sigma}{d^3p}\right|_{AB \to CX} \sim p_T^{-8} f(t/s) \tag{14.33}$$

This is in agreement with the observed power dependence for $pp \to \pi +$ anything (section II.B of Sivers *et al.*, 1977).

There seems to be little motivation for this model other than to fit the faster power dependence by replacing one pointlike quark by a composite meson. What is not clear is why one should restrict oneself to elastic scattering and not also $qM \to qM^*$. If the latter is allowed then the subprocess $ab \to cd$ will be the inelastic $qM \to q +$ anything in which event one will presumably again find s^{-2} and p_T^{-4} (by analogy with $ep \to e +$ anything having the slower dependence on Q^2 then $ep \to ep$, i.e. the inelastic scatter again sees the quark structure of the meson M and so effectively $qq \to qq$ reappears).

The particle ratios in the final state will be similar in this model and the $qq \to qq$.

14.4.3 $q\bar{q} \to M\bar{M}$: QUARK FUSION

$q\bar{q} \to \gamma \to q\bar{q}$ will have s^{-2} behaviour like $e^+e^- \to \gamma \to \mu^+\mu^-$. Generalise this to gluon instead of photon and $q\bar{q} \to$ gluon $\to q\bar{q}$ is still s^{-2}. The quark–lepton counting also gives s^{-2}. These results are trivial since $q\bar{q} \to q\bar{q}$ is just like $qq \to qq$ turned on its side ($s \leftrightarrow t$) (cf. $e\mu \to e\mu$ and $e^+e^- \to \mu^+\mu^-$, equations 14.7 and 14.8).

Similarly $q\bar{q} \to M\bar{M}$ will be expected to have an s^{-4} behaviour (like $qM \to qM$ and so again will be a viable candidate for the observed p_T^{-8} behaviour of $(E \, d\sigma/d^3p)|_{AB \to C...}$. This mechanism has been investigated in detail by Landshoff and Polkinghorne (1973a,b,c, 1974).

A similar criticism applies here as to the previous example of $qM \to qM$: Why does $q\bar{q} \to q\bar{q}$ not assert itself and generate p_T^{-4} ? One possibility is that the subprocess is still at relatively low energies on the quark scale and "coherent" $qM \to qM$, or $q\bar{q} \to M\bar{M}$, are at present dominant whereas at (much) higher energies the basic (incoherent) quark scattering will show up and p_T^{-4} will emerge.

A process where $q\bar{q}$ annihilation can indeed be isolated is when a lepton pair is produced. The basic subprocess here is $q\bar{q} \to \mu\bar{\mu}$ and the

s^{-2} of this shows up as an m^{-4} ($s \equiv m^2_{\mu\mu}$) dependence in the mass distribution of the lepton pairs. This is known as the Drell–Yan process and is of particular interest in that it can be explicitly calculated in principle (Drell and Yan, 1971).

14.5 An example of $q\bar{q}$ fusion: the Drell–Yan process $pp \rightarrow \mu^+ \mu^- \ldots$

In the production of lepton pairs at high p_T with invariant mass away from the prominent vector mesons one is presumably studying the large p_T production of massive photons. The basic parton subprocess is

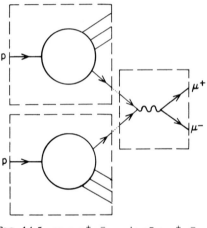

FIG. 14.5. $pp \rightarrow \mu^+ \mu^- \ldots$ via $q\bar{q} \rightarrow \mu^+ \mu^-$.

hypothesised to be $q\bar{q} \rightarrow \gamma \rightarrow \mu^+ \mu^-$ (Fig. 14.5). The cross-section (equation 14.4) is very simple to write down

$$\frac{d\sigma}{dm^2}\bigg|_{AB \rightarrow \mu^+ \mu^-} = \sum_a \int dx \, dy \, f_a^A(x) f_{\bar{a}}^B(y) \frac{d\sigma}{dm^2}(q\bar{q} \rightarrow \mu^+ \mu^-)$$

$$(14.34)$$

The cross-section for $q\bar{q} \rightarrow \mu^+ \mu^-$ will be e_q^2 times that for $e^+ e^- \rightarrow \mu^+ \mu^-$ (equation 14.9)

$$\frac{d\sigma}{dm^2}(q\bar{q} \rightarrow \mu^+ \mu^-) = \frac{4\pi\alpha^2}{3m^2} e_a^2 \delta(m^2 - s) \qquad (14.35)$$

Hence

$$\frac{d\sigma}{dm^2}(AB \to \mu^+\mu^- \ldots)$$

$$= \sum_a \int_0^1 dx \, dy \, f_a^A(x) f_{\bar{a}}^B(y) \frac{4\pi\alpha^2}{3m^2} e_a^2 \delta(m^2 - xyz)$$

$$\equiv \sum_a \int dx \, dy \frac{f_a^A(x)}{x} \frac{f_{\bar{a}}^B(y)}{y} \frac{4\pi\alpha^2 e_a^2}{3m^4} \delta\left(\frac{1}{xy} - \frac{s}{m^2}\right) \qquad (14.36)$$

We therefore have two levels of prediction:

i. The cross-section is predicted to scale as

$$m^4 \frac{d\sigma}{dm^2} \left(\text{or } m^3 \frac{d\sigma}{dm} \right) \sim f(m^2/s) \qquad (14.37)$$

ii. If A and B are protons then the quark distribution functions $f_a^p(x)$ can be obtained from analysis of the inelastic lepton scattering data. Hence the absolute size and energy dependence of $(d\sigma/dm^2)(\text{pp} \to \mu^+\mu^- \ldots)$ can be predicted.

The J/ψ discovery has stimulated experimental studies of pp $\to$ $l^+l^- \ldots$ at large s with large m^2 and one can begin to compare the results with small m^2 small s data, hence testing the scaling prediction (equation 14.37). An analysis by Lederman and Pope (1977) suggests that the data are consistent with this (see also Antreasyan et al., 1977).

With better data one may eventually hope to extract the sea antiquark distributions $f_{\bar{a}}(x)$ since the valence quarks' $f_a(x)$ is well determined from lepton scattering data (at least for $x \geqslant 0.3$). Also one could envisage making an absolute computation of the cross-section (equation 14.36). In this connection one should note the role that colour plays. Inclusion of colour decreases the predicted cross-section relative to that without colour. This is because

$$F_2^{\gamma P} \sim \sum_i f_i(x)$$

and

$$\frac{d\sigma}{dm^2} \sim \sum_i f_i(x) f_{\bar{i}}(y) \sim F_2(x) F_2(y) \qquad (14.38)$$

without colour whereas if there are N colours these would read

$$F_2^{\gamma P} \sim N \sum_i f_i(x) \quad \text{and} \quad \frac{d\sigma}{dm^2} \sim N \sum_i f_i(x)f_{\bar{i}}(y) \sim \frac{1}{N} F_2(x) F_2(y) \quad (14.39)$$

The crucial feature is that in the production of $q(\bar{q})$ there is a factor of N for colours at each vertex which is already included in F_2 measured in deep inelastic lepton scattering. However, only the N combinations where flavour *and* colour match can contribute to the $q\bar{q} \to \mu^+\mu^-$, hence the overall depletion in coloured quark models relative to uncoloured.

14.5.1 $p-p$ COLLISIONS AS $q-\bar{q}$ COLLIDING BEAMS

The Drell–Yan process $pp \to \mu^+\mu^- \ldots$ is an example of annihilation of a quark in one proton with an antiquark of the other. Can high energy pp collisions simulate e^+e^- annihilation physics due to the $q\bar{q}$ annihilation mechanism? If the proton beams each have energy E in the centre of mass then quarks in the beam have energies between 0 and E GeV ($0 \leqslant x \leqslant 1$) with $\langle x \rangle \simeq 0.2$ to 0.3, and hence $\langle E \rangle \simeq E/4$. The antiquarks are dominantly at small x, say $\langle x \rangle \leqslant 0.1$, so $\langle E \rangle \leqslant E/10$. Hence the $q\bar{q}$ collision is centred on $(E/10)(\bar{q}) + (E/4)(q)$ so that in the $q\bar{q}$ centre of mass the $q\bar{q}$ beam energy is $1/\sqrt{40} \simeq 1/6$ of the proton beam energy.

Violations of scaling, such as are found in deep inelastic lepton scattering, may have profound implications for ultra-high energy $pp \to l^+l^- \ldots$ predictions. As $\langle Q^2 \rangle$ increases the structure functions $F_2(x)$ die at large x and grow at small x so that $\langle x \rangle$ decreases (section 9.4). This will affect predictions for large Q^2 Drell–Yan production. This is particularly relevant to $pp \to W^+W^- \ldots$ searches if $m_W \simeq 60$ GeV. Discussion of scaling violations in this process have recently attracted some theoretical interest in the literature (Hinchliffe and Llewellyn Smith, 1977).

14.5.2 SINGLE LEPTON PRODUCTION $pp \to \mu^+ X$

There are many sources of leptons at large p_T. One particular source of current interest is charmed particle production in pp collisions with subsequent semileptonic decays of the charmed states. A "background" to this will be those leptons which are one of a Drell–Yan pair, the other being not detected.

The basic subprocess $q\bar{q} \to \mu\bar{\mu}$ has cross-section

$$\frac{d\sigma}{dt}(q\bar{q} \to \mu\bar{\mu}) = 2\pi\alpha^2 \frac{t^2 + u^2}{s^4} e_q^2$$

or equivalently

$$E \frac{d\sigma}{d^3p}\bigg|_{q\bar{q}\to\mu\bar{\mu}} = \frac{s^{q\bar{q}}}{\pi} \delta[s + t + u - \Sigma m^2] 2\pi\alpha^2 \frac{t^2 + u^2}{s^4} \qquad (14.40)$$

The inclusive cross-section for $q\bar{q} \to \mu X$ is identical in form with this, where Σm^2 now includes m_X^2 instead of m_μ^2. If we neglect all masses and insert the subprocess into the full diagram we can immediately calculate

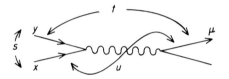

FIG. 14.6. stu scaling of subprocesses in $q\bar{q}$ fusion.

the cross-section for $pp \to \mu^+ +$ anything. Note that the s, t, u of the subprocess have scaled with respect to the overall s, t, u as (Fig. 14.6)

$$\left. \begin{array}{l} s = 2p_q \cdot p_{\bar{q}} = xys \\[4pt] t = -2p_q \cdot \mu = yt \\[4pt] u = -2p_{\bar{q}} \cdot \mu = xu \end{array} \right\} \qquad (14.41)$$

This leads to the cross-section

$$\frac{E \, d\sigma}{d^3p}(pp \to \mu^+ X)_{\text{Drell–Yan}} = \sum_a \int dx \, dy \, f_a(x) f_{\bar{a}}(y) \frac{xys}{\pi} \delta(xys + yt + xu)$$

$$\times 2\pi\alpha^2 \frac{y^2 t^2 + x^2 u^2}{(xys)^4} \qquad (14.42)$$

Rewrite the δ function as $(1/s)\delta[xy + y(t/s) + x(u/s)]$; note that t/s and u/s are functions only of x_T, θ and that $p_T^2 = tu/s \equiv sg(x_T, \theta)$, then we can rearrange equation (14.42) into the form

$$\frac{E \, d\sigma}{d^3p}(pp \to \mu^+ X)_{DY} = \frac{1}{p_T^4} F(x_T, \theta) \tag{14.43}$$

and so $p_T^4 (E \, d\sigma/d^3p)$ is predicted to scale (i.e. a function of x_T and angle only).

15.5.3   DRELL–YAN ANNIHILATION WITH MESON BEAMS

A rather direct way of investigating if it is $q\bar{q}$ annihilation that indeed generates the massive muon pairs is to compare the rates for dimuon production with incident π^+ and π^- beams. If one uses an isoscalar target (e.g. carbon) then the $q\bar{q}$ annihilation predicts that for large dimuon masses

$$\frac{\pi^+ C \to \mu^+ \mu^- \ldots}{\pi^- C \to \mu^+ \mu^- \ldots} \xrightarrow{m^2/s \to 1} \frac{1}{4} \tag{14.44}$$

This is essentially because $\pi^+(u\bar{d})$ produces photons by annihilation of the $\bar{d}$ with a d quark in the target while $\pi^-(d\bar{u})$ produces it by $\bar{u}$ annihilating with u. The ratio of the squared charges is $1:4$ and in an isoscalar target d and u quarks are equiprobable; hence the ratio of $1:4$ for the production.

More quantitatively one can predict the $x \equiv M^2/s$ dependence of this ratio. Since $X_q X_{\bar{q}} = M^2/s$ then for small M^2/s quarks and antiquarks from the sea (in either π or nucleon) will give the dominant contribution and so the π^+ and π^- beams will have equal probabilities to produce the photon; hence

$$\frac{\pi^+ C \to \mu^+ \mu^- \ldots}{\pi^- C \to \mu^+ \mu^- \ldots} \xrightarrow{m^2/s \to 0} 1 \tag{14.45}$$

For large $x (\geq 0.1)$ the valence quarks will dominate and the ratio of $1:4$ will obtain. As x is increased from zero to moderate values a gradual transition in the ratio from 1 to $\frac{1}{4}$ should be seen if the $q\bar{q}$ annihilation into a photon is the operative mechanism.

The data are shown in Fig. 14.7. There is some indication that this mechanism is indeed operative. Notice that at the masses where prominent vector mesons are produced (e.g. J/ψ) the ratio becomes unity, showing that here it is the meson rather than a photon which produces the muon pair.

With this indication of support for the $q\bar{q}$ annihilation mechanism one can calculate the production ratio with $\pi^{\pm}$ beams on arbitrary nuclei. Consider a nucleus with α protons and β neutrons. As usual $d^P(x)$ represents the probability for a down quark in the proton etc.

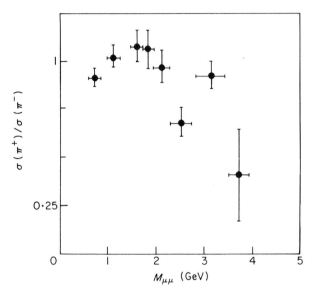

FIG. 14.7. Ratio of cross-sections for $\pi^+ C \rightarrow \mu^+\mu^-.../\pi^- C \rightarrow \mu^+\mu^-...$ (Smith, 1976).

Then since $\pi^+(u\bar{d})$ selects out d quarks while $\pi^-(d\bar{u})$ selects u,

$$\frac{\pi^+(\alpha P + \beta N) \rightarrow \mu^+\mu^- ...}{\pi^-(\alpha P + \beta N) \rightarrow \mu^+\mu^- ...}(x) = \frac{\alpha d^P(x) + \beta d^N(x)}{4(\alpha u^P(x) + \beta u^N(x))}$$

$$\equiv \frac{\alpha d^P(x) + \beta u^P(x)}{4(\alpha u^P(x) + \beta d^P(x))} \quad (14.46)$$

To say more requires knowledge of $d/u(x)$. Since this ratio can lie anywhere from zero to infinity one has a range of values $\beta/4\alpha$ (when $d = 0$) to $\alpha/4\beta$ (when $u = 0$) for the π^+/π^- production rates. For isoscalar nuclei $\alpha = \beta$ and so the 1:4 obtains independent of $d/u(x)$. For other nuclei the ratio will depend upon $d/u(x)$.

As an example consider hydrogen ($\alpha = 1, \beta = 0$). Then as $x \to 1$ where $d/u(x) \to 0$ (Chapter 11) we expect the ratio will tend to zero (the π^+ cannot annihilate) while as $x \to \frac{1}{3}$, where $2d \simeq u$, the ratio will be $1:8$.

14.6   Constituent interchange

A contribution to the scattering cross-section for the interaction of two composite systems will be the exchange between the systems of one or more of their constituents. Examples include interchange of electrons in atom–atom collisions and possibly interchange of quarks in hadron collisions. Blankenbecler *et al.* (1972) originally hypothesised that the different normalisation of the cross-sections for large angle pp $\to$ pp and p$\bar{\text{p}}$ $\to$ p$\bar{\text{p}}$ might be accounted for if quark interchange was the dominant mechanism in large-angle scattering. Encouraging support for the model came from the discovery that the angular structure of K$^+$p $\to$ K$^+$p and pp $\to$ pp appears to be consistent with the model calculation if the proton is treated as a quasi two-body system comprising a quark and an elementary diquark "core" with spin 1 and coupling to the quark with a γ_μ vertex (i.e. the proton–quark–vector core vertex conserves helicity asymptotically). If a scalar core had been used then the angular distribution is altered but the s^{-N} dependence at fixed angle stays the same.

The diagrams calculated by Blankenbecler *et al.* are shown in Fig. 14.8. Matveev *et al.* (1974) employed dimensional counting and argued

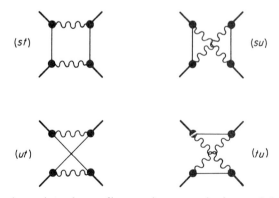

FIG. 14.8 Constituent interchange diagrams in pp $\to$ pp in the quark (——) plus core ($\sim\!\sim\!\sim$)model.

that the (ut) graph dominates the interchange amplitude because the (st) and (su) graphs are absent in leading order (where no antiquarks are present in the proton). Hence their angular dependence is different from that of Blankenbecler *et al.*

In general these models predict that the energy and angular dependences factorise

$$\frac{d\sigma}{dt}(pp \rightarrow pp) \sim s^{-N}G(\cos \theta)$$

A detailed description of these models is given in section V of Sivers *et al.* (1977). This work is also recommended as a detailed introduction to large transverse momentum processes and contains an extensive bibliography of both the phenomena and theoretical works. The application of the nonperturbative covariant parton model to large p_T physics is described in the earlier review of Landshoff and Polkinghorne (1972).

PART 3
FIELD THEORIES OF WEAK-
ELECTROMAGNETIC AND STRONG
INTERACTIONS, NEW PARTICLES
AND RECENT DEVELOPMENTS

15 Introduction to the Phenomenology of Non-Abelian Gauge Theories

In recent years two dramatic developments have taken place in high-energy physics, both of the first order of magnitude in significance and, moreover, apparently intimately linked.

First of all, on the experimental side, has been the discovery of a new spectroscopy of hadrons associated with the existence of a fourth flavour of quark (charmed quark). Initially, in November 1974, two groups independently announced the discovery of a massive metastable meson called J or ψ (Aubert et al., 1974; Augustin et al., 1974). This was the first clear evidence for the charmed quark, and is a $^3S_1$ state of $c\bar{c}$. Within a few months a spectroscopy of these states with hidden charm ($c\bar{c}$) had been revealed. In the middle of 1976 the first evidence was found conclusively proving the existence of hadrons with manifest charm and the spectroscopy of charmed particles is now slowly emerging.

What makes these discoveries even more dramatic is the fact that theoretical developments in the preceding years had anticipated and indeed required them. In order to appreciate some of the features of the charmed particles whose discovery and properties are described in the following sections we will first make a brief and superficial survey of the theoretical developments that bear on this topic. These are the discoveries that non-Abelian gauge field theories appear to play a crucial role in high energy physics ('t Hooft, 1971a,b, 1972; Gross and Wilczek,

1973a,b; Politzer, 1973), both in connection with unifying the weak and electromagnetic interactions and also the possibility that coloured quarks might interact by exchanging coloured gluons and hence generate a field theory of strong interactions (quantum chromodynamics or QCD) (Fritzsch *et al.*, 1972; Weinberg, 1973; Gross and Wilczek, 1973b).

Having made the theoretical introduction to the non-Abelian field theories we shall discuss the phenomenological consequences of them with particular regard to hadron spectroscopy and the emerging data on the charmed particles.

15.1 Non-Abelian gauge theories of weak-electromagnetic interactions

A major theoretical development recently has been the realisation that non-Abelian gauge field theories may play a role in high-energy physics.

As a first example we cite the discovery that a unified field theory of weak and electromagnetic interactions could be formulated (Glashow, 1961; Salam and Ward, 1964; Weinberg, 1967; Salam, 1967, 1968) which satisfied requirements of gauge invariance, renormalisability etc. ('t Hooft, 1971a,b). The phenomenological support for such an idea is now very impressive and will be described in the first part of this section.

Feynman and Gell-Mann (1958) successfully hypothesised that the vector part of the charged weak interaction ($W^{\pm}$ emission or absorption)[1] is related by an isospin rotation to the isovector piece of the electromagnetic interaction. In a world of four quarks and four leptons the left-handed components of these fermions form doublets of this "weak isospin" ($SU(2)_{wk}$)

$$\begin{pmatrix} \nu_e \\ e^- \end{pmatrix}_L \begin{pmatrix} \nu_\mu \\ \mu^- \end{pmatrix}_L \begin{pmatrix} u \\ d_\theta \end{pmatrix}_L \begin{pmatrix} c \\ s_\theta \end{pmatrix}_L \qquad (15.1)$$

which characterise their weak couplings. Here c is a fourth flavour of quark originally invented to introduce a symmetry between the leptons

---

[1] The idea of $W^{\pm}$ vector partners of the photon appears to be due to Schwinger (1957).

and quarks. Furthermore

$$d_\theta \equiv d \cos \theta_c + s \sin \theta_c$$
$$s_\theta \equiv s \cos \theta_c - d \sin \theta_c$$

(15.2)

are the eigenquarks of weak isospin, in contrast to s, d which are the eigenquarks of strong interactions, the angle θ_c being the Cabibbo angle.

The interactions with charged W bosons (which are the $I_3 = \pm 1$ members of a weak isotriplet) then cause left-handed transitions between ν_e and e^-, ν_μ and μ^- in the lepton sector and between u and d or s (amplitudes $\cos \theta_c$ or $\sin \theta_c$ respectively) in the quark sector. With the hypothesised charmed quark, then, the orthogonal combination of d and s quarks also takes part in the weak interactions, the d and s coupling to the c quark with amplitudes $-\sin \theta_c$ and $\cos \theta_c$ respectively.

Glashow (1961) extended the idea of weak isospin by introducing a U(1) degree of freedom called weak hypercharge. The resulting model has an $SU(2)_{wk} \otimes U(1)$ structure (Glashow, 1961; Salam and Ward, 1964). The leptons and quarks are distinguished by their weak hypercharge, related to electrical charge by

$$Q = Y_{wk} + I_{3,wk}$$

(15.3)

in the left-handed doublets.

The multiplet structure of the right-handed fermions is controversial. There is good evidence against the charged weak interactions having a right-handed component connecting u and d or s quarks and $\nu_{e,\mu}$ with e^-, μ^- leptons. If there are only four leptons and quarks then it appears that the right-handed fermions are scalars under weak isospin with hypercharge and charge

$$Q = Y_{wk}$$

(15.4)

However, if further heavy leptons or quark flavours exist, then it is possible that these could exist in doublets containing u, d, s in the right-handed sector (e.g. Fritzsch *et al.*, 1975). This is a question to be settled experimentally.

The model of Glashow (1961), Salam and Ward (1964), developed by Salam (1967, 1968) and Weinberg (1967), postulated that the weak and electromagnetic interactions occur through coupling of the fermions to weak isospin and weak hypercharge gauge vector bosons W^+, W^-, W^0 and B^0; specifically

$$\mathscr{L}_{wk} = g\left(\sum_i \psi_L^i \frac{\boldsymbol{\tau}}{2} \psi_L^i\right) \cdot \mathbf{W} + g'\left(\sum_i \psi^i Y_{wk} \psi^i\right) B \tag{15.5}$$

where g and g' are *a priori* arbitrary coupling strengths.

If all interactions are invariant under weak isospin and hypercharge gauge transformations and all particles and gauge bosons are massless then the theory will be renormalisable. The resulting field theory is called "non-Abelian" because the gauge fields do not commute (cf. section 2.2.1.4) (Yang and Mills, 1954)

$$[W_i, W_j] = i\varepsilon_{ijk} W_k$$

The empirical problem is that $m_w \neq 0$ since it is not seen in neutron β-decay nor in K decay. More recently the non-observation of significant scaling violations in weak interactions (Chapter 11) suggests that the W is more than 10 GeV in mass.

The above model's possible relevance as a physical *theory* came about when 't Hooft (1971a,b) proved the conjecture of Weinberg (1967) and Salam (1967, 1968) that if the particles and bosons gained masses by the mechanism of spontaneous symmetry breaking (Guralnik *et al.*, 1964; Higgs, 1964a,b; Brout and Englert, 1964) then the theory remained renormalisable. After spontaneous symmetry breaking the $\mathscr{W}_3$ and B become mixed yielding states

$$A = \cos \theta_W B + \sin \theta_W \mathscr{W}_3$$

$$Z^0 = \cos \theta_W \mathscr{W}_3 - \sin \theta_W B$$

where θ_W is known as the Weinberg angle. The fact that A couples to the charge, and that the charge is conserved, implies that A remains massless and hence this is the physical photon. In contrast the currents (charges) to which the Z couples are not conserved and the Z gains a mass, probably of the order 70 GeV (Weinberg, 1974).

We will formally write the neutral piece in $\mathscr{L}_{wk}$ (equation 15.5) as

$$g\mathscr{W}_3 \frac{\tau_3}{2} + g'BY \tag{15.8}$$

Then substituting in equations (15.6) and (15.7) for the fields A, Z and constraining the couplings by

$$g' = g \tan \theta_W \qquad (15.9)$$

the neutral interaction becomes

$$Qg \sin \theta_W \cdot A + \frac{g}{\cos \theta_W} Z\left(\frac{\tau_3}{2} - Q \sin^2 \theta_W\right) \qquad (15.10)$$

Rewriting $g \sin \theta_W \equiv e$ then this becomes

$$e\hat{Q}A + \frac{2e}{\sin 2\theta_W} Z\left(\frac{\tau_3}{2} - Q \sin^2 \theta_W\right) \qquad (15.11)$$

The first term is the familiar electromagnetic interaction. The second term is a new neutral current predicted by the theory.

At the very least one has for the first time a renormalisable field theory of weak interactions. The observed weakness of the weak charged interaction relative to the electromagnetic strength is understood if the W have masses of order 75 GeV ($G \sim g^2/m_W^2$ with $g^2 = 4\pi\alpha$). The masses of Z and $W^\pm$ are related by

$$m_W^2 = m_Z^2 \cos^2 \theta_W \qquad (15.12)$$

and so the Z is predicted to be more massive than the $W^\pm$ (Weinberg, 1974).

The immediate prediction of the theory is that weak neutral currents should exist, e.g. $\nu p \to \nu p$ as partner of the familiar $\nu n \to \mu^- p$. The discovery at CERN in 1973 that such neutral currents exist in Nature was the first hint that such a theory is relevant to the real world (Hasert et al., 1973; Benvenuti et al., 1974). The second feature of these neutral currents is that they empirically conserve strangeness. The status of neutral and charged weak currents is summarised below

	$\Delta S = 0$	$\Delta S \neq 0$
$W^\pm$ charged	$\nu n \to \mu^- p$	$K^\pm \to \mu^\pm \nu$
Z^0 neutral	$\nu p \to \nu p$	$K^0 \not\to \mu^+ \mu^-$

The weak neutral current connecting d_θ with d_θ appears to give rise to unwanted strangeness changing neutral currents[1] of form $ds \sin \theta_c \cos \theta_c$. However, if the u and c quarks are degenerate in mass, the neutral current connecting s_θ with s_θ exactly cancels the unwanted $ds \sin \theta_c \cos \theta_c$ term and no effective $d \leftrightarrow s$ transition can be induced in any order. The observed suppression of strangeness changing neutral currents $K_L \to \mu^+ \mu^-$, $\pi \nu \bar\nu$ etc. requires only that the u, c mass splitting is small on the scale of W boson masses (Gaillard and Lee, 1974).

The suppression of strangeness changing neutral currents requires the existence of the charmed quark whose weak interaction is[2]

$$c \leftrightarrow s \cos \theta - d \sin \theta \qquad (15.14)$$

(Glashow et al., 1970). The consequences of this are that a whole new spectroscopy of particles will exist containing one or more charmed quarks. The lightest of these must be limited to being at most a few GeV in mass in order to understand the suppression of strangeness changing neutral currents.

The discovery of the ψ particle in 1974 was the first evidence for a fourth flavour of quark, specifically the ψ being a $c\bar c$ combination. Before entering into the phenomenology of charmed particles and the evidence for them we will give the theoretical background to the second place where we believe that non-Abelian gauge field theories may be relevant, namely in the strong interactions. This will then provide the complete theoretical background against which we may compare the emerging phenomenology of the new particles.

15.2 Quantum chromodynamics: a non-Abelian gauge field theory of strong interactions

The gauge field theory uniting the weak and electromagnetic interactions is non-Abelian due to the fact that the gauge bosons carry

---

[1] The problem of strangeness changing neutral currents in the $SU(2) \times U(1)$ model with nonzero Cabibbo angle was commented upon by Salam and Ward (1964).

[2] Gell-Mann had already suggested that a fourth fundamental field might exist and have a weak current of the form in equation (15.14) (see footnote 3 in Gell-Mann, 1964). This idea was developed by Hara (1964). However, these fields were postulated to have charges of zero or minus one and the parallelism with the leptons was commented upon. Bjorken and Glashow (1964) built a model based on these four fields and invented the name "charm". This has little relation with the modern concept of charm. Their fields had integer values of baryon number and hypercharge and charm was carried by *three* of the four fields. This gave rise to integer charges through the modified Gell-Mann, Nishijima and Nakano relation which became $Q = I_3 + \frac{1}{2}(Y + C)$.

flavour and are coupled to the weak isospin and hypercharge of an $SU(2)_{wk} \times U(1)$ group structure. This non-Abelian gauge field theory has been evocatively named "Quantum Flavour Dynamics", an obvious extension of the Quantum Electrodynamics contained within it. The QED coupling to electrical charge is a $U(1)$ coupling, e. In QFD the $SU(2) \times U(1)$ structure involves also 2×2 matrices, τ, describing the couplings of the $SU(2)$ portion of the flavour flow (in direct analogy to the way the Pauli τ matrices appear in the isospin exchange in nuclear forces). This was seen explicitly in the Langrangian (equation 15.5).

The idea that each flavour of quark appears in any of three colours has been exploited to formulate a non-Abelian gauge field theory of strong interactions. The gauge bosons carry colour and if the three quark colours generate an $SU(3)_{colour}$ group, then the gauge bosons couple via 3×3 λ matrices (analogues of the 2×2 τ matrices in the $SU(2)_{isospin}$ example). The resulting theory is called quantum chromodynamics (QCD). If the $SU(3)_{colour}$ is an exact symmetry so that the three colours of any particular quark flavour are degenerate, and if the coloured vector gluons are massless then the theory is renormalisable. It is hoped that quark confinement will be a consequence of this theory and hence part of its potential significance.

Before developing the theory we will survey the evidence suggesting that quarks indeed come in three colours.

i. The large magnitude of the cross-section for e^-e^+ annihilation suggests that the quarks are coloured. Specifically below charm production threshold

$$R \equiv \frac{\sigma^{e^+e^- \rightarrow hadrons}}{\sigma^{e^+e^- \rightarrow \mu^+\mu^-}_{QED}} = \sum_i e_i^2 = \begin{cases} \frac{2}{3} \text{ uds uncoloured} \\ 2 \text{ uds in 3 colours} \end{cases} \qquad (15.15)$$

Upon crossing charm production threshold one expects an increase in R of $\frac{4}{9}$ for uncoloured charmed quarks or $\frac{4}{3}$ if they have three colours. The observed increase is of order 2 to 2·5, one unit of which is due to pair production of a new heavy lepton $\tau^{\pm}$ of mass $\simeq 2$ GeV (Perl *et al.*, 1977). There is no evidence for a second heavy lepton threshold in this region, so the remaining increase of 1 to 1·5 is presumably due to the charm degree of freedom. This is consistent with $\frac{4}{3}$ and hence a charmed quark with charge $\frac{2}{3}$ (squared being $\frac{4}{9}$) in three colours.

ii. There is a theorem (Adler, 1969; Bell and Jackiw, 1969) which states that the rate for π^0 to decay to two photons can be calculated exactly by coupling two photons and an axial current to a quark loop.[1] The amplitude is then proportional to $\sum_i I_3^i e_i^2$, the I_3 being due to the isovector axial current and the squared charge arising from the coupling of the two photons. The sum is over all quarks (in general all fundamental fermion fields) in the triangle. The data require this sum to be of order $\frac{1}{2}$. For uncoloured ud quarks we have

$$\sum_{u,d} I_3 e^2 = \frac{1}{2}(\frac{4}{9} - \frac{1}{9}) = \frac{1}{6} \qquad (15.16)$$

whereas for three colours this becomes $\frac{1}{2}$ due to the sum now being over each and every colour.

iii. Hadroproduction of lepton pairs (the Drell–Yan process pp $\rightarrow$ $\mu^+\mu^- + \cdots$, section 14.5) has a cross-section whose magnitude can be related to the $F_2(x)$ measured in deep inelastic lepton scattering (Chapter 9). The lepton pair is the decay product of a massive timelike photon produced when a quark in the proton annihilates with an antiquark of the same flavour in the other proton. The cross-section will depend in particular upon the x distribution of the antiquarks in the proton and this explicit behaviour is controversial. Modulo this uncertainty one can calculate the cross-section.

If the q and $\bar{q}$ each can exist in N colours then the probability that they produce a photon (hence that they have the same colour) will be $1/N$ times that if no colour degree of freedom were present. Thus the predicted cross-section will be $1/N_c$ smaller with N_c colours than if the quarks were uncoloured.

Whether or not $N_c = 3$ fits better than $N_c = 1$ will not be entirely clear until the $\bar{q}(x)$ distribution and also nuclear physics effects are better understood (most experiments have involved nuclear targets).

iv. In spectroscopy of baryons we have met (sections 5.1 and 8.1) the classical reason for inventing the coloured degree of freedom for quarks. The nucleons are fermions and hence require antisymmetric wavefunctions, yet in an uncoloured quark model their wavefunctions are symmetric. This is readily seen by considering the Δ^{++} made of three up quarks in a total state of spin $\frac{3}{2}$ along the z-axis. This state clearly has a wavefunction which is totally symmetric and by the SU(6) symmetry

---

[1] Note that the coupling of f_π of π^0 to the divergence of the axial current $\partial_\mu A_\mu$ is a *measured* number (Gasiorowicz, p. 580 *et seq.*). Hence there is no $1/\sqrt{N}$ colour normalisation factor from the π^0 wavefunction present explicitly in the amplitude and so the rate is proportional to N^2. Contrast with $\eta_c \rightarrow \gamma\gamma$ (G. Kane, 1977, section 16).

this implies that the nucleon has a totally symmetric wavefunction. In order that the nucleon should have an antisymmetric wavefunction one invents a further degree of freedom for the quarks, namely colour. With three colours available for each quark one can form a totally antisymmetric wavefunction for the Δ^{++} as follows

$$\Delta^{++}(u^{\uparrow}_R u^{\uparrow}_B u^{\uparrow}_Y)\varepsilon_{RBY}$$

v. Further hints that colour may be relevant come from the observation of semileptonic decays of charmed mesons which appear to be of the order of 10 per cent each into $e^- \nu_e$ and $\mu^- \nu_\mu$ (Brandelik *et al.*, 1977b). Within the framework of a quark model if the weak decays are triggered as follows

$$c \leftrightarrow s(e\bar{\nu}, \mu\bar{\nu}, u_R \bar{d}_R, u_B \bar{d}_B, u_G \bar{d}_G) \qquad (15.17)$$

and furthermore if the amplitude for each of these is the same, then one expects that the semileptonic decays to electron and muon are equal and are each one fifth of the total. The nonleptonic decays will be 60 per cent due to the three colour degrees of freedom. The data suggest that the strong interactions enhance the nonleptonic decays somewhat leading to 80 per cent of the total.

There is also rather good evidence that a heavy lepton, τ, exists with mass of order of 2 GeV and with leptonic decays of order 20 per cent into electron and neutrinos and again 20 per cent into muon and neutrinos (Perl *et al.*, 1977). This is as expected with three colours of quark since the decay is triggered as in Fig. 15.1(a).

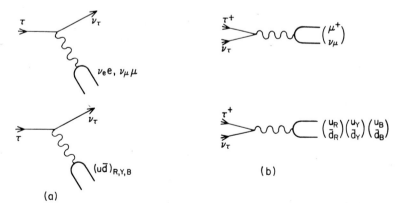

FIG. 15.1. (a) Heavy lepton decays $\tau \to \nu + $ leptons, $\tau \to \nu + q\bar{q}$. (b) $\tau^+ \nu \to$ hadrons, $\tau^+ \nu \to$ leptons.

If one twists this Figure round, forming Fig. 15.1(b), then we can imagine it as $\tau^+\nu \to W^+ \to$ all hadrons and so

$$``R" \equiv \frac{\tau^+\nu \to \text{hadrons}}{\tau^+\nu \to e^+\nu} = N_c = 3 \qquad (15.18)$$

is the analogue of $R^{e^+e^- \to \text{hadrons}} = N_c \sum_i e_i^2$.

vi. Colour provides a way of motivating why the interquark forces saturate the three-quark system so that qqq are low-lying states whereas $qqqq$ etc. are not seen. This is described in more detail later.

15.2.1 COLOUR SINGLET HADRONS

When coloured quarks are viewed on a short time scale the effective quark–gluon coupling tends asymptotically to zero (asymptotic freedom). Conversely, when quarks are separated from each other by large distances the interquark force grows. It has been widely speculated that this may provide permanent confinement of quarks.

Heuristically a coloured quark polarises the vacuum about it with emission and absorption of coloured gluons and quark pairs. From this many-body soup the colour singlet combinations crystallise out with low masses, the other states having large masses, probably infinite, and hence do not appear as observables in the laboratory. We will attempt to illustrate these ideas in a rough and ready way while bearing in mind that no one has yet satisfactorily proven a mechanism for confinement.

If the quarks form the fundamental triplet representation of an SU(3) group of colour then the following colour representations will ensue for the particular combination of quarks and antiquarks listed.

$$q : \mathbf{3}$$

$$q\bar{q} : \mathbf{3} \otimes \bar{\mathbf{3}} = \mathbf{1} \oplus \mathbf{8}$$

$$qq : \mathbf{3} \otimes \mathbf{3} = \mathbf{6} \oplus \bar{\mathbf{3}}$$

$$qq\bar{q} : \mathbf{3} \otimes \mathbf{3} \otimes \bar{\mathbf{3}} = \mathbf{3} \oplus \bar{\mathbf{6}} \oplus \bar{\mathbf{3}} \oplus \mathbf{15}$$

$$qqq : \mathbf{3} \otimes \mathbf{3} \otimes \mathbf{3} = \mathbf{1} \oplus \mathbf{8} \oplus \mathbf{8} \oplus \mathbf{10} \qquad (15.19)$$

Notice that only $q\bar{q}$ and qqq contain SU(3) colour singlets. If we invented a rule that only colour singlets have low masses then we could easily understand why qq and qqq occur in Nature and why q, qq, $qq\bar{q}$, . . . do not. Of course, what we are doing is replacing one puzzle with another: namely the puzzle of quark confinement has been replaced with the puzzle of colour confinement.

A possible clue as to why colour singlets may lie low in mass comes from nuclear physics (Lipkin, 1973). If one takes two nucleons, then of the three possibilities nn, np, pp only the isoscalar (SU(2) singlet) combination is bound—the deuteron. The binding of this state and pushing up in energy of the $I = 1$ states comes about as a consequence of isospin exchange between the nucleons. Analogously the exchange of coloured gluons between the quarks can yield low-lying colour singlets.

First of all we will recall why the isospin exchange selects the $I = 0$ combination from the two nucleon system. The electromagnetic interaction in hydrogen is proportional to the charges $e_1 e_2$ of electron and proton. This product is negative and hydrogen is bound. By analogy the exchange of isospin between two fermions has $\mathcal{H}_I \sim \mathbf{I}_1 . \mathbf{I}_2$. In order to calculate the expectation value of this isospin interaction we note that

$$2\mathbf{I}_1 . \mathbf{I}_2 \equiv (\mathbf{I}_1 + \mathbf{I}_2)^2 - \mathbf{I}_1^2 - \mathbf{I}_2^2$$

$$\rightarrow \langle 2\mathbf{I}_1 . \mathbf{I}_2 \rangle \equiv I_{\text{tot}}(I_{\text{tot}} + 1) - \tfrac{1}{2}\tfrac{3}{2} - \tfrac{1}{2}\tfrac{3}{2} \rightarrow \begin{cases} -\tfrac{3}{2}, & I = 0 \\ +\tfrac{1}{2}, & I = 1 \end{cases} \tag{15.20}$$

Consequently the interaction for isospin zero is bound whereas for isospin one it is unbound and so the isospin zero, or SU(2) singlet nucleon combination in the nucleus, lies lower in mass than the unwanted SU(2) triplet, or isospin 1, combination.

By analogy we can consider the exchange of the coloured gluon between the coloured quarks. The interaction in this case will be of the form

$$\mathcal{H}_I \sim \mathbf{F}_1 . \mathbf{F}_2 (\mathbf{F} \equiv \tfrac{1}{2}\boldsymbol{\lambda} \equiv \text{colour SU(3) } 3 \times 3 \text{ matrices}) \tag{15.21}$$

and by analogy with the previous example we can calculate the expectation value and it gives (cf. equation 5.45)

$$\langle 2\mathbf{F}_1 . \mathbf{F}_2 \rangle \equiv \lambda_{\text{tot}}^2 - \lambda_q^2 - \lambda_{\bar{q}}^2 \equiv \lambda_{\text{tot}}^2 - 2\lambda_q^2 \equiv \lambda_{\text{tot}}^2 - \tfrac{2}{3} \tag{15.22}$$

(the λ^2 is the Casimir operator for SU(3) and for a triplet or antitriplet has the magnitude $\tfrac{4}{3}$ whereas for a singlet it is 0). For an N body system by analogy the interaction has the form

$$\mathcal{H}_I = \sum_{i \neq j} \mathbf{F}_i . \mathbf{F}_j \tag{15.23}$$

and one in this case finds

$$\langle \mathscr{H}_I \rangle \to \lambda_{\text{tot}}^2 - \tfrac{4}{3}N \qquad (15.24)$$

The energy of a system of quarks will then have a contribution from the quark masses and a contribution from this gluon exchange potential and this total energy will be given by

$$E = N m_q + V \langle \mathbf{F}_i \cdot \mathbf{F}_j \rangle \qquad (15.25)$$

Then, inserting the expectation values of the colour gluon interaction, we obtain for the energy of the system the following expression:

$$E = N(m_q - \tfrac{4}{3}V) + V\lambda_{\text{tot}}^2 \qquad (15.26)$$

The energy of the system is, of course, nothing more or less than the mass of the hadron built from these quarks.

Now imagine that the quark's mass tends to infinity (which will of course kill quarks as observables). Furthermore let

$$V = \tfrac{3}{4}m_q \xrightarrow{\;m_q \to \infty\;} \infty \qquad (15.27)$$

then this will give the possibility of finite masses for quark–antiquark bound states due to the infinite potential cancelling the infinite quark masses. The energy of the system is then found from equations (15.26) and (15.27) to be given by

$$E = \tfrac{3}{4}\lambda_{\text{tot}}^2 m_q \qquad (15.28)$$

λ^2 is 0 for a colour singlet state but is greater than 0 for any colour nonsinglet state. Consequently the quark mass being infinity means that the energy of the system will be infinity for any colour nonsinglet state. Only for the case of colour singlets will the energy go to zero. Hence in this rather simple example we see that one can send certain unwanted states to infinite masses, while keeping the wanted colour singlet states with zero (i.e. finite) mass.

If we return to equation (15.27) and let the potential now be given by

$$V = \tfrac{3}{4}(m_q - \varepsilon) \qquad (15.29)$$

then the energy of the system will be given instead by the expression

$$E = N\varepsilon + \tfrac{3}{4}\lambda_{\text{tot}}^2 (m_q - \varepsilon) \qquad (15.30)$$

As before the colour nonsinglets will have infinite energy but the colour singlet energies will now be given by

$$E = N\varepsilon \tag{15.31}$$

Phenomenologically, the quantity ε has the following magnitudes for different flavours of quarks

$$\varepsilon_{ud} \sim 350 \text{ MeV}; \qquad \varepsilon_s \sim 500 \text{ MeV}; \qquad \varepsilon_c \sim 1500 \text{ MeV} \tag{15.32}$$

These quantities are effectively the energy that the quarks carry inside the colour singlet hadrons or "effective masses" of the quarks (see also Chapters 17 and 18).

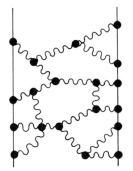

FIG. 15.2. A QCD topology which does not factorise into $\lambda_i \cdot \lambda_j$.

Clearly there is no reason to believe that this example is in any obvious way relevant to Nature since the binding of quarks will involve contributions from diagrams like Fig. 15.2 which are not like those covered in the above argument. However, we have at least found a model example where the colour degree of freedom can push certain unwanted states up to infinite masses while retaining other wanted states with low masses.

Note in particular that the above required the interquark forces to be infinite in order to send the nonsinglet states to infinity. If we had done this in a purely *ad hoc* way then we would have been at a loss to explain why one can at the same time have quasi-free quarks in deep inelastic phenomena. The beauty of the asymptotic freedom in the non-Abelian colour gauge field theory is that we can naturally reconcile these two disparate effects.

15.2.2. COLOUR AND HYPERFINE SPLITTINGS IN HADRON SPECTROSCOPY

We noted in the previous section that in colour singlet hadrons the quarks have effective masses ε and hence we would expect that typical meson masses will be of order 700 MeV and baryon masses 1050 MeV when all quarks are in their ground states.

In hydrogen the spin dependent, or magnetic, interaction between the fermion constituents generates a spin-dependent contribution to the energy which manifests itself in the hyperfine splitting of $^3S_1$ and $^1S_0$ levels. The former is higher than the latter and photoemission in the transition between them yields light of 21 cm wavelength.

Analogously the quark–vector-gluon exchange *including colour* generates both the correct sign and relative sizes of meson and baryon spin–spin splittings. In particular we shall see that in the absence of colour the baryon mass splittings would have the wrong sign and also the $\Delta - N$ magnitude would be larger than $\rho - \pi$.

First we will see how the relative magnitudes of baryon and meson splittings are unsatisfactory in the absence of colour. The spin-dependent interaction in the case of mesons has the form

$$\langle 2\mathbf{S}_1 . \mathbf{S}_2 \rangle = S(S+1) - \tfrac{3}{2}$$

$$\Rightarrow \Delta E_M = +\tfrac{1}{2}\Delta : 1^-$$

(15.33)

$$-\tfrac{3}{2}\Delta : 0^-$$

and hence the pseudoscalar and vector are separated by a quantity 2Δ (where Δ has dimensions of energy whose magnitude is not yet specified). Similarly for the baryons the form of the interaction is

$$\langle 2(\mathbf{S}_1 . \mathbf{S}_2 + \mathbf{S}_1 . \mathbf{S}_3 + \mathbf{S}_2 . \mathbf{S}_3) \rangle \equiv \langle (\mathbf{S}_1 + \mathbf{S}_2 + \mathbf{S}_3)^2 - \mathbf{S}_1^2 - \mathbf{S}_2^2 - \mathbf{S}_3^2 \rangle$$

$$= S(S+1) - \tfrac{9}{4}$$

$$\Rightarrow \Delta E_M = +\tfrac{3}{2}\Delta : \tfrac{3}{2}^+$$

$$-\tfrac{3}{2}\Delta : \tfrac{1}{2}^+$$

(15.34)

and so $\frac{3}{2}^+$ and $\frac{1}{2}^+$ are predicted to be separated by an amount 3Δ. Empirically the baryon splittings are smaller than the mesons whereas the above is predicting the opposite.

A second problem concerns the signs of the splitting. The calculation above took no account of the fact that in the meson case $q - \bar{q}$ were involved whereas in the baryon it was $q - q$. Heuristically we can see that the sign is correctly predicted in the meson case and wrongly for baryons. Recall that in hydrogen the dipole–dipole electromagnetic interaction is written

$$\mathcal{H}_{s.s} \sim -\boldsymbol{\mu}_1 . \boldsymbol{\mu}_2 \sim -\frac{e_1 e_2}{m_1 m_2} \mathbf{S}_1 . \mathbf{S}_2 \qquad (15.35)$$

Here $e_{1,2}$ are the electric charges of the particles and hence the strength of their couplings to the vector photon. For two identical particles $e_1 e_2 = e_1^2 > 0$ while for particle–antiparticle $e_1 e_2 = -e_1^2 < 0$. Hence the interaction becomes

$$E_{s.s} \sim -\mathbf{S}_1 . \mathbf{S}_2 \quad \text{for particle–particle}$$
$$E_{s.s} \sim +\mathbf{S}_1 . \mathbf{S}_2 \quad \text{for particle–antiparticle} \qquad (15.36)$$

For this reason the $^3S_1$ is more energetic than $^1S_0$ in hydrogen.

By immediate analogy the vector-gluon interaction will have the form

$$E_{s.s} \sim -\mathbf{S}_1 . \mathbf{S}_2 \quad \text{for } q - q$$
$$+\mathbf{S}_1 . \mathbf{S}_2 \quad \text{for } q - \bar{q} \qquad (15.37)$$

and so the meson phenomenology was satisfactory in sign but for the baryons one is now predicting that $\frac{1}{2}^+$ is more massive than $\frac{3}{2}^+$.

Consequently we see that there are two problems with the hyperfine splitting if colour is not included. Firstly, the relative magnitude of baryon and meson is not satisfactory and secondly the sign of the baryon splitting is not correct. Both of these problems are remedied when the colour degree of freedom is introduced. The splitting pattern in the data with uncoloured gluons and coloured gluons is summarised in Fig. 15.3. The reason that coloured gluons give satisfactory phenomenology is entirely due to that fact that the coupling between *coloured quarks* in a colour singlet baryon is of the *same sign* as between *coloured quark* and *coloured antiquark* and half of the magnitude. This is the crucial feature and we shall now prove this.

In the electromagnetic case the $\mathbf{S}_1 \cdot \mathbf{S}_2$ interaction was multiplied by $e_1 e_2$ which took account of the electrical charge in the photon exchange. In the case of quantum chromodynamics the interaction will instead have $\mathbf{F}_1 \cdot \mathbf{F}_2$ to take account of the "colour charge" in the colour gluon exchange. We must therefore calculate the expectation value of

$$2\mathbf{F}_1 \cdot \mathbf{F}_2 \equiv (\mathbf{F}_1 + \mathbf{F}_2)^2 - \mathbf{F}_1^2 - \mathbf{F}_2^2$$

first of all for $q\bar{q}$ in a colour singlet (i.e. meson) and secondly for qq in a coloured antitriplet (i.e. a baryon). In a colour singlet qqq state, any qq pair must be in $\bar{3}$ of colour if they are to form a $\mathbf{1}$ on combining with the third quark.

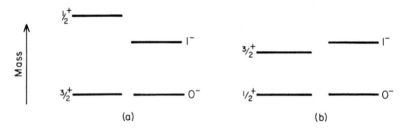

FIG. 15.3. Hyperfine splittings of mesons and baryons (a) without colour, and (b) with colour.

In the colour singlet meson we have that

$$\langle 2\mathbf{F}_1 \cdot \mathbf{F}_2 \rangle = 0 - \lambda_{(3)}^2 - \lambda_{(\bar{3})}^2$$

$$= -2\lambda^2$$

$$= -\tfrac{8}{3} \qquad\qquad (15.38)$$

where λ^2 is the Casimir operator for an $SU(3)$ triplet or antitriplet and is equal to $\tfrac{4}{3}$ in either case.

In the case of a baryon we have almost the same thing as before except that in this case the quark pair in the baryon are in the coloured antitriplet and hence

$$\langle 2\mathbf{F}_1 \cdot \mathbf{F}_2 \rangle = \lambda_{(\bar{3})}^2 - 2\lambda_{(3)}^2$$

$$= -\lambda^2$$

$$= -\tfrac{4}{3} \qquad\qquad (15.39)$$

Hence

$$\langle \mathbf{F}_1 . \mathbf{F}_2 \rangle_{qq \text{ in baryon}} = +\tfrac{1}{2} \langle \mathbf{F}_1 . \mathbf{F}_2 \rangle_{q\bar{q} \text{ in meson}} \qquad (15.40)$$

in contrast to the uncoloured case where the qq and $q\bar{q}$ had equal magnitude but opposite sign. The coloured degree of freedom therefore gives satisfactory phenomenology, namely $\tfrac{3}{2}^+$ more massive than $\tfrac{1}{2}^+$ but split by less than 1^- and 0^- mesons (Fig. 15.3).

From this discussion we have seen that the hyperfine splitting patterns of baryons and mesons are satisfactorily dealt with in quantum chromodynamics. A quantitative discussion of this phenomenon, in particular the relative magnitudes of the hyperfine splittings in the nonstrange, strange and charmed spectroscopies, will be given in Chapter 17 after the charmed particles have been introduced.

16 The New Particles

16.1 Charmonium

The dramatic discovery (Aubert *et al.*, 1974; Augustin *et al.*, 1974) of the J/ψ vector meson with mass 3095 MeV in November 1974 opened a new chapter in high-energy physics. This particle was seen as a distinct enhancement in the e^+e^- mass spectrum in $pBe \rightarrow e^+e^- +$ anything at Brookhaven and in $e^+e^- \rightarrow$ hadrons at SPEAR. Apart from its high mass it had the remarkable property of being metastable, its width to e^+e^- being $4 \cdot 8 \pm 0 \cdot 6$ keV (which is typical of vector mesons), whereas its total width is only 69 ± 15 keV (for a 3 GeV state one would have *a priori* anticipated a width of some hundreds of MeV).

Within a few days a heavier vector particle, $\psi'(3684)$, was discovered in e^+e^- annihilation. It had similar properties to the $\psi(3095)$, namely conventional leptonic width ($\Gamma^{e^+e^-} = 2 \cdot 1 \pm 0 \cdot 3$ keV) and narrow hadronic width ($\Gamma_{\psi'} = 225 \pm 56$ keV). Moreover about 50 per cent of the ψ' decays contain ψ in the decay products ($\psi' \rightarrow \psi\pi\pi$, $\psi\eta$) and some 20 per cent are radiative decays to heavy metastable states χ of masses $3 \cdot 4$–$3 \cdot 55$ GeV.

These states have all the properties expected of the $c\bar{c}$ spectroscopy associated with a fourth flavour of quark (charm). The ψ and ψ' are $1\,{}^3S_1$ and $2\,{}^3S_1$ states of $c\bar{c}$ while the χ states are probably $1\,{}^3P_{0,1,2}$.

In such a spectroscopy one naturally expects that $\eta_c(1\,{}^1S_0)$ will exist below the ψ. This state may have been found in $\psi \rightarrow \gamma\eta_c \rightarrow \gamma\gamma\gamma$. The observed states between $2 \cdot 7$ and $3 \cdot 8$ GeV are shown in Fig. 16.1 together with their possible J^{PC} and $c\bar{c}$ assignments. These will be discussed in more detail in subsequent sections. Experimental details and extensive references can be found in the bibliography cited at the end of this chapter.

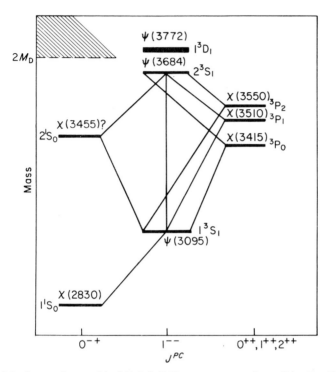

FIG. 16.1. States observed in 2·7–3·8 GeV mass range and possible $c\bar{c}$ assignments.

16.1.1   THE OKUBO–ZWEIG–IIZUKA (OZI) RULE

The lightest charmed particles $c\bar{u}$ have masses of order 1·8–2 GeV (sections 16.4 and 16.5). The ψ'', ψ''' ($c\bar{c}$) states above 4 GeV in mass can pair produce charmed mesons (Fig. 16.2(a)). Such states have widths of several MeV, typically hadronic in magnitude. The states ψ, ψ' however lie below threshold for charm pair production so that the ψ decays into uncharmed hadrons by disconnected quark diagrams (Fig. 16.2(b)). This situation is analogous to the ϕ ($s\bar{s}$) which readily decays to strange pairs $K\bar{K}$ but whose decay to 3π, by disconnected quark diagrams, appears to be suppressed.

Motivated by the suppression[1]

$$\frac{\Gamma(\phi \to 3\pi)}{\Gamma(\phi \to K\bar{K})} \simeq \frac{1}{5} \tag{16.1}$$

a rule was invented (OZI rule, Okubo, 1963; Zweig, 1964b; Iizuka, 1966). Transitions are forbidden if they are described by a diagram that can be cut by a line which does not cross any quark line and originates outside the hadrons. The OZI rule is not exact since ϕ does decay to 3π and the ψ does decay into hadrons. However, the rule does appear to be better satisfied in the case of the ψ than the ϕ (probably due to its greater mass). Also the ψ, χ, ψ' system is very useful as a laboratory for studying the breaking of the OZI rule since all of these states lie below charm production threshold.

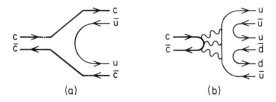

FIG. 16.2. (a) $\psi'' \to D\bar{D}$. (b) $\psi \to$ uncharmed hadrons.

As we will see in section 16.5 the properties of the charmed quark c in the $\psi(c\bar{c})$ system and in the charmed particles D(c$\bar{u}$) etc. are in accord with those anticipated in the non-Abelian gauge theory of weak and electromagnetic interactions (quantum flavour dynamics). This has in turn excited interest in the possibility of non-Abelian gauge theories of the *strong* interactions (quantum chromodynamics). Several features of the charmed spectroscopy fit in with this latter theory. In turn QCD provides a well-defined mechanism for violating the OZI rule. Quarks and antiquarks in a meson can annihilate producing gluons which in turn produce $q\bar{q}$ pairs, e.g.

$$c\bar{c} \to \text{gluons} \to u\bar{u}, d\bar{d}, s\bar{s} \tag{16.2}$$

The way in which this happens for η_c, ψ and χ decays can be quantified in QCD if the charmonium system is similar to positronium.

[1] Note that phase space favours $\phi \to 3\pi$.

Qualitatively we can describe the QCD approach to the OZI rule as follows. The processes $\phi \to \rho\pi$ and $\psi \to \rho\pi$ are described by the diagram of Fig. 16.2(b) where solid lines are coloured quarks and wiggly lines are coloured gluons. A diagram with n hard gluons will have rate proportional to $[\alpha_s(s)]^n$ where $\alpha_s(s)$ is the quark–gluon coupling constant which is a function of s, the squared mass running through the gluons. If $\alpha(s)$ is small then the amplitude will be small and so $\phi \to \rho\pi$ and $\psi \to \rho\pi$ will be suppressed. In QCD the asymptotic freedom property implies that $\alpha_s(s)$ decreases as s increases. Hence $\alpha_s(s = m_\psi^2)$ will be smaller than $\alpha_s(s = m_\phi^2)$ and the greater inhibition of $\psi \to \rho\pi$, or in general $\psi \to$ ordinary hadrons, can be understood.[1] Empirically we shall see that $\alpha(s = m_\phi^2) \simeq 0.5$ while $\alpha(s = m_\psi^2) \simeq 0.2$.

The same remarks can be applied to the hadronic decays of η_c, χ states and ψ'. In the latter case phase space allows the decay $\psi' \to \psi\pi\pi$ to occur. This process is Zweig rule violating and in the quark–gluon field theory can proceed by a diagram like Fig. 16.3. The dipion system has an invariant mass that is less than 600 MeV and so the gluons that produce them will be soft. Consequently $\alpha_s(s)$ is small and this process should not be strongly inhibited. This appears to be qualitatively true, the width of this process being larger than the total width of $\psi' \to$ ordinary hadrons.

A similar contribution to ψ' decay is $\psi' \to \eta\psi$. This process has a width of about 10 keV (4 per cent of all ψ' decays) which appears large given that phase space inhibits it (P-wave) and that η is dominantly octet.[2] It may be that η contains some $c\bar{c}$ and so this decay takes place by

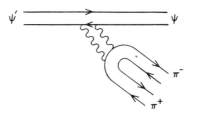

FIG. 16.3. Quark–gluon diagram for $\psi' \to \psi\pi\pi$.

[1] See, however, the footnote number 27 in Appelquist and Politzer (1975a) where it is noted that if the $c\bar{c}$ annihilates into a large number of gluons then each may be soft enough that α_s is large and the coupling is in the strong regime. The perturbation approach and dominance of the three gluon mode is therefore questionable. No completely satisfactory resolution of this question yet exists in the literature. It may be possible to justify the three-gluon dominance by appealing to Kinoshita's theorem (Kinoshita, 1962).

[2] The $\psi(c\bar{c})$ and $\psi'(c\bar{c})$ are SU(3) singlets.

an OZI-allowed diagram without suppression. The possibility that η and η' contain $c\bar{c}$ will be discussed again in section 17.6.

These singly disconnected topologies are known as "first forbidden". Examples in ψ decays include ωf and $\phi f'$ in addition to $\pi\rho$ etc. Diagrams which are twice disconnected are known as "doubly forbidden" (Fig. 16.4). Examples include $\psi \to f\phi, f'\omega$ (the $\omega\phi ff'$ system being ideally mixed, viz. f' and ϕ are $s\bar{s}$ while ω and f contain only $u\bar{u}$ and $d\bar{d}$).

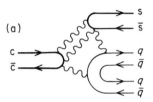

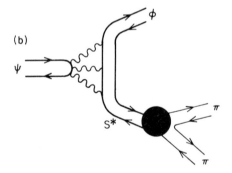

FIG. 16.4. (a) Doubly forbidden processes $\psi \to f\phi, f \to \pi\pi$. (b) $\psi \to \phi S^*, S^* \to \pi\pi$.

Are doubly forbidden decays suppressed relative to singly forbidden? The data bearing on this are shown in Fig. 16.5. The data for $\psi \to \phi\pi\pi$ or $\phi K\bar{K}$ are shown as a function of the $\pi\pi$ and $K\bar{K}$ invariant masses. A clear f peak is seen in $\psi \to \omega\pi\pi$ (i.e. $\psi \to \omega f$) but none is seen in $\psi \to \phi\pi\pi$ (i.e. $\psi \not\to \phi f$). Similarly in $\psi \to \phi K\bar{K}$ and f' peak is seen but no such enhancement occurs in the $\psi \to \omega K\bar{K}$ mode. Hence $\psi \to \phi f', \not\to \omega f'$.

One might conclude that $\psi \to \phi\pi\pi$ will in general be suppressed relative to $\omega\pi\pi$ since the former would be doubly forbidden. However some care is needed. Imagine that one produces a state A which is not ideally mixed. If we write u$\bar{\text{u}}$ to denote u$\bar{\text{u}}$ + d$\bar{\text{d}}$ then in production

$$\psi \rightarrow \phi(s\bar{s}) + A(s\bar{s})$$

$$\psi \rightarrow \omega(u\bar{u}) + A(u\bar{u}) \tag{16.3}$$

The A is a physical eigenstate:

$$A = \cos\theta \; s\bar{s} + \sin\theta \; u\bar{u} \tag{16.4}$$

and so can decay into K$\bar{\text{K}}$ or $\pi\pi$. Consequently $\psi \rightarrow \phi\pi\pi$ or ωK$\bar{\text{K}}$ can be quite significant if the $m_{\pi\pi}$ or m_{KK} is at a resonance mass.

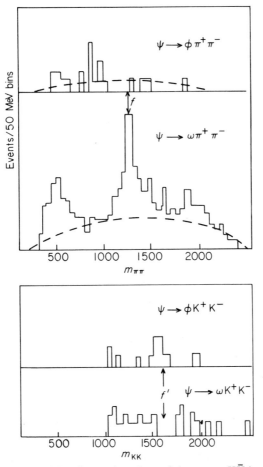

FIG. 16.5. $\psi \rightarrow \phi\pi\pi$, ϕKK data as functions of the $\pi\pi$ or K$\bar{\text{K}}$ invariant masses.

From the data one can hope eventually to extract information on the $s\bar{s}/u\bar{u}$ content of states A. As an example we cite the S*(990) which is clearly seen (?) in $\phi\pi\pi$ and hardly seen at all in $\omega\pi\pi$. This suggests that S* has significant $s\bar{s}$ content. The possibility of extracting information on the quark content or existence of $\varepsilon(1200)$ also raises itself. Much more data are needed before such a programme can be systematically carried out. One begins to see here the possibility of using the ψ as a laboratory for learning about old-fashioned meson spectroscopy.

Finally there is the possibility of OZI-allowed decays $\psi' \to \gamma\chi, \chi \to \gamma\psi$ etc. The decays $\psi' \to \gamma\chi$ have a sum total of around 60 keV which is about 25 per cent of all ψ' decays. The decays $\psi \to \eta\gamma, \eta'\gamma$ and $\pi^0\gamma$ have also been seen. The widths

$$\Gamma(\psi \to \eta\gamma) = 55 \pm 12 \text{ eV}$$
$$\Gamma(\psi \to \eta'\gamma) = 152 \pm 117 \text{ eV}$$

(16.5)

for individual two-body channels are quite sizeable. This might support the notion that η and η' contain some $c\bar{c}$, the η' being more dominantly singlet having the greater amount of $c\bar{c}$, which accounts for its greater production. Compare with $\pi^0\gamma$ which has width

$$\Gamma(\psi \to \pi^0\gamma) = 5 \pm 3 \text{ eV}$$

(16.6)

This is in fair agreement with the naive vector-meson dominance prediction

$$\Gamma(\psi \to \pi^0\gamma) \simeq \tfrac{1}{300}\Gamma(\psi \to \pi^0\rho) \simeq 1 \text{ eV}$$

(16.7)

and hence is consistent with Zweig single disconnection. The $\eta\gamma$ and $\eta'\gamma$ are 50 to 150 times more abundant than $\pi^0\gamma$. This may be due to $c\bar{c}$ in the η, η', or due to SU(3) singlet photon couplings being enhanced relative to the SU(3) octet. This latter seems unlikely since $\psi' \to \gamma\chi$ have canonical widths (0(keV)), and involve SU(3) singlet photon couplings.

We now will make a quantitative investigation into some of the preceding qualitative remarks. We begin with the ψ, χ decays via coloured gluons in singly forbidden OZI topologies.

16.2　Charmonium decays in quantum chromodynamics

The production of noncharmed hadrons in the decays of $c\bar{c}$ states like ψ, η_c, χ will be radiative (e.g. $\psi \to \gamma \to$ hadrons) or violate Zweig's rule.

The mechanism proposed in quantum chromodynamics is that the $c\bar{c}$ first annihilate into gluons, the gluons then produce noncharmed quarks and antiquarks which in turn comprise the familiar hadrons.

It is supposed that the gluons are quasi-free (i.e. propagate as if Abelian, no three gluon vertices being present) and fragment into hadrons with unit probability. If, in addition, the quark–gluon coupling α_s is small then lowest-order perturbation theory can be used and the matrix elements will be related to the analogous QED annihilation of positronium into two or more photons (Appelquist and Politzer, 1975b).

We collect here the positronium annihilation formulae into photons for $L = 0$ singlet and triplet states, mass M, and $R(0)$ is the radial wavefunction of the fermion–antifermion state at the origin.

$$\Gamma({}^1S_0 \to \gamma\gamma) = \frac{4\alpha^2}{M^2}|R_s(0)|^2 \tag{16.8}$$

$$\Gamma({}^3S_1 \to \gamma\gamma\gamma) = \frac{16}{9\pi}(\pi^2 - 9)\frac{\alpha^3}{M^2}|R_s(0)|^2 \tag{16.9}$$

The reader interested in the derivation of these formulae should consult Chapter 12 of Jauch and Rohrlich (1955). For the P-waves the derivative of the wavefunction at the origin occurs in the positronium annihilation

$$\Gamma({}^3P_0 \to \gamma\gamma) = \frac{256}{3}\frac{\alpha^2}{M^4}|R'_P(0)|^2 \tag{16.10}$$

$$\Gamma({}^3P_2 \to \gamma\gamma) = \frac{4}{15}\Gamma({}^3P_0 \to \gamma\gamma)\left(\frac{M_0}{M_2}\right)^4 \tag{16.11}$$

The derivation of these results can be found in Alekseev (1958). The ${}^3P_1$ and ${}^1P_1$ discussion will be deferred.

First we study the S-wave states of $c\bar{c}$, the ${}^1S_0(\eta_c)$ and ${}^3S_1(\psi)$. If the charge of the quark in the ψ is e_q then for N colours of quark

$$\Gamma(\psi \to e^+e^-) = \frac{4\alpha^2 e_q^2 N}{3M^2}|R_s(0)|^2 \tag{16.12}$$

The decay of $\eta_c \to \gamma\gamma$ is given immediately from equation (16.8) and for N colours of quark[1] with charge e_q

$$\Gamma(\eta_c \to \gamma\gamma) = \frac{4\alpha^2 e_Q^4 N}{M^2}|R_s(0)|^2 \tag{16.13}$$

(the η_c and ψ have the same radial wavefunction). Then independent of the number of colours we have the prediction

$$\frac{\Gamma(\eta_c \to \gamma\gamma)}{\Gamma(\psi \to e^+ e^-)} = 3e_Q^2 \left(\frac{M_\psi}{M_{\eta,c}}\right)^2 \tag{16.14}$$

If $e_q = \frac{2}{3}$ (charmed quark) and $m_{\eta,c} \lesssim m$ then

$$\Gamma(\eta_c \to \gamma\gamma) \gtrsim \tfrac{4}{3}\Gamma(\psi \to e^+ e^-) \tag{16.15}$$

If $m_{\eta,c} \simeq 2 \cdot 8$ GeV then this predicts some 8 keV for this decay mode. The data on $\eta_c \to \gamma\gamma$ will provide a good test for e_q's magnitude, or the extent to which η_c contains more than just $c\bar{c}$.

The decay $\eta_c \to$ hadrons proceeds by the intermediary of two gluons. In QCD with α_s the quark–gluon coupling strength at the η_c mass, the prescription is to make the replacement in equation (16.13)

$$\alpha^2 e_q^4 \to \tfrac{2}{9}\alpha_s^2 \tag{16.16}$$

and with $N = 3$ in QCD we obtain

$$\Gamma(\eta_c \to gg) = \frac{8}{3}\frac{\alpha_s^2(m^2)}{M^2}|R_s(0)|^2 \tag{16.17}$$

(equation 16.16 is derived later in equation 16.40). If the gluons turn into hadrons with unit probability then this formula gives the total hadronic width of the η_c. Comparing with equation (16.13) we find

$$\frac{\Gamma(\eta_c \to \text{hadrons})}{\Gamma(\eta_c \to \gamma\gamma)} = \frac{2}{9}\frac{\alpha_s^2(m^2)}{\alpha^2 e_q^4} \equiv \frac{9}{8}\left(\frac{\alpha_s}{\alpha}\right)^2\left(\frac{2/3}{e_q}\right)^4 \tag{16.18}$$

From this we can predict the η_c width once α_s is known, or conversely, knowing the width, then $\alpha_s(m_{\eta,c}^2)$ is obtained.

The decay mode $\psi \to$ hadrons requires a three-gluon intermediate state (section 16.2.1.1). The $^3S_1 \to 3$ gluons decay is obtained from

---

[1] A $c\bar{c}$ state made from uncoloured quarks will become $1/\sqrt{N}(c_i\bar{c}_i)$ with N colours. It is this $1/\sqrt{N}$ normalisation which causes the widths to be proportional to N and not N^2. Compare and contrast the $\pi^0 \to \gamma\gamma$ in Chapter 15 of Kane (1977).

equation (16.9) by the replacement

$$\alpha^3 \to \frac{5\alpha_s^3}{18} \tag{16.19}$$

(derived in equation 16.45). If the gluons create hadrons with unit probability then

$$\Gamma(^3S_1 \to \text{hadrons}) = \frac{40}{81\pi}(\pi^2 - 9)\frac{\alpha_s^3}{M^2}|R_s(0)|^2 \tag{16.20}$$

Compare this with the width for $^3S_1 \to e^+e^-$ (equation 16.12) and we have

$$\frac{\Gamma(\psi \to \text{hadrons})}{\Gamma(\psi \to e^+e^-)} = \frac{5}{18\pi}(\pi^2 - 9)\frac{\alpha_s^2(m_\psi)}{\alpha^2}\left(\frac{2/3}{e_q}\right)^2 \tag{16.21}$$

The data give

$$\Gamma(\psi \to e^+e^-) = 4 \cdot 8 \pm 0 \cdot 6 \text{ keV}$$

$$\Gamma(\psi \to \text{hadrons}) = 69 \pm 15 \text{ keV}$$

and so, if $e_q = \frac{2}{3}$, we find from equation (16.21) that

$$\alpha_s(m_\psi \approx 3 \text{ GeV}) \approx 0 \cdot 2 \tag{16.22}$$

This quark–gluon coupling is indeed small enough that our lowest-order perturbation theory is justifiable. In passing it is interesting to compare this result with the ϕ meson. Clearly from equation (16.21) we can obtain

$$\frac{\Gamma(\phi \to \text{nonstrange hadrons})}{\Gamma(\psi \to \text{hadrons})} \cdot \frac{\Gamma(\psi \to e^+e^-)}{4\Gamma(\phi \to e^+e^-)} = \left|\frac{\alpha_s(m_\phi)}{\alpha_s(m_\psi)}\right|^3 \tag{16.23}$$

(the $1:4$ being the ratio of strange and charmed quarks squared charges). The ψ leptonic width is roughly four times the ϕ and so the ratio of hadronic widths is directly a measure of $[\alpha_s(m_\phi)/\alpha_s(m_\psi)]^3$. This yields

$$\alpha_s(m_\phi \approx 1 \text{ GeV}) \approx 0 \cdot 5 \tag{16.24}$$

Comparing with equation (16.22) we see a beautiful realisation of the asymptotic freedom in the theory, the coupling constant at 3 GeV being only 40 per cent of that at 1 GeV. Hence it is argued that the asymptotic freedom of the quantum chromodynamics theory is responsible for the OZI rule becoming better at high masses, and so the narrow width of the ψ emerges.

As a test of this theory of the OZI rule one can compare the widths of η_c and ψ. From equations (16.17) and (16.20) we have that

$$\frac{\Gamma(\eta_c \to \text{hadrons})}{\Gamma(\psi \to \text{hadrons})} = \frac{27\pi}{5(\pi^2 - 9)} \frac{\alpha_s^2(\eta_c)}{\alpha_s^3(\psi)} \tag{16.25}$$

Inserting $\alpha_s(m_\psi) \simeq \frac{1}{5}$ and taking $\alpha_s(m_{\eta,c}) \gtrsim \alpha_s(m_\psi)$ then

$$\Gamma(\eta_c \to \text{hadrons}) \gtrsim 80\Gamma(\psi \to \text{hadrons}) \simeq 7 \text{ MeV} \tag{16.26}$$

Consequently the η_c is much broader than the ψ but still much smaller than "typical" hadronic widths of order hundreds of MeV at this sort of mass. Comparing equations (16.15) and (16.26) (or inserting $\alpha_s \simeq 0 \cdot 2$ into equation 16.18) we expect that

$$\frac{\Gamma(\eta_c \to \gamma\gamma)}{\Gamma(\eta_c \to \text{hadrons})} \simeq 10^{-3} \tag{16.27}$$

A final feature of S-wave decays concerns the radial excitations ψ' and η_c'. Since production of ordinary hadrons and e^+e^- both involve annihilation of the $c\bar{c}$ at the origin then

$$\frac{\Gamma(\psi' \to \text{ordinary hadrons})}{\Gamma(\psi \to \text{ordinary hadrons})} = \frac{\Gamma(\psi' \to e^+e^-)}{\Gamma(\psi \to e^+e^-)} \simeq 0 \cdot 4 \tag{16.28}$$

Common channels in ψ and ψ' decays do appear to be consistent with this.

The annihilation of P-wave positronium into two photons has been studied by Tumanov (1953) and Alekseev (1958). Making the substitution $\alpha^2 \to \alpha^2 e_q^4$ then the two photon decays of the P-wave $c\bar{c}$ system are

$$\Gamma(\chi_2 \to \gamma\gamma) = \frac{4}{15} \Gamma(\chi_0 \to \gamma\gamma) = \frac{1024}{45} \frac{\alpha^2 e_q^4}{M^4} |R_P'(0)|^2 \tag{16.29}$$

where $|R_P'(0)|$ is the derivative of the P-wave radial wavefunction at the origin. With the colour substitution (equation 16.40) of

$$\frac{\Gamma(2g)}{\Gamma(2\gamma)} = \frac{2\alpha_s^2}{9\alpha^2 e_Q^4} \tag{16.30}$$

then

$$\Gamma(\chi_2 \to gg) = \frac{4}{15}\Gamma(\chi_0 \to gg) = \frac{128}{5}\frac{\alpha_s^2}{M^4}|R_P'(0)|^2 \qquad (16.31)$$

which is the result found by Barbieri *et al.* (1976a). The reader interested in the detailed derivation should consult the work of Jackson (1976b). The $4:15$ ratio is a combination of a spin-orbital Clebsch coupling ($\frac{1}{2}$) and $\sin^4\theta$ versus isotropic decay distributions ($\frac{8}{15}$).

The decay width of these $^3P_{2,0}$ states depends upon $|R_P'(0)|^2$ which has dimensions of mass to the fifth power and so is strongly dependent upon the explicit details of the potential. Typically magnitudes of the order 0.05 to 0.1 $(GeV)^5$ have been estimated (Eichten *et al.*, 1975; Barbieri *et al.*, 1976a; Jackson, 1976b). These predict widths of the order of 0.5 to 1 MeV for $\Gamma(^3P_2)$ and 2 to 4 MeV for the $^3P_0$.

A final question in the P-wave concerns the decay of the axial vector mesons $J^{PC} = 1^{+-}$ and 1^{++}.

A state with $J^{PC} = 1^{++}$ cannot decay into two massless vectors (in particular two photons). This is known as the theorem of Landau (1948) or Yang (1950). This in turn has implications for the two-gluon decay of the $^3P_1$ (contrast the $^3P_{0,2}$). The two photons (gluons) have polarisation vectors $\varepsilon_{1,2}$ and the relative momentum of them is $\mathbf{k}$. Any possible final state wavefunction will be linear in ε_1 and ε_2 and transform as a vector if the total state has spin 1. The only three possibilities are

i. $\varepsilon_1 \times \varepsilon_2$
ii. $(\varepsilon_1 \cdot \varepsilon_2)\mathbf{k}$ $\qquad (16.32)$
iii. $\mathbf{k} \times (\varepsilon_1 \times \varepsilon_2)$

The first is antisymmetric in the two photons and so is the second (since $\mathbf{k} \to -\mathbf{k}$). The final possibility satisfies Bose–Einstein statistics. However the transversality condition $\mathbf{k} \cdot \varepsilon = 0$ kills it and so the two photon annihilation is forbidden. If gluons are massless then $^3P_1 \to$ two on-shell gluons is forbidden.

The decay of $^1P_1(J^{PC} = 1^{+-})$ will be into three gluons symmetrically coupled in colour (d_{ijk}) and the $^3P_1(J^{PC} = 1^{++})$ will decay by three gluons antisymmetrically coupled (f_{ijk}).[1] Hence naively one expects the total widths to be much smaller than the $^3P_{0,2}$ brothers (which only

---

[1] A $c = -$ state decays into three photons by charge conjugation. The photons are trivially symmetric in colour, hence the gluons must also be. Conversely the $c = +$ decays require antisymmetry in colour.

required two gluons). However the $^3P_1$ decay *can* take place into *two* gluons if one is on and the other *off* mass shell (or if both are off shell). Barbieri *et al.* (1976b) find that this process has a logarithmic divergence. Similarly the three-gluon annihilation diverges logarithmically. Their results are for the annihilation contribution to the axial widths

$$\Gamma\binom{1^{++}}{1^{+-}} = \frac{1}{3\pi}\left(\frac{128}{320/3}\right)\frac{\alpha_s^3}{M^4}|R_P'(0)|^2 \log\frac{4m^2}{4m^2 - M^2} \qquad (16.33)$$

where M is the system's mass and $2m$ the fermion constituents' total mass. In the zero binding limit there is a logarithmic divergence. Comparing those results with the calculation of the $^3P_0$ decay yields for the annihilation contributions

$$\frac{\Gamma(^3P_1)}{\Gamma(^3P_0)} = \frac{4}{9}\frac{\alpha_s}{\pi} \log\frac{4m^2}{4m^2 - M^2} \qquad (16.34)$$

and

$$\frac{\Gamma(^1P_1)}{\Gamma(^3P_0)} = \frac{10}{27}\frac{\alpha_s}{\pi} \log\frac{4m^2}{4m^2 - M^2} \qquad (16.35)$$

Similar logarithmic effects may be expected from nonleading graphs in the $^3P_0$ and $^3P_2$ decays.

This brings into focus some of the questions that need further study. If quarks have large masses m (which is not the case for u, d, s but might be all right for charm) then the lowest bound states' characteristics may be well approximated by the short distance Coulomb type of force. In a Coulomb potential (e.g. hydrogen) the binding energy in the Nth quantum state is

$$E_B^{(N)} = -\frac{2m\alpha^2}{8N^2} \qquad (16.36)$$

For Coulombic binding in the $q\bar{q}$ system $\alpha \to \frac{4}{3}\alpha_s$ (compare sections 15.2 and 17.3, in particular equations 15.38 and 17.34) and so

$$M_N = 2m\left(1 - \frac{1}{8N^2}\left(\frac{4}{3}\alpha_s\right)^2\right) \qquad (16.37)$$

If $N = 1, 2$ are the $\psi(^3S_1)$ and $\psi'(2\,^3S_1)$ then either one has a small α_s and large $2m$, or small $2m$ and large α_s. In the former case one has a large binding, which violates the weak binding assumption. In the latter case the large α_s violates the Coulombic bound state picture and no hope of perturbation theory can be realised. Therefore it seems that the charmed system is not yet massive enough for the Coulomb picture to apply. However it may be valid for heavier quarks (b, t ...) and gives some predictions for ψ and χ annihilations which can be tested empirically.

16.2.1   RELATION BETWEEN HADRON AND PHOTON DECAYS

The $q\bar{q}$ annihilation into photons is a function of $\alpha \equiv e^2/4\pi$ and the quark charge e_q. The annihilation into coloured gluons involves $\alpha_s \equiv g^2/4\pi$ and the Gell-Mann 3×3 matrices λ corresponding, at each gluon emission vertex, to the $SU(3)_{\text{colour}}$ representation of the particular gluon emitted. The relative rates for N photon and N gluon emission will then be given by contraction over the $SU(3)$ λ indices.

16.2.1.1 $N = 2$

The diagrams for $c\bar{c} \to 2\gamma$ or 2 gluons are shown in Fig. 16.6. The colour labels i, j = 1, 2, 3 (red, blue, yellow) and the initial $c\bar{c}$ being a colour singlet requires the colour label of c and $\bar{c}$ to be the same.

Photo-emission does not change the colour of the quark, hence the δ^i_j colour space contribution at the upper and δ^j_i at the lower vertex. Gluon emission in general can change the quark colour. A member of the gluon

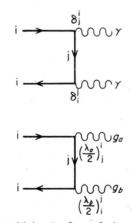

FIG. 16.6. $c\bar{c} \to 2\gamma$ or 2 gluons.

octet ($a = 1 \ldots 8$) will change colour i to colour j and couple with a factor $(\lambda_a/2)^i_j$, one of the 3×3 matrices of Chapter 2.

Apart from the overall scale of the gluon–quark coupling strength (α_s) to that of the photon–quark ($e_q^2 \alpha$) the only differences in the two matrix elements are the above colour factors. Hence

$$\frac{M(2g)}{M(2\gamma)} = \frac{\alpha_s}{e_q^2 \alpha} \frac{(\lambda_a/2)^i_j (\lambda_b/2)^j_i}{\delta^i_j \delta^j_i} = \frac{\alpha_s}{e_q^2 \alpha} \frac{\frac{1}{2}\delta_{ab}}{3} \qquad (16.38)$$

where in the last step we used

$$\text{Tr}\left(\frac{\lambda_a}{2} \frac{\lambda_b}{2}\right) = \frac{1}{2}\delta_{ab} \qquad (16.39)$$

which follows from the commutation properties of the λ matrices. (Add together equations 2.45 and 2.46 and remember that the λ are traceless.)

To compare the rates one must also include the crossed diagrams (Fig. 16.6(b)). Due to the δ_{ab} the gluons are identical particles just like the two photons. Hence all counting of states is the same in the two cases and so the ratio of rates can be immediately computed by squaring equation (16.38). Since $\sum_{ab} \delta_{ab}\delta^{ab} = 8$ then

$$\Gamma(2g)/\Gamma(2\gamma) = \frac{2}{9} \frac{\alpha_s^2}{\alpha^2 e_q^4} \qquad (16.40)$$

Notice that this result depends only on the colour indices and not upon angular momentum properties etc. Hence it applies equally well to $^1S_0$ as to P-waves etc. Finally we note that states with $c = +$ can decay into 2γ and hence also into 2 gluons (except if $J^P = 1^+$ or 1^-, section 16.2). A $c = +$ state can also decay to 3 gluons if any pair is coupled antisymmetrically (f_{ijk}) in the colour labels (this degree of freedom is absent for colour singlet photons).

Two colour-octet gluons cannot couple to a colour-singlet state with $c = -$. This is most easily seen by considering G parity (section 4.1.1) generalised to SU(3). The two gluons will necessarily have positive G parity and hence the colour singlet state must also have positive G parity if it is to couple. For a (colour) singlet then the (colour) isospin is clearly zero and hence the charge conjugation must be positive if its G parity is to be positive. Hence the $c = +$ colour-singlet can decay into two octet gluons, whereas $c = -$ needs at least three.

The analogous argument in flavour SU(3) forbids the singlet ψ to decay to the octet K^+K^- but allow KK^* since K and $K^*(890)$ have opposite generalised G parity. The ϕ can decay to K^+K^- due to its octet component.

16.2.1.2 $N = 3$

States with $c = -$ decay into three gluons coupled symmetrically in colour (analogous to the fact that the decay into three photons is trivially symmetric in colour). Hence, in the analogous notation to the $N = 2$ example, we have

$$\frac{M(3g)}{M(3\gamma)} = \frac{\alpha_s^{3/2}}{e_q^3 \alpha^{3/2}} \frac{[(\lambda_a/2)_j^i (\lambda_b/2)_k^j (\lambda_c/2)_i^k]_{sym}}{\delta_j^i \delta_k^j \delta_i^k} \qquad (16.41)$$

Symmetrising the ab labels in the colour case we have

$$\text{Tr}\left(\frac{\lambda_a}{2} \frac{\lambda_b}{2} \frac{\lambda_c}{2}\right) = \frac{1}{2} \text{Tr}\left(\left\{\frac{\lambda_a}{2}, \frac{\lambda_b}{2}\right\} \frac{\lambda_c}{2}\right)$$

and since trace $\lambda \equiv 0$ then from the commutation relations in equation (2.46) we find that

$$\frac{1}{2}\text{Tr}\left(\left\{\frac{\lambda_a}{2}, \frac{\lambda_b}{2}\right\} \frac{\lambda_c}{2}\right) \equiv \frac{1}{8}\text{Tr}\, d_{abc}\lambda_b\lambda_c \equiv \frac{1}{4}d_{abc}\delta_{bc} \qquad (16.42)$$

Insert this into equation (16.41), square it, use the fact that

$$\sum_{abc} (d_{abc})^2 = \tfrac{40}{3} \qquad (16.43)$$

(which follows from Table 2.1) and verify that

$$\frac{\Gamma(3g)}{\Gamma(3\gamma)} = \frac{5}{54} \frac{\alpha_s^3}{\alpha^3 e_q^6} \qquad (16.44)$$

The $^3S_1 \to 3\gamma$ width is given by equation (16.9) multiplied by $3e_q^6$ (the 3 is for colour). Hence the three-gluon decay width is obtained from equation (16.9) by the replacement

$$\alpha^3 \to \frac{5\alpha_s^3}{18} \qquad (16.45)$$

as asserted at equation (16.19).

16.3 The charmonium spectroscopy

16.3.1 THE χ PARTICLES

If $\psi(3\cdot1 \text{ GeV})$ and $\psi'(3\cdot7 \text{ GeV})$ are respectively the $1\,^3S_1$ and $2\,^3S_1$ states of a $c\bar{c}$ system, then one expects that $c\bar{c}$ states $1\,^3P_{0,1,2}$ should exist with masses less than or of the order of the ψ' mass (e.g. (i) the spectroscopy of uds flavours suggests that P-wave $q\bar{q}$ states are of order 500 MeV heavier than the S-wave; (ii) a harmonic oscillator model would predict the P-wave states roughly midway between the 1S and 2S levels) (see also section 16.3.3). The $^3P_{0,1,2}$ states have positive charge conjugation and hence can be produced in radiative decays of $\psi'(3\cdot7 \text{ GeV})$ if they are below 3·7 GeV in mass. Up to four narrow states in the mass range 3·4

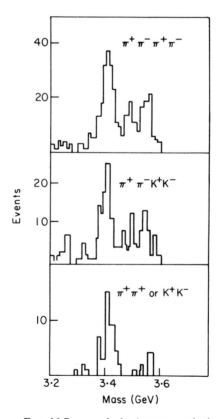

FIG. 16.7. χ peaks in $\psi \to \gamma\chi$, $\chi \to$ hadrons.

to 3·6 GeV have been found in ψ' radiative decays and one therefore immediately wonders if they are indeed the P-wave states 0^{++}, 1^{++}, 2^{++} and possibly also the $2\,^1S_0(\eta_c')$.

Three of these states ($\chi(3410)$, $\chi(3500)$, $\chi(3550)$) have been seen as narrow peaks in the modes (Fig. 16.7)

$$\psi' \to \gamma\chi$$
$$\,\raisebox{.5ex}{$\llcorner$}\!\to 2\pi, 4\pi, \pi\pi K\bar{K} \ldots \qquad\qquad (16.46)$$

Evidence for them is also seen in

$$\psi' \to \gamma\chi$$
$$\,\raisebox{.5ex}{$\llcorner$}\!\to \gamma\psi \qquad\qquad\qquad (16.47)$$

The invariant masses of the $\gamma\psi$ systems are shown in Fig. 16.8 which appears to show the three previous states as well as a new state $\chi(3460)$. How do we know which photon came first and which is to be combined with the ψ to form the χ mass? The ψ' is produced at rest in the

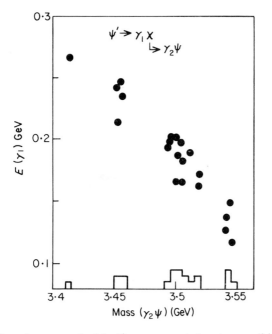

FIG. 16.8. Invariant mass of $\gamma\psi$ in $\psi' \to \gamma\chi$, $\chi \to \gamma\psi$ showing possible evidence for four χ states.

laboratory (e^+e^- centre of mass) and will produce a photon of well-defined energy in association with a given χ state. The χ state is now in motion and on decaying into $\gamma\psi$ produces a photon whose frequency will be Doppler shifted depending upon whether the χ was moving towards or away from the detector. Hence for a χ of zero width, the first photon will have a fixed energy while the second will be spread over an energy range. This effect is already visible in Fig. 16.8 (in particular the $\chi(3460)$ state).

The four χ states are also visible as enhancements in the spectrum of photon energies measured in $\psi' \to \gamma + \text{anything}$. From these data the branching ratios for $\psi' \to \gamma\chi$ have been determined (Feldman, 1976; Whitaker *et al.*, 1976):

$$\text{B.R.}(\psi' \to \gamma\chi(3410)) \simeq (7\cdot5 \pm 2\cdot5) \text{ per cent}$$

$$\text{B.R.}(\psi' \to \gamma\chi(3460)) \lesssim (0\cdot8 \pm 0\cdot4) \text{ per cent}$$

$$\text{B.R.}(\psi' \to \gamma\chi(3500)) \simeq (9 \pm 3) \text{ per cent} \qquad (16.48)$$

$$\text{B.R.}(\psi' \to \gamma\chi(3550)) \simeq (8 \pm 3) \text{ per cent}$$

The existence and quantum numbers of $\chi(3460)$ are still unclear (Chanowitz and Gilman, 1976) and we shall not discuss it further here. The properties of the other χ states are consistent with their assignments as $0^{++}(3410)$, $1^+(3500)$, $2^{++}(3550)$.[1] We will describe the evidence for this and refer the reader to recent reviews for further details (see the Bibliography for this chapter).

The 0^{++}, 1^{++}, 2^{++} states can each be produced by E1 (electric dipole) radiative transitions from the $^3S_1$ ψ'. The phase space for such transitions is proportional to $|\mathbf{p}|^3$ and hence *increases* as the χ mass *decreases*. The width for $\psi' \to \gamma\chi$ is also proportional to the spin J of the χ as follows:

$$\Gamma(\psi' \to \gamma\chi) \propto (2J_x + 1)|\mathbf{p}|^3 \qquad (16.49)$$

Since the widths to each χ state are consistent with being equal, then this suggests that the lowest mass (largest p^3) has the lowest spin etc. and hence the ordering $0^{++}(3410)$, $1^{++}(3500)$, $2^{++}(3550)$. The angular distributions of the photon in the $\psi' \to \gamma\chi$ transitions are also consistent with this assignment (section 16.3.2).

---

[1] Compare with the mass splittings of light mesons, section 5.3.1.

The $\chi(3410)$ and possibly also $\chi(3550)$ are seen to decay into $\pi\pi$ or K$\bar{\text{K}}$ (Fig. 16.7). Parity forbids a state $J^{PC} = 1^{++}$ to decay into two pseudoscalars, so these states are not 1^{++}. The angular distribution of $e^+e^- \to \gamma + \chi(3550)$ appears inconsistent with 0^+ and so we again find consistency with $0^+(3410)$, $2^+(3550)$. The $\chi(3500)$ is not seen decaying into $\pi\pi$ or K$\bar{\text{K}}$ and is consistent with a $J^{PC} = 1^{++}$ assignment for this state.

16.3.2 $e^+e^- \to \psi' \to \gamma\chi$

The angular distribution of the χ production in $e^+e^- \to \psi' \to \gamma\chi$ gives information on their spins and parities.

Returning to equations (12.50) to (12.53) we have

$$\frac{d\sigma}{d\theta} \sim \sum_{m'} |d^1_{1m'}(\theta)|^2 |A_{m'}|^2 \tag{16.50}$$

where $A_{m'}$ is the amplitude for the $\gamma\chi$ system to have net projection of spin $m' = \pm 1, 0$ along the z'-axis ($\gamma\chi$ axis). Invoking parity conservation,

$$\frac{d\sigma}{d\theta} \sim \tfrac{1}{2}(1 + \cos^2\theta)\{|A_1|^2 + |A_{-1}|^2\} + \sin^2\theta |A_0|^2 \tag{16.51}$$

It will be useful to rewrite this in terms not of the total m' but instead in terms of the χ helicity. Since the γ is real then it has helicity ± 1 so we can construct a Table of the helicity amplitudes.

m	λ_γ	λ_χ	Amplitude
1	1	0	$A_0^\uparrow$
	-1	2	$A_2^\downarrow$
-1	1	-2	$A_{-2}^\uparrow$
	-1	0	$A_0^\downarrow$
0	1	-1	$A_{-1}^\uparrow$
	-1	1	$A_1^\downarrow$

$$\tag{16.52}$$

The angular distributions may be rewritten

$$\frac{d\sigma}{d\theta} \sim \tfrac{1}{2}(1+\cos^2\theta)[|A_0^\uparrow|^2 + |A_2^\downarrow|^2 + |A_0^\downarrow|^2 + |A_{-2}^\uparrow|^2]$$

$$+ \sin^2\theta\,[|A_1^\downarrow|^2 + |A_{-1}^\uparrow|^2] \tag{16.53}$$

Then from parity $|A_{-m'}^\uparrow| = |A_{+m'}^\downarrow|$, so finally

$$\frac{d\sigma}{d\theta} \sim (1+\cos^2\theta)[|A_0|^2 + |A_2|^2] + 2\sin^2\theta\,|A_1|^2 \tag{16.54}$$

which describes the angular distribution for $\psi' \to \gamma\chi$ in terms of the amplitudes for χ to have helicity 0, 1 or 2.

For χ with $J = 0$ only A_0 exists so that:

$$0^{++}: \frac{d\sigma}{d\theta} \sim 1 + \cos^2\theta \tag{16.55}$$

For $J_\chi = 1, 2$ we must know the relative importance of the different helicity amplitudes. This is a question of dynamics.

It is stated widely that in the charmonium scheme the resulting angular distributions are

$$
\begin{aligned}
0^{++}&: \quad 1 + \cos^2\theta \\
1^{++}&: \quad 1 - \tfrac{1}{3}\cos^2\theta \\
2^{++}&: \quad 1 + \tfrac{1}{13}\cos^2\theta
\end{aligned}
\tag{16.56}
$$

We shall describe the derivation of these results below. It is interesting to compare these with the data. If $d\sigma/d\theta \sim 1 + A\cos^2\theta$ then (Feldman, 1976, and Perl, 1977)

$$
\begin{aligned}
\chi(3410)&: \quad A = 1\cdot4 \pm 0\cdot4 \\
\chi(3500)&: \quad A = 0\cdot26 \pm 0\cdot5 \\
\chi(3550)&: \quad A = 0\cdot22 \pm 0\cdot4
\end{aligned}
\tag{16.57}
$$

Hence $\chi(3410)$ is consistent with $J^{PC} = 0^{++}$. The 3500 and 3550 are in turn consistent with the 1^{++} and 2^{++} expectations.

In the calculation of $d\sigma/d\theta$ in the charmonium model the essential feature is the assumption that

$$2\,^3S_1(q\bar{q}) \to 1\,^3P_{0,1,2}(q\bar{q}) + \gamma \tag{16.58}$$

where $q \to q\gamma$ or $\bar{q} \to \bar{q}\gamma$ by flipping the L_z (orbital angular momentum projection) of the quark. From this the angular distributions follow at once.

Consider the emission of a photon with $J_{z'} = +1$ from ψ' with $J_{z'} = \pm 1, 0$ (it has $J_z = \pm 1$ only but along z' it can have $J_{z'} = 0$). Then the matrix element, where subscripts denote $J_{z'}$, can be formally written

$$\langle \chi_{0,1,2} | j_+^{\text{e.m.}} | \psi'_{-1,0,1} \rangle \tag{16.59}$$

The quark spin structure contained in this is

$$\langle 1 \; S_{z'} | 0 \; 0 | 1 \; S_{z'} \rangle \tag{16.60}$$

since the photon is not flipping the spin. In the orbital space we have the structure

$$\langle 1 \; 1 | 1 \; 1 | 0 \; 0 \rangle \tag{16.61}$$

since the photon flips L_z and carries $L = 1$ to go from S- to P-wave $q\bar{q}$. Hence the χ has $\langle JJ_{z'} | LL_{z'}; SS_{z'} \rangle$ as

$$\langle J, 1 + S_{z'} | 1 \; 1; 1 \; S_{z'} \rangle \tag{16.62}$$

and so the amplitudes $A_{1+S,z'}^J$ have relative magnitudes given by the ratios of these Clebsch–Gordan coefficients. Consequently squaring these Clebsch–Gordans we have

$$\text{for } J = 2, \quad |A_2|^2 : |A_1|^2 : |A_0|^2 = 6 : 3 : 1 \tag{16.63}$$

$$\text{for } J = 1, \quad |A_1|^2 : |A_0|^2 = 3 : 3 \tag{16.64}$$

Substituting into $d\sigma/d\theta$, the $(1 + A \cos^2 \theta)$ with $A = -\frac{1}{3}$ and $\frac{1}{13}$ arise immediately.

There is no general reason why the L_z flip should dominate. A general approach can be based on the assumptions:

a. $\psi'(2\,^3S_1)\chi(1\,^3P_{0,1,2})$ of $q\bar{q}$,
b. q has spin $\frac{1}{2}$,
c. $q(\bar{q}) \to q(\bar{q})\gamma$ triggers the transition.

The assumption (c) with (b) means that the $\gamma \leftrightarrow q\bar{q}$ has spin 0 or 1 only. Hence (section 7.3)

$$j_+^{\text{e.m.}} = AL_+ + BS_+ + CS_z L_+ + DS_- L_+ L_+ \tag{16.65}$$

is the most general form of $j_+^{\text{e.m.}}$ (S_+, L_+ refer to $\Delta S_z = 1$, $\Delta L_z = 1$ respectively). Since we are interested in S to P transitions then the last term cannot contribute since it has $\Delta L_z = 2$. A, B, C are singlet operators under the spin group and depend upon spatial wavefunctions, potentials, quark masses etc. and are supposed to be essentially the same for all the χ states. The Clebsch–Gordan coefficients for the AL_+ term have been calculated above. By analogous procedure the Clebsch–Gordan coefficients for BS_+ and $CS_z L_+$ can be calculated.

The resulting structure of the helicity amplitudes becomes:

	A_2	$6(A-C)$
$J = 2$	A_1	$3(A-B)$
	A_0	$(A-2B-C)$
$J = 1$	A_1	$3(A+B)$
	A_0	$3(A-C)$
$J = 0$	A_0	$2(A+B-C)$

Without further knowledge of A, B, C we can say very little.

To go beyond these discussions and calculate the absolute widths $\psi' \to \gamma\chi$ requires commitment to a potential model. The calculations will then parallel those of $N^* \to N\gamma$ (Chapter 7). For explicit discussion in the harmonic oscillator potential see Jackson (1976b). A linear potential has been investigated by Eichten *et al.* (1975, 1976) and Barbieri *et al.* (1976c).

16.3.3 THE $c\bar{c}$ POTENTIAL

At short distances when r (the $q\bar{q}$ separation) tends to zero one gluon exchange (OgE) is assumed to be dominant and generate a Coulomb–Fermi–Breit potential. The phenomenological consequences of this are discussed in sections 17.2 and 17.3. At large distances multiple gluon exchange and gluon splittings (Fig. 16.9) are believed to generate an effective confining potential $V_c(r)$. In a world of one space and one time dimension (2D QCD) then $V_c(r) \sim r$, a linearly rising potential. Whether this result also obtains in the real world is not yet known. What can we learn about $V_c(r)$ by studying the spectrum of states?

If $V_c \equiv 0$ then the familiar Coulomb spectrum (Fig. 1.1) would emerge. The phenomenological absence of a continuum (free q or $\bar{q}$ states) shows that $V_c \neq 0$. The harmonic oscillator potential $V_c(r) \sim r^2$ is much loved by quark model devotees (e.g. Chapter 7) and its spectroscopy is shown in Fig. 1.1. Clearly the ψ, χ spectrum is rather similar to this (1p at 3·4–3·6 GeV about midway between 1s at 3·1 and 2s at 3·7 GeV: the 1d at 3·77 is almost degenerate with the 2s ψ'). The 1p and 1d are both slightly higher than in the oscillator spectrum and this may be due to the residual effect of the Coulomb OgE potential (Fig. 1.1) or to nonharmonic $V_c(r)$.

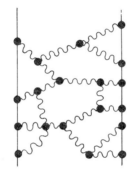

FIG. 16.9. Gluon exchanges which generate a non-Coulombic (confining?) potential.

The ordering of the low-lying levels in a potential $V_c(r)$ has been studied by Martin (1977) and by Grosse (1977). In general $E(1p) > E(1s)$ due to the repulsive centrifugal barrier for $L = 1$ in the Schrödinger equation. $E(1p) < E(2s)$ and $E(1d) < E(2p)$ if

$$\text{i.} \quad \frac{d^3}{dr^3}(r^2 V_c) > 0 \quad \text{for all } r \tag{16.66}$$

$$\text{ii.} \quad \lim_{r \to 0}\left[2rV_c + r^2 \frac{dV_c}{dr} \right] = 0 \tag{16.67}$$

and so follow for $N > 0$ if $V(r) \sim r^N$. The Coulomb potential ($N = -1$) has $E(1p) = E(2s)$ and $E(1d) = E(2p)$.

We have noted that it phenomenologically appears that $E(2s) \lesssim E(1d)({}^3S_1(3684)$ and ${}^3D_1(3772))$. This follows if, in addition to the constraints in equations (16.66) and (16.67), one also has

$$\frac{d}{dr}\left[\frac{1}{r}\frac{d}{dr}\left(2V_c+r\frac{dV_c}{dr}\right)\right]<0 \quad \text{for all } r \qquad (16.68)$$

If $V \sim r^N$ then this requires $0 < N < 2$. The harmonic oscillator potential with $N = 2$ has the 1d and 2s degenerate.

The spectroscopies of the harmonic oscillator and linear potentials are compared with the emerging data in Fig. 16.10. In this Figure the

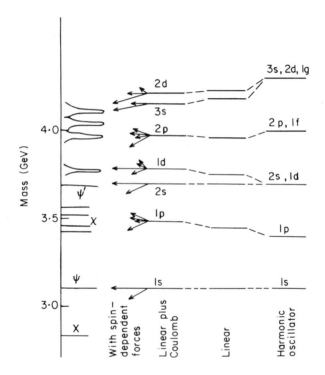

FIG. 16.10. Comparison of harmonic oscillator and linear potentials with the emerging spectra of $c\bar{c}$ states.

levels have been scaled to give the observed $\psi' - \psi$ mass difference for the $2s - 1s$ interval. It is clear that only above 4 GeV does a significant difference emerge among the potentials. When we take the linear plus Coulomb potentials and incorporate the spin-dependent splittings (Chapter 17) the pattern of levels will be as indicated by the arrows.

We have followed the evocative procedure of Jackson (1977) by employing long arrows to represent $J^{PC} = 1^{--}$ states (directly produced in e^+e^- annihilation), the middle length arrows to represent states with $C = +1$ that can be reached by single photon emission from the 1^{--} states and the short arrows to indicate $C = -1$, $J^P \neq 1^-$ states (which are difficult to observe in e^+e^- annihilation).

The states indicated by the long and medium arrows in the energy range 2·7 to 3·8 GeV appear to have been found (see this chapter's bibliography for details), namely $1\ ^1S_0(2·8?)$, $1\ ^3S_1(3·1)$, $1\ ^3P_0(3·41)$, $1\ ^3P_1(3·50)$, $1\ ^3P_2(3·55)$, $2\ ^1S_0(3·46??)$, $2\ ^3S_1(3·68)$, $1\ ^3D_1(3·77)$. In this entire mass range it is possible that all but the $1\ ^1P_1$ state have been discovered[1] within less than three years of the initial discovery of the J/ψ (3·1 GeV). Possible ways that one might hope to isolate the $1\ ^1P_1$ state are discussed in Segre and Weyers (1976).

The $1\ ^1D_2$ state may also be expected to exist in or near to this mass range. If its mass is greater than 3·8 GeV then it will decay strongly into charmed hadrons and will not easily be visible. It has been suggested that a large spin–spin splitting might have caused this state to exist as low as 3·46 GeV and hence to be the $\chi(3·46)$ (Harari, 1976b). Possible problems with this interpretation of $\chi(3·46)$ have been raised by Jackson (1977).

Information on the potential *at the origin* can be obtained in principle from a study of the leptonic widths (equation 16.12). In practice this will be difficult since the ψ at 3772 MeV and at 3684 MeV are so near in mass that one expects them to be mixtures of $2\ ^3S_1$ and $1\ ^3D_1$. Since the wavefunction at the origin vanishes for a D-wave state then naively one expects the $^3D_1$ to decouple from the lepton decay

$$\Gamma(^3D_1 \to e^+e^-) \simeq 0$$

The leptonic width of 3772 MeV is about 0·4 keV in contrast to the 2·1 keV for $\psi(3684)$. These suggest that the $\psi(3684)$ is dominantly $2\ ^3S_1$ with a small $1\ ^3D_1$ admixture. The conclusions are complicated, however, by the fact that the threshold for producing charmed mesons (section 16.4) is at 3730 MeV and so these states will also couple to $D\bar{D}$ (charmed mesons) which affects the $\Gamma^{e\bar{e}}$ calculation (Eichten *et al.*, 1976).

---

[1] Problems with a $^1S_0$ assignment for $\chi(3·46)$ are raised by Chanowitz and Gilman (1976).

In addition to these complications one must also recognise that the
$^3D_1$ can couple to e^+e^- through a nonvanishing *second derivative* of the
wavefunction at the origin (Jackson, 1977)

$$\Gamma(^3D_1 \to e^+e^-) = \frac{4\alpha^2 e_q^2}{M^2} \left| \frac{R_D''(0)}{M^2} \right|^2$$

(compare equation 16.12 with three colours).

16.3.31 *Mass dependence of the potential and leptonic widths*

The leptonic decay of a vector meson, $V \to l^+l^-$, occurs as a result of the
q and $\bar{q}$ annihilating and producing a virtual photon. The annihilation
amplitude will be proportional to $R_s(0)$, the $q\bar{q}$ wavefunction at zero
separation, and also the quark charge.

First consider the latter contribution. For mesons containing a single
flavour of quark then

$$\langle q_i \bar{q}_i | \gamma | 0 \rangle \propto e_i = \begin{cases} \frac{2}{3} : \psi(c\bar{c}) \\ -\frac{1}{3} : \phi(s\bar{s}) \end{cases}$$

For isospin states like the ρ and ω

$$\left\langle \frac{u\bar{u} \pm d\bar{d}}{\sqrt{2}} \middle| \gamma \middle| 0 \right\rangle = \frac{1}{\sqrt{2}} \left(\frac{2}{3} \pm \left(-\frac{1}{3} \right) \right) = \begin{cases} \dfrac{1}{3\sqrt{2}} : \omega \\ \dfrac{1}{\sqrt{2}} : \rho \end{cases}$$

Hence if we regard the mesons as containing effective quark charges e_q
the leptonic width will be proportional to $(e_q)^2$ so that $\frac{4}{9}, \frac{1}{9}, \frac{1}{18}, \frac{1}{2}$ for ψ, ϕ, ω
and ρ.

If we neglect the masses of the leptons then the width will be

$$\Gamma(V \to l^+l^-) = 4\alpha^2 e_q^2 \frac{|R_s(0)|^2}{M_V^2}$$

That this expression is plausible may be seen as follows. The $|R_s(0)|^2$ is a
probability density and therefore has dimensions of $(\text{mass})^3$. The width
has dimensions of mass, hence an $[M^{-2}]$ factor is needed. The vector
meson mass is the only mass scale in the problem and hence $[M^{-2}] = M_V^{-2}$.

If ρ, ω, ϕ, ψ, all had the same mass then their leptonic widths would be in the ratio of the quark charges and hence $9:1:2:8$. As their masses are in fact rather different, their leptonic widths will be a function of the dependence of $R_s(0)$ on the quark masses. This will be a function of the potential at the origin, in particular how the potential's short-range behaviour causes the wavefunction to scale as a function of quark mass.

For a potential $V(r) \sim r^N$ then $|R_s(0)|^2$ scales with the quark mass as $m^{3/(2+N)}$. Taking the quark mass to be proportional to the vector meson mass:

$$\frac{\Gamma(V \to l^+ l^-)}{e_q^2} \propto \left[\frac{1}{M_V}\right]^{(2N+1)/(N+2)} \tag{16.69}$$

Before confronting this with the data we should derive it in order to exhibit the implicit assumptions.

The (coloured) quark is the source of a (chromodynamic) field. We shall suppose that the resulting interquark potential is independent of the masses and velocities and is a function only of their spatial separation. Consequently this is a nonrelativistic picture. The Schrödinger equation may be written

$$\left\{\frac{\nabla^2}{2m} + V(r)\right\} \psi(r) = E\psi(r)$$

where $V = Ar^N$ and A is a constant, independent of mass. Now scale all lengths $r \to \lambda r$ and hence

$$\left\{\frac{\nabla^2}{2m\lambda^2} + A\lambda^N r^N\right\} \psi(\lambda r) = E\psi(\lambda r)$$

or equivalently

$$\{\nabla^2 + A2m\lambda^{N+2} r^N\}\psi(\lambda r) = 2m\lambda^2 E\psi(\lambda r)$$

Imagine that we had already solved this equation for some mass m_1 and wished now to solve it for another mass m_2. We could do this immediately by scaling out the explicit mass dependence. Thus from the potential we can see that the distances scale as $m^{-1/(2+N)}$ and

$$\{\nabla^2 + Ar^N\}\psi(\lambda r) \sim 2m^{N/(2+N)} E\psi(\lambda r)$$

which implies that energies scale as $m^{-N/(2+N)}$, hence like m for a Coulomb, $m^{-1/3}$ for linear and $m^{-1/2}$ for harmonic oscillator potentials. If the spectra of Fig. 1.1 were normalised to agree with the $\psi' - \psi$

separation of about $600\,\text{MeV}$ where the $m_c \simeq 1 \cdot 5 (\text{GeV})$, then the separation for $\Upsilon' - \Upsilon (m_b \simeq 4 \cdot 5\,\text{GeV} \simeq 3 m_c)$ will be given by this same figure but with the abscissa scaled by $3, 3^{-1/3}, 3^{-1/2}$ respectively.

There is some evidence to suggest that $\Upsilon(b\bar{b}?)$ exists with a mass approximately three times that of the ψ and also that $\delta m (\Upsilon' - \Upsilon) \simeq \delta m (\psi' - \psi)$ (Innes et al., 1977). Equal spacing independent of mass can be achieved if one has a combination of Coulomb and linear (or oscillator) potentials, or if one has a logarithmic potential. That a logarithmic potential gives energy levels whose separations are independent of the mass scale can be seen immediately since $r \to \lambda r$ in the potential just shifts the logarithm:

$$V(\lambda r) = \log (\lambda r) = \log r + \log \lambda = V(r) + \text{constant}$$

and so the energy levels are shifted by a constant amount when m_1 is replaced by m_2.

Since distances scale as $m^{-1/(2+N)}$ in the r^N potential then the probability density $|R_s(0)|^2 \sim (\text{length})^{-3}$ will scale as $m^{3/(2+N)}$. Therefore equation (16.69) follows immediately.

Since we have explicitly neglected mass-dependent potentials (such as the magnetic contributions in the Fermi–Breit Hamiltonian—equation 17.35) then a test of this equation needs a nonrelativistic situation. This may have some chance of success in the case of massive quarks such as are relevant in the ψ and Υ spectroscopies. However it is almost certainly not applicable to the light quarks present in the $\rho \omega$ and ϕ. If one ignores this caveat and applies it to $\rho \omega \phi$ and ψ then the data suggest that $\Gamma^{e^+e^-}/e_q^2$ is almost mass independent (a best fit suggests $N \sim 0$ to $-\frac{1}{2}$ in $V \sim r^N$ at short distances).

16.4   Charmed particles

The three quark flavours uds formed a fundamental triplet representation of SU(3) whose weight diagram in $I_3 - Y$ space was a triangle (Fig. 2.3). Incorporation of a fourth quark, c, with charge $\frac{2}{3}$, isospin and strangeness zero and carrying one unit of charm, generates the fundamental quartet representation of SU(4). The weight diagram is now a pyramid (Fig. 16.11).

In SU(2) the weight diagram is simply a line with two ends. The three sides of the SU(3) triangle show the three SU(2) subgroups

(I, U, V) contained within the fundamental SU(3) representation. Similarly the four triangles forming the sides of the pyramid show the SU(3) subgroups (uds), (udc), (dsc), (usc) contained in the basic SU(4).

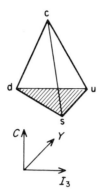

FIG. 16.11. Fundamental representation of SU(4).

16.4.1 MESONS

In SU(3) the mesons formed from $q\bar{q}$ fell into nonets (singlets and octets):

$$\mathbf{3} \otimes \bar{\mathbf{3}} = \mathbf{1} \oplus \mathbf{8}$$

With the new quark generating SU(4) we now have

$$\mathbf{4} \otimes \bar{\mathbf{4}} = \mathbf{1} \oplus \mathbf{15}$$

and so seven new states should exist. Clearly these are:

i. Three states with charm $+1$ forming a $\bar{\mathbf{3}}$ of SU(3)

$$\left.\begin{array}{ll} c\bar{u} & (D^0) \text{ or } (D^{0*}) \\ c\bar{d} & (D^+) \text{ or } (D^{+*}) \\ c\bar{s} & (F^+) \text{ or } (F^{+*}) \end{array}\right\} \tag{16.70}$$

ii. Three states with charm -1 forming a $\mathbf{3}$ of SU(3)

$$\left.\begin{array}{ll} \bar{c}u & (\bar{D}^0) \text{ or } (\bar{D}^{0*}) \\ \bar{c}d & (D^-) \text{ or } (D^{-*}) \\ \bar{c}s & (F^-) \text{ or } (F^{-*}) \end{array}\right\} \tag{16.71}$$

iii. A state with charm zero ("hidden charm") which is a singlet of SU(3):

$$c\bar{c} \quad (\eta_c) \text{ or } (\psi) \tag{16.72}$$

This latter state could be mixed in with the SU(3) singlet components of η and η' (or ω and ϕ in the vector case) or it can exist separately. In the former case η or η' (ω or ϕ) would necessarily have some $c\bar{c}$ content themselves.

The meson multiplets are shown diagrammatically for 0^- and 1^- in Fig. 16.12.

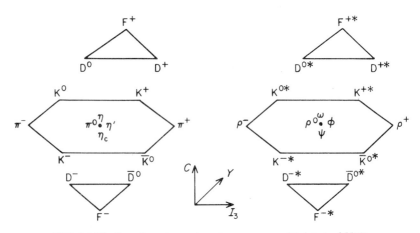

FIG. 16.12. Pseudoscalar and vector meson multiplets in SU(4).

16.4.2 BARYONS

The qqq baryons in SU(3) formed multiplets with dimensions

$$3 \otimes 3 \otimes 3 = 1 \oplus 8 \oplus 8 \oplus 10$$

The Young diagram techniques immediately show that in SU(4) the dimensions will be

$$4 \otimes 4 \otimes 4 = 4 \oplus 20 \oplus 20 \oplus 20 \tag{16.73}$$

Hence the octets gain twelve charmed partners while the symmetric decuplet gains another ten. There are now four antisymmetric possibilities in place of the one in SU(3). The nature of these new states can be readily seen by explicit construction.

With charm of one there are six possibilities, three of which are manifestly symmetric in the two noncharmed quarks and three which can be either symmetric or antisymmetric in that pair (compare Table 3.6). The symmetric states form the set in the first column of Table 16.1. There are three states with charm of two units and one state with charm three (analogue of Ω^- in the strange quark world).

TABLE 16.1
Charmed $C = +1$ baryons

Σ_c^{++}	cuu			
Σ_c^+	c(ud + du)	Λ_c^+	c(ud − du)	strangeness 0
Σ_c^0	cdd			
S^+	c(us + su)	A^+	c(us − su)	strangeness −1
S^0	c(ds + sd)	A^0	c(ds − sd)	
T^0	css			strangeness −2
	6		$\bar{3}$	

As we have already discussed in Chapter 3, any state can be written in a totally symmetric fashion. Hence there are 10 symmetric possibilities to be added to the old symmetric uncharmed decuplet and so there results a 20plet of states in SU(4). This forms a pyramid with four floors, charm zero, 1, 2, 3. These floors contain triangles with respectively ten members (charm zero), six (charm 1), three (charm 2) and one (charm 3). Hence

$$20_S = 10 \oplus 6 \oplus 3 \oplus 1 \qquad (16.74)$$

is the SU(3) decomposition of the symmetric 20plet of SU(4). The lowest mass examples of these states will be the $\frac{3}{2}^+$ partners of the Δ.

In Tables 3.2 and 3.7 we have noted that a mixed symmetry wavefunction can be written for a three-body system if at least one label is distinct from the other two. This is true of the charm = +1 sextet and of the charm = +2 triplet. Three of the charm = +1 states have all three labels distinct and hence give three further mixed states. Hence

$$20_M = 8 \oplus (6 \oplus \bar{3}) \oplus 3 \qquad (16.75)$$

is the SU(3) decomposition of the mixed symmetry 20plet of SU(4) which contains the familiar octet representation of SU(3) in the uncharmed states. Note that the **6** and **$\bar{3}$** are both charmed $+ 1$; there is no charm $+ 3$ state here.

At present there is no universally accepted nomenclature for these particles. One system due to Gaillard *et al.* (1974) is for the strangeness zero:

$$\text{cuu, c(ud + du), cdd are } C_1^{++}, C_1^+, C_1^0$$

$$\text{c(ud - du) is } C_0^+$$

Here C stands for charm and the subscript denotes the isospin of the u, d diquark system. An alternative system is to call these states Σ_c^{++}, Σ_c^+, Σ_c^0 and Λ_c^+ by analogy with their strange counterparts. We use this scheme in this book.

For the strangeness $- 1$ states with charm $+ 1$ we have S^+S^0 and A^+A^0 for the symmetric or antisymmetric noncharmed pair respectively. Finally the strangeness minus two is called T^0.

The three states with $c = +2$ are named X_u^{++}, X_d^+, X_s^+, the subscript denoting the flavour of the uncharmed quark. The former pair are sometimes named Ξ_c^{++} and Ξ_c^+ having two charmed quarks and being analogues of the familiar strangeness minus two Ξ states of noncharmed spectroscopy.

Pictorially these states are shown in Fig. 16.13. The familiar octet hexagon of (uds) SU(3) is exhibited and charm increases vertically in

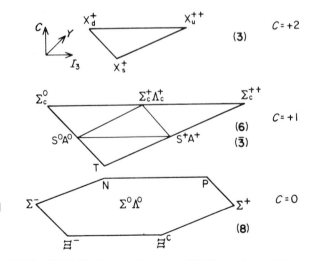

FIG. 16.13. SU(4) 20plet containing the SU(3) noncharmed baryon octet.

the three-dimensional Figure. With charm $+1$ we have the **6** and $\bar{\mathbf{3}}$ immediately above the octet and highest with charm $+2$ is the triplet (X_{uds}).

We now complete the Figure to form the truncated tetrahedron in the three-dimensional plot of I_z, Y and C (weight diagram) which is essentially Fig. 16 of Gaillard *et al.* (1975). This is displayed in Fig. 16.14.

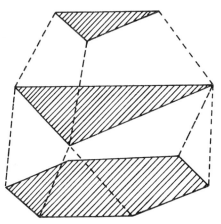

FIG. 16.14. Truncated tetrahedron weight diagram for the 20plet of SU(4) containing the nucleon octet.

The truncated tetrahedron has four hexagonal faces each one representing one of the four SU(3) subgroups of SU(4) formed by uds, udc, usc, cds. In Fig. 16.15 we explode the truncated tetrahedron to manifestly expose these hexagons, the three-quark subsystem associated with each one also being shown.

Finally the three charm $+1$ states with three different quark labels can combine with the charm zero uds to form four totally antisymmetric states. The lowest mass examples of these will be the $\frac{1}{2}^-$, $\frac{3}{2}^-$ partners of $\Lambda(1405$ and $1520)$. The weight diagram is in Fig. 16.16.

16.5   Weak decays of charmed hadrons

We noted in section 15.1 that the weak interaction charged current involving the charmed quark triggered transitions

$$c \leftrightarrow s \cos \theta_c - d \sin \theta_c \tag{16.76}$$

in contrast to the analogous transitions for up quarks

$$u \leftrightarrow d \cos \theta_c + s \sin \theta_c \qquad (16.77)$$

where θ_c is the Cabibbo angle $\simeq 0\cdot23$ radians. We can refer to the $\cos \theta_c$ amplitude as Cabibbo allowed and the $\sin \theta_c$ as Cabibbo forbidden since

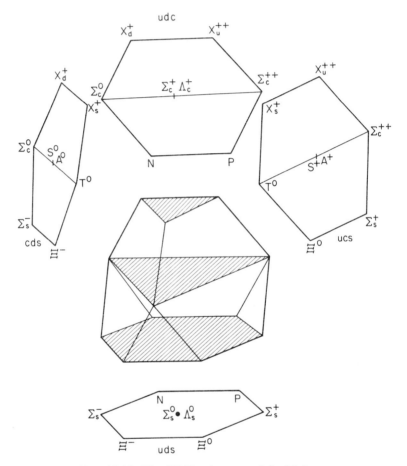

FIG. 16.15. The SU(3) subgroups of the 20plet.

in the rates $\tan^2 \theta_c \simeq \frac{1}{25}$. We see then that the hadronic selection rules for the charm changing transitions are:

i. Cabibbo allowed $c \leftrightarrow s$:

$$\Delta Q = -\Delta S = \Delta C; \qquad \Delta I = 0 \qquad (16.78)$$

ii. Cabibbo forbidden $c \leftrightarrow d$:

$$\Delta Q = \Delta C, \qquad \Delta S = 0; \qquad \Delta I = \tfrac{1}{2} \qquad (16.79)$$

Since the dominant decay of a charmed quark is into a strange quark, then strange particles should be an important feature of the charmed

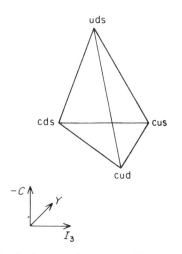

FIG. 16.16. Antisymmetric quartet of baryons in SU(4).

decay products. For example

$$D^0(c\bar{u}) \xrightarrow{\cos\theta} K^-(s\bar{u}) + \pi^+ \qquad (16.80)$$

will dominate over the Cabibbo suppressed

$$D^0(c\bar{u}) \xrightarrow{\sin\theta} \pi^-(d\bar{u}) + \pi^+ \qquad (16.81)$$

Indeed a clear peak in the $K^-\pi^-$ mass distribution has been seen at SPEAR in $e^+e^- \to K^-\pi^+ +$ hadrons with a mass of $1\cdot863$ GeV. This state may also have been seen decaying into $\pi^-\pi^+$ ($6\cdot5 \pm 4$) per cent as often as into $K^-\pi^+$. In the above theory of the charmed weak interactions one expects

$$\frac{\Gamma(D^0 \to \pi^+\pi^-)}{\Gamma(D^0 \to K^-\pi^+)} = \tan^2\theta_c \times \text{phase space} \simeq 4 \text{ or } 5 \text{ per cent}$$

$$(16.82)$$

This neutral state can also decay

$$D^0(c\bar{u}) \xrightarrow{\cos\theta} K^0(s\bar{d}) + \pi^-(d\bar{u}) + \pi^+ \tag{16.83}$$

and this decay has also been observed. The observation of D^0 decaying into two pseudoscalars ($K\pi$) and also three pseudoscalars ($K\pi\pi$) shows that parity is violated in the decay (analogous to the old $\theta - \tau$ puzzle in $K \to 2\pi$ and 3π). This confirms the weak interaction nature of the decay mode.

A very clear pointer to the weak current equation (16.76) comes from the decay of the charged partner of the D^0. Here

$$D^+(c\bar{d}) \xrightarrow{\cos\theta} K^0(s\bar{d}) + \pi^+ \tag{16.84}$$

or alternatively

$$D^+(c\bar{d}) \xrightarrow{\cos\theta} K^-(s\bar{u}) + \pi^+(u\bar{d}) + \pi^+ \tag{16.85}$$

This latter decay pattern is particularly remarkable in that a negatively charged K is produced from a positively charged initial state. This is quite distinct from the decay of a K^{*+} which would produce a K^+. Experimentally

$$D^+ \to K^-\pi^+\pi^+ \tag{16.86}$$

has been clearly seen with a mass of 1.868 GeV. The similar final state $K^+\pi^+\pi^-$ shows no enhancement at this mass and these observations all point at the veracity of the $c \leftrightarrow s$ transition.

The above data and further studies (Nguyen *et al.*, 1977) support these as being the 0^- mesons

$$\begin{aligned}
D^0(c\bar{u}) \quad & 1863.3 \pm 0.9 \text{ MeV} \\
D^+(c\bar{d}) \quad & 1868.3 \pm 0.9 \text{ MeV}
\end{aligned} \tag{16.87}$$

By studying the data on $e^+e^- \to D + $ other hadrons, evidence for the possibility of the vector partners has been found. Their masses are (Perruzi *et al.*, 1977)

$$\begin{aligned}
D^{0*} \quad & 2006 \pm 1.5 \text{ MeV} \\
D^{+*} \quad & 2009 \pm 1 \text{ MeV}
\end{aligned} \tag{16.88}$$

The separation in mass of D^* and D is some 140 MeV, noticeably less than K^* and K or ρ and π. This phenomenon is expected in the hyperfine splittings that arise in quantum chromodynamics (Chapter 17).

To complete the charmed meson picture we need the states $F^+(c\bar{s})$ and $F^{*+}(c\bar{s})$. Having replaced a $\bar{u}$ or $\bar{d}$ by an $\bar{s}$ one anticipates that these states will be about 150 MeV more massive than the D and D^* and hence about 2·0 and 2·1 GeV in mass. Possible evidence for $F(2·03 \pm 0·06)$ decaying into $\eta\pi$ has been reported by Brandelik *et al.* (1977a). These authors also claim evidence for $F^*(2·14 \pm 0·06)$.

Bibliography

This is a very active experimental and theoretical field at present and is likely to remain so for several years to come. Some phenomenological reviews with detailed experimental references are Feldman and Perl (1975, 1977), Schwitters and Strauch (1976), Wiik and Wolf (1977).

17   Hadron Mass Shifts in Field Theories

17.1   Fine and hyperfine structure in atomic and quark physics

In the hydrogen atom the massive proton sits at the centre of mass. The coupling of the electron spin with its orbital angular momentum about the centre of mass is called spin–orbit or $\mathbf{S} \cdot \mathbf{L}$ coupling. The strength is proportional to the inverse square of the electron's mass and generates *fine structure* splittings of the energy levels (see any standard atomic physics or quantum mechanics textbook).

The coupling of the electron spin with the proton spin is known as spin–spin or $\mathbf{S} \cdot \mathbf{S}$ coupling. This strength is proportional to $(m_e m_p)^{-1}$ and so is much smaller than the strength of $\mathbf{L} \cdot \mathbf{S}$ couplings. The resulting splittings in the energy levels are called *hyperfine structure*.

In the S-wave $\langle \mathbf{S} \cdot \mathbf{L} \rangle = 0$ and the fine structure is absent. The electron and proton couple their spins to $S = 0$ or 1 and the $\langle \mathbf{S} \cdot \mathbf{S} \rangle$ in these two states differs giving two energy levels, $^3S_1$ and $^1S_0$. Radiative transitions between these levels yield the famous 21 cm wavelength signal.

In higher waves, P, D, F . . . , both fine and hyperfine structure occur leading to a rich spectrum.

In positronium the $\langle \mathbf{S} \cdot \mathbf{S} \rangle$ and $\langle \mathbf{L} \cdot \mathbf{S} \rangle$ have comparable magnitudes since there is no massive proton suppressing the former relative to the latter. Similarly in quarkonium (the $q\bar{q}$ mesons) one may expect analogous splittings to occur. We have made some phenomenological discussion on the spin–orbit splittings in section 5.3; here we will concentrate on the quantitative aspects of the hyperfine or spin–spin splittings.

The splittings arising from quantum electrodynamics, i.e. photon exchange between the quarks, will be at most a few MeV. These electromagnetic effects will generate mass differences between $\pi^{\pm} - \pi^{0}$, $K^{+} - K^{0}$ etc. and are discussed in section 17.5. However, they cannot explain the large separation in mass of π and ρ. If the quantum chromodynamics picture is correct then vector–gluon exchange will generate hyperfine splittings proportional to the quark–vector-gluon coupling. If this coupling is much stronger than the electromagnetic, then the $\mathbf{S} \cdot \mathbf{S}$ splittings could easily be tens or hundreds of MeV. Indeed, typical $\mathbf{S} \cdot \mathbf{S}$ splittings ($\rho - \pi$, $K^{*} - K$) do appear to be quite sizeable, and we may hypothesise that they have their origin in the vector–gluon exchange.

In section 15.2 we saw that the qualitative behaviours of the $^{3}S_{1} - {}^{1}S_{0}$ and $\frac{3}{2}^{+} - \frac{1}{2}^{+}$ splittings of mesons and baryons could be understood in quantum chromodynamics, and noticed in particular the crucial role played by the colour. We will now study some of the quantitative aspects of this model.

17.2 Spin-dependent mass splittings for baryons

17.2.1 THE $\Delta - N$ SYSTEM

The Δ and nucleon N are made of three quarks $q(\equiv u, d)$ in overall S-wave with spins coupled to $\frac{3}{2}$ or $\frac{1}{2}$ respectively. The mass separation of 300 MeV between these states is hypothesised to be a manifestation of a quantum chromodynamic hyperfine splitting. The splitting will be proportional to the product of the quarks' colour-magnetic moments defined in analogy to their electromagnetic moments. As discussed in section 15.2 the interaction (with correct sign) has then the form

$$\mathcal{H}_{S.S} = -c^{2} \sum_{j \neq k} \mathbf{F}_{j} \cdot \mathbf{F}_{k} \mathbf{S}_{j} \cdot \mathbf{S}_{k} / \varepsilon_{j} \varepsilon_{k} \tag{17.1}$$

where $\mathbf{F}, \mathbf{S}$ are $SU(3)$ colour and $SU(2)$ spin matrices respectively, and ε the effective mass of the quark. c^{2} is a positive constant with dimensions mass cubed, related to wavefunction overlap (see section 17.3), and will not enter further into our discussion.

If we set $\varepsilon_d \equiv \varepsilon_u$ then for the $\Delta - N$ system we have

$$\mathcal{H}_{S.S} = \frac{2c^2}{3\varepsilon_u^2}(\mathbf{S}_1 . \mathbf{S}_2 + \mathbf{S}_1 . \mathbf{S}_3 + \mathbf{S}_2 . \mathbf{S}_3) \qquad (17.2)$$

since $\langle \mathbf{F}_j . \mathbf{F}_k \rangle = -\frac{2}{3}$ (equation 15.39) for each and every j, k. By exploiting

$$\mathbf{S}_1 . \mathbf{S}_2 + \mathbf{S}_1 . \mathbf{S}_3 + \mathbf{S}_2 . \mathbf{S}_3 \equiv \tfrac{1}{2}\{(\mathbf{S}_1 + \mathbf{S}_2 + \mathbf{S}_3)^2 - \mathbf{S}_1^2 - \mathbf{S}_2^2 - \mathbf{S}_3^2\}$$

we have that

$$\mathbf{S}_1 . \mathbf{S}_2 + \mathbf{S}_1 . \mathbf{S}_3 + \mathbf{S}_2 . \mathbf{S}_3 \equiv \tfrac{1}{2}(S_t(S_t+1) - \tfrac{9}{4}) = \left\{ \begin{array}{l} \tfrac{3}{4} : S_t = \tfrac{3}{2} \\ -\tfrac{3}{4} : S_t = \tfrac{1}{2} \end{array} \right. \qquad (17.3)$$

where S_t is the total spin of the three-quark system. Hence

$$\Delta E_M \equiv \langle \mathcal{H}_{S.S} \rangle_{\frac{3}{2}^+ - \frac{1}{2}^+} = \frac{c^2}{\varepsilon_u^2} \equiv \mu_u^2 \text{(definition)} \qquad (17.4)$$

17.2.2 THE $\Lambda\Sigma\Sigma^*$ SYSTEM

We will now compare this $\Delta - N$ splitting with the $\Lambda\Sigma\Sigma^*$ splittings. If we replace one of the quarks in the Δ by a quark of a different flavour, i, where $i = s, c \ldots$ then we obtain $\Sigma_i^*(iqq)$. The nucleon has the quarks pairwise in either $I = 1$ or 0, and upon replacing the third quark by i yields respectively $\Sigma_i(iqq)$ or $\Lambda_i(iqq)$ states.

For the choice $i \equiv s$ we have the familiar states

$$\Sigma_s^*(sqq) \,(1385); \qquad \Sigma_s(sqq) \,(1193); \qquad \Lambda_s(sqq) \,(1115) \quad (17.5)$$

We see that the act of substituting a strange quark for a u or d quark has:

1. Increased the mass of the three-quark system by around 150–200 MeV.
2. Decreased the $\frac{3}{2}^+ - \frac{1}{2}^+$ mass splittings.
3. Split the $\Lambda(I = 0)$ and $\Sigma(I = 1)$ states.

It is easy to see qualitatively how these three effects are related. For the strange quark

$$\varepsilon_s \simeq \varepsilon_u + (150–200 \text{ MeV}) \qquad (17.6)$$

thus the increase in mass relative to N or Δ. Since the hyperfine splittings are inversely proportional to the quark masses, then the $\Sigma^* - (\Sigma\Lambda)$ will be in turn less split than the $\Delta - N$. Finally, the overall spin–unitary spin symmetry of the wavefunction (section 4.2 and Table 4.3) requires that the nonstrange pair in Λ have spin 0 and in the Σ have spin 1. This will lead to different spin–spin expectation values for the two states and hence to different masses.

We will now study this quantitatively. For ease of notation we write $\mu_i \equiv c/\varepsilon_i$ for any quark, and hence

$$\mathscr{H}_{S.S} = -\sum_{j \neq k} \mathbf{F}_j \cdot \mathbf{F}_k \mathbf{S}_j \cdot \mathbf{S}_k \mu_j \mu_k \tag{17.7}$$

Since $\langle \mathbf{F}_j \cdot \mathbf{F}_k \rangle = -\frac{2}{3}$ for any pair of quarks, we will study

$$\mathscr{H}_{S.S} = +\frac{2}{3} \sum_{j \neq k} \mu_j \mu_k \mathbf{S}_j \cdot \mathbf{S}_k \tag{17.8}$$

Imagine the $\Sigma_i \Sigma_i^*$ or Λ_i made of a quark flavour $i = s, c \ldots$ and a pair of quarks ("diquark") of flavours u or d. Taking quark number 1 to be that with flavour i then the spin–spin interaction may be written

$$\tfrac{3}{2}\mathscr{H}_{S.S} = \mu_q^2 \mathbf{S}_2 \cdot \mathbf{S}_3 + \mu_q \mu_i \mathbf{S}_1 \cdot (\mathbf{S}_2 + \mathbf{S}_3) \tag{17.9}$$

where μ_q denotes either of μ_u or μ_d which we shall assume to be equal by isospin symmetry of the vector gluon–quark interaction. To evaluate the expectation values of these products of spin operators we write

$$\mathbf{S}_{tot} \equiv \mathbf{S}_1 + \mathbf{S}_2 + \mathbf{S}_3 \tag{17.10}$$

and square it. This gives

$$\mathbf{S}_{tot}^2 \equiv \mathbf{S}_1^2 + (\mathbf{S}_2 + \mathbf{S}_3)^2 + 2\mathbf{S}_1 \cdot (\mathbf{S}_2 + \mathbf{S}_3) \tag{17.11}$$

and so

$$2\mathbf{S}_1 \cdot (\mathbf{S}_2 + \mathbf{S}_3) \equiv \mathbf{S}_{tot}^2 - \mathbf{S}_1^2 - \mathbf{S}_{diquark}^2 \tag{17.12}$$

The desired expectation value is now immediately found to be

$$\langle 2\mathbf{S}_1 \cdot (\mathbf{S}_2 + \mathbf{S}_3) \rangle = S_{tot}(S_{tot}+1) - \tfrac{1}{2}(\tfrac{1}{2}+1) - S_d(S_d+1) \tag{17.13}$$

where S_{tot} and S_d are the magnitudes of the total and diquark spin angular momenta.

The expectation value of $\mathbf{S}_2 \cdot \mathbf{S}_3$ is similarly obtained by noting that

$$\mathbf{S}_d^2 \equiv (\mathbf{S}_2 + \mathbf{S}_3)^2 = \mathbf{S}_2^2 + \mathbf{S}_3^2 + 2\mathbf{S}_2 \cdot \mathbf{S}_3 \qquad (17.14)$$

Hence

$$2\langle \mathbf{S}_2 \cdot \mathbf{S}_3 \rangle = S_d(S_d + 1) - 2 \cdot \tfrac{1}{2}(\tfrac{1}{2} + 1) \qquad (17.15)$$

We now combine equations (17.13) and (17.15) with equation (17.9) to obtain the total spin–spin contribution to the energy (mass) of the system. It reads

$$\frac{3}{2} E_{\text{s.s}} = \mu_q^2 \left(\frac{S_d(S_d + 1)}{2} - \frac{3}{4} \right) + \tfrac{1}{2}\mu_q\mu_i \left(S_t(S_t + 1) - S_d(S_d + 1) - \frac{3}{4} \right) \qquad (17.16)$$

where $S_t \equiv S_{\text{total}} = \tfrac{1}{2}$ or $\tfrac{3}{2}$ for Σ, Λ or Σ^* respectively and $S_d \equiv S_{\text{diquark}} = 0$ or 1 for Λ or Σ, Σ^* respectively (recall that the spin–isospin symmetry of the wavefunctions—Chapter 4—requires the $I = 0(1)$ diquark to have spin $= 0(1)$ in the totally symmetric 56plet and so the Σ and Λ have differing diquark spins). Hence we have

$$\tfrac{3}{2} E_{\text{s.s}} \left\{ \begin{matrix} \Sigma^* \\ \Sigma \\ \Lambda \end{matrix} \right\} = \mu_q^2 \left\{ \begin{matrix} \tfrac{1}{4} \\ \tfrac{1}{4} \\ -\tfrac{3}{4} \end{matrix} \right\} + \mu_q\mu_i \left\{ \begin{matrix} \tfrac{1}{2} \\ -1 \\ 0 \end{matrix} \right\} \qquad (17.17)$$

The mass splitting are therefore

$$\Sigma_i^* - \Sigma_i = \mu_q\mu_i \qquad (17.18)$$

$$\Sigma_i - \Lambda_i = \tfrac{2}{3}\mu_q(\mu_q - \mu_i) \qquad (17.19)$$

(where the particle symbols here refer also to their masses), while the combination

$$\frac{2\Sigma_i^* + \Sigma_i}{3} - \Lambda_i = \tfrac{2}{3}\mu_q^2 \qquad (17.20)$$

is independent of μ_i (Federman et al., 1966).

Note in particular that the observed difference in the masses of Σ_s and Λ_s forces $\mu_s \neq \mu_q$. To quantify these splittings we should compare with the $\tfrac{3}{2}^+ - \tfrac{1}{2}^+$ splittings when $i \equiv q$, that is the splitting $\Delta - N$. From the above example one can set $\mu_i \equiv \mu_q$ and recover our earlier result (equation 17.4)

$$\Delta - N = \mu_q^2 \qquad (17.21)$$

This has finally brought us to our desired formulae with which we can compare the known masses. We have

$$\frac{\Sigma_i^* - \Sigma_i}{\Delta - N} = \frac{\mu_i}{\mu_q} \tag{17.22}$$

$$\frac{\Sigma_i - \Lambda_i}{\Delta - N} = \frac{2}{3}\left(\frac{\mu_q - \mu_i}{\mu_q}\right) \tag{17.23}$$

$$\frac{2\Sigma_i^* + \Sigma_i}{3} - \Lambda_i = \tfrac{2}{3}(\Delta - N) = \text{independent of i} \tag{17.24}$$

That $\Sigma_s^* - \Sigma_s$ is smaller than $\Delta - N$ requires $\mu_s < \mu_q$ (equation 17.22). This in turn causes the Σ_s to be heavier than Λ_s (equation 17.23), as is empirically observed. The magnetic moment of quark i being inversely proportional to the quark mass, then it is reasonable that $\mu_s < \mu_q$ correlates with $\varepsilon_s > \varepsilon_q$ and hence $\Sigma_s^* > \Delta$.

A diagrammatic representation of these formulae is shown in Fig. 17.1. If $\varepsilon_s \sim \varepsilon_q$ then Σ_s and Λ_s are nearly degenerate. For $\varepsilon_c \to \infty$ however it will be Σ_c and Σ_c^* that become degenerate.

The pattern of the splittings in the $(\Lambda\Sigma\Sigma^*)_i$ system indeed appears to be realised in the data and is an example of the insights into quark

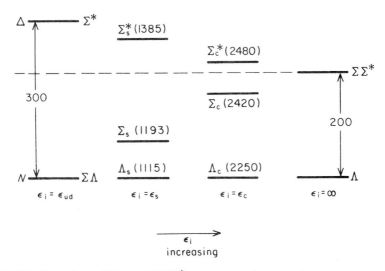

FIG. 17.1. Hyperfine splittings of $(\Lambda\Sigma\Sigma^*)_i$ system as ε_i increases from exact symmetry $(\varepsilon_i \equiv \varepsilon_{u,d})$ to infinity.

dynamics that are emerging with the discovery of heavy flavours $i =$ strange and charm. If further heavier flavours are discovered then the $\Sigma^*\Sigma$ degeneracy and 200 MeV splitting from the associated Λ will be a clear-cut prediction to be investigated.

What is even more interesting is that, on top of the above qualitative successes, there is a significant amount of quantitative consistency in this picture.

17.2.3 QUANTIFYING THE SPIN–SPIN SPLITTING AND ITS EFFECT ON BARYON SPECTROSCOPY

This vector gluon exchange picture has been investigated in some detail by de Rujula *et al.* (1975). For the mass of the u, d quarks they choose 336 MeV, this being motivated by the idea that the anomalous magnetic moment of the quark is small and hence that

$$\varepsilon_u = \frac{m_{\text{proton}}}{2\cdot 79} \simeq 336 \text{ MeV} \tag{17.25}$$

A possible meaning of this effective mass will be discussed in Chapter 18 on bag models (see also section 15.2.1, equation 15.32).

The observed ratio of $\Delta - N$ to $\Sigma_s^* - \Sigma_s$ masses yields, by equation (17.22), that

$$\varepsilon_s \simeq \tfrac{3}{2}\varepsilon_u \tag{17.23}$$

and hence that

$$\varepsilon_s \simeq 510 \text{ MeV} \tag{17.24}$$

We see that this gives

$$\varepsilon_s - \varepsilon_u \simeq 170 \text{ MeV}$$

which is *of the order* of the mass splittings between the $\Delta^- - \Sigma^{*-} - \Xi^- - \Omega^-$ members of the decuplet. Furthermore the SU(3) prediction

$$\frac{\mu(\Lambda)}{\mu(P)} = -\frac{1}{3} \tag{17.25}$$

is modified to read

$$\frac{\mu(\Lambda)}{\mu(P)} = -\frac{1}{3}\frac{\varepsilon_q}{\varepsilon_s} = -0\cdot 22 \tag{17.26}$$

which is in much better agreement with the data (-0.24 ± 0.02) (equation 17.20) than without the inclusion of the quark mass effect. As a final check on the consistency we can compute the expected $\Sigma - \Lambda$ mass difference. The above quark masses predict

$$\Sigma - \Lambda \simeq 99 \text{ MeV} \tag{17.27}$$

to be compared with the observed 75 to 80 MeV.

For the charmed states $(\Sigma \Lambda \Sigma^*)_c$ one inserts

$$\varepsilon_c \simeq 1.5 \text{ GeV} \tag{17.28}$$

into the formulae. This magnitude is consistent with the $\psi(c\bar{c})$ at 3.1 GeV, the observation of charmed mesons $D(c\bar{q})$ with masses around 1.85 to 2 GeV $(\simeq 1.5 + 0.34$ for c and $\bar{q} \equiv \bar{u}, \bar{d})$ and also with the charmed baryon $\Lambda_c(cqq)$ with mass of 2.26 GeV $(\simeq 1.5 + 0.34 + 0.34$ for c and $qq)$ (Knapp et al., 1976).

Having parametrised ε_c we find the important ratios of the different flavour masses to be

$$\frac{\varepsilon_u}{\varepsilon_c} \simeq \frac{1}{5}, \frac{\varepsilon_s}{\varepsilon_c} \simeq \frac{1}{3} \tag{17.29}$$

and hence

$$\frac{\mu_u}{\mu_c} \simeq 5; \frac{\mu_s}{\mu_c} \simeq 3 \tag{17.30}$$

These ratios are all that we need to quantify the spin–spin mass splittings of the charmed $(\Sigma \Lambda \Sigma^*)_c$ compared to their strange analogues $(\Sigma \Lambda \Sigma^*)_s$. We have from equations (17.18) and (17.19) that

$$(\Sigma^* - \Sigma)_c = \frac{\mu_c}{\mu_s}(\Sigma^* - \Sigma)_s \simeq 60 \text{ MeV} \tag{17.31}$$

$$(\Sigma - \Lambda)_c = \frac{\mu_u - \mu_c}{\mu_u - \mu_s}(\Sigma - \Lambda)_s \simeq 2(\Sigma - \Lambda)_s \simeq 160 \text{ MeV} \tag{17.32}$$

This pattern is quite different from the strange system and is consistent with the emerging data (Knapp et al., 1976; Cazzoli et al., 1976; Dalitz, 1976).

17.2.4   THE $\Xi^* - \Xi$ SYSTEM

If $q = $ u or d then states with two quarks of the same flavour, but differing from u or d will form a system $\Xi_i^* \Xi_i (\mathrm{ii}q)$. If $\mathrm{i} \equiv \mathrm{s}$ then the familiar strange baryons $\Xi_\mathrm{s}(1320)$ and $\Xi_\mathrm{s}^*(1535)$ are relevant. If $\mathrm{i} \equiv \mathrm{c}$ then Ξ_c^{++}(ccu), Ξ_c^+(ccd) occur (called sometimes X_u^{++}, X_d^+).

We can calculate the hyperfine splittings by using the techniques of the previous section and the wavefunctions of Table 3.7. It is quicker to note that Ξ_s^- is in the same U-spin multiplet as Σ_s^- and so these states are related by replacing d $\leftrightarrow$ s or $q \leftrightarrow$ i in general. Then on making this replacement in equation (17.18) we have

$$\Xi_i^* - \Xi_i = \mu_i \mu_q \equiv \Sigma_i^* - \Sigma_i \qquad (17.33)$$

which is interesting because naively one might have expected the Ξ split to be smaller due to the extra strange (massive) quark.

For the strange states this relation is satisfied to within 10 per cent, which is all we should expect as we have ignored electromagnetic effects, quark kinetic energies etc. which can contribute a few MeV to the masses and may be sufficient to explain the 10 per cent mismatch. It will be interesting to see how well this prediction fares in the doubly charmed baryons.

17.2.5   MASSES OF OCTET AND DECUPLET BARYONS

To summarise the phenomenology of baryon masses we present here the contributions to the masses of the 56plet $\mathbf{8}\,\tfrac{1}{2}^+$ and $\mathbf{10}\,\tfrac{3}{2}^+$ baryons arising from the u, d and s quark masses and the spin–spin couplings for coloured vector–gluon exchange between coloured quarks. We write

TABLE 17.1
Octet and decuplet masses

N: $3\varepsilon - g^2\mu_q^2$	Δ: $3\varepsilon + g^2\mu_q^2$
Σ^0: $3\varepsilon + \Delta m - \tfrac{1}{3}g^2(4\mu_q\mu_\mathrm{s} - \mu_q^2)$	Σ^*: $3\varepsilon + \Delta m + \tfrac{1}{3}g^2(2\mu_q\mu_\mathrm{s} + \mu_q^2)$
Λ^0: $3\varepsilon + \Delta m - g^2\mu_q^2$	
Ξ: $3\varepsilon + 2\Delta m - \tfrac{1}{3}g^2(4\mu_q\mu_\mathrm{s} - \mu_s^2)$	Ξ^*: $3\varepsilon + 2\Delta m + \tfrac{1}{3}g^2(2\mu_q\mu_\mathrm{s} + \mu_s^2)$
	Ω: $3\varepsilon + 3\Delta m + g^2\mu_\mathrm{s}^2$

$\varepsilon_s = \varepsilon + \Delta m$ where $\varepsilon \equiv \varepsilon_{u,d} \simeq 335\,\mathrm{MeV}$. If g is proportional to the quark–gluon coupling strength then, from equations (17.3), (17.4), (17.17) and (17.33), we have for the masses the expressions in Table 17.1.

If $\Delta m = 0$ and $g = 0$ then the supermultiplet would be mass degenerate (exact $SU(6)$) with a mass of 3ε, and thus around $1100\,\mathrm{MeV}$. This is illustrated in Fig. 17.2(a).

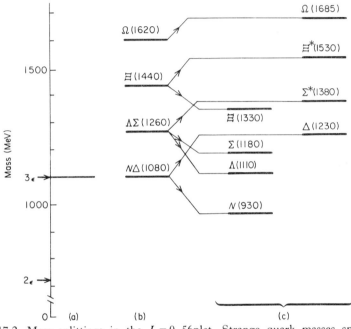

FIG. 17.2. Mass splittings in the $L = 0$ 56plet. Strange quark masses split the 1100 MeV supermultiplet (a) into four levels distinguished by strangeness (b). In (c) hyperfine splitting separates the $J = \frac{3}{2}$ **10** and $J = \frac{1}{2}$ **8** and also Σ^0 and Λ.

Now switch on the mass difference between strange and nonstrange quarks, $\Delta m \simeq 180\,\mathrm{MeV}$. This separates the states of different strangeness by equal amounts (Fig. 17.2(b)). We are not making an attempt to find a "best fit" to masses; our intention is only to illustrate schematically what is going on. Hence, for example, the masses used here are not exactly the same as elsewhere; instead they are chosen as nice round numbers to facilitate rapid assimilation.

If $\varepsilon_{ud} \sim 360\,\mathrm{MeV}$ then $\varepsilon_s \sim 540\,\mathrm{MeV}$ and so we take $\mu_s/\mu_{ud} = \varepsilon_{ud}/\varepsilon_s \simeq \frac{2}{3}$. Now switch on the spin–spin splitting by having $g \neq 0$. We have chosen for simplicity $g^2\mu_q^2 = 150$ MeV and hence $g^2\mu_s^2 = 66$ MeV, $g^2\mu_N\mu_S = 100$ MeV. The $\mathbf{10}\ \tfrac{3}{2}^+$ states are pushed up while $\mathbf{8}\ \tfrac{1}{2}^+$ come down in mass and the $\Sigma\Lambda$ are split.

17.3 Fermi–Breit Hamiltonian in QCD

With the simple picture developed above, the pattern of masses is already well reproduced. One could clearly improve on this by introducing further parameters, e.g. optimising μ_s/μ_{ud}, including kinetic energy effects $\sim 1/m_i^2$ etc. Within the philosophy of quantum chromodynamics the fermion–vector gluon Hamiltonian will be of the same form as that familiar in QED.

This has been investigated in some detail by de Rujula *et al.* (1975) who write the strong and electromagnetic part of the Hamiltonian in the form

$$\mathcal{H} = L(\mathbf{r}_1, \mathbf{r}_2 \ldots) + \sum_i \left(m_i + \frac{\mathbf{p}_i^2}{2m_i} + \cdots\right) + \sum_{i>j} (\alpha Q_i Q_j + k\alpha_s) S_{ij}$$

(17.34)

In this Hamiltonian L is the universal interaction binding the quarks whose positions, masses, momenta are respectively $\mathbf{r}_i$, m_i, $\mathbf{p}_i$. S_{ij} is the two-body interaction which de Rujula *et al.* take to be Coulombic. The electromagnetic piece involves α and is proportional to the quark charges Q_i while the colour gluon exchange has coupling strength $k\alpha_s$ where $k = -\tfrac{4}{3}$ for mesons and $-\tfrac{2}{3}$ for baryons (compare equations 15.38 and 15.39).

Neglecting relativistic corrections the two-body Fermi–Breit and Coulomb interaction S_{ij} has the form

$$S_{ij} = \frac{1}{|\mathbf{r}|} - \frac{1}{2m_i m_j}\left(\frac{\mathbf{p}_i \cdot \mathbf{p}_j}{|\mathbf{r}|} + \frac{\mathbf{r} \cdot (\mathbf{r} \cdot \mathbf{p}_i)\mathbf{p}_j}{|\mathbf{r}|^3}\right) - \frac{\pi}{2}\delta^3(\mathbf{r})\left(\frac{1}{m_i^2} + \frac{1}{m_j^2} + \frac{16\mathbf{s}_i \cdot \mathbf{s}_j}{3m_i m_j}\right)$$
$$-\frac{1}{2|\mathbf{r}|^3}\left\{\frac{1}{m_i^2}\mathbf{r}\times\mathbf{p}_i \cdot \mathbf{s}_i - \frac{1}{m_j^2}\mathbf{r}\times\mathbf{p}_j \cdot \mathbf{s}_j + \frac{1}{m_i m_j}\left[2\mathbf{r}\times\mathbf{p}_i \cdot \mathbf{s}_j - 2\mathbf{r}\times\mathbf{p}_j \cdot \mathbf{s}_i\right.\right.$$
$$\left.\left. -2\mathbf{s}_i \cdot \mathbf{s}_j + \frac{6(\mathbf{s}_i \cdot \mathbf{r})(\mathbf{s}_j \cdot \mathbf{r})}{|\mathbf{r}|^2}\right]\right\}$$

(17.35)

Here $\mathbf{r} = \mathbf{r}_i - \mathbf{r}_j$ and $\mathbf{s}_i$ is the i-th quark's spin.

The hadron masses are the expectation values of this Hamiltonian placed between the hadron SU(6) wavefunctions $\Psi(r_1 r_2 r_3)$. For S-wave states the $\mathbf{L} . \mathbf{S}(\mathbf{r} \times \mathbf{p} . \mathbf{s})$ interactions give no contribution. Writing the expectation values of the spatial operators of kinetic energy:

$$a = \left\langle \Psi_0 \left| \frac{p_i^2}{2} \right| \Psi_0 \right\rangle \tag{17.36}$$

Coulomb interaction,

$$b = \left\langle \Psi_0 \left| \frac{1}{|\mathbf{r}_{12}|} \right| \Psi_0 \right\rangle \tag{17.37}$$

Darwin–Breit interaction,

$$c = \tfrac{1}{2} \left(\Psi_0 \left[\frac{\mathbf{r}_{12}^2 \mathbf{p}_1 . \mathbf{p}_2 + \mathbf{r}_{12} . (\mathbf{r}_{12} . \mathbf{p}_1)\mathbf{p}_2}{|\mathbf{r}_{12}|^3} \right] \Psi_0 \right) \tag{17.38}$$

and finally the point interaction,

$$d = \frac{\pi}{2} \langle \Psi_0 | \delta^3(\mathbf{r}_{12}) | \Psi_0 \rangle \tag{17.39}$$

we have the first-order mass formula for the S-wave baryons,

$$
\begin{aligned}
M = M_0 + \sum_i & \left[\Delta m_i + a \left(\frac{1}{m_i} - \frac{1}{m_{ud}} \right) \right] \\
+ \sum_{i>j} & (\alpha Q_i Q_j - \tfrac{2}{3}\alpha_s) \left[b - \frac{c}{m_i m_j} - d \left(\frac{1}{m_i^2} + \frac{1}{m_j^2} + \frac{16 \mathbf{S}_i . \mathbf{S}_j}{3 m_i . m_j} \right) \right]
\end{aligned} \tag{17.40}
$$

where M_0 is the eigenstate of

$$M_0 \equiv \Sigma L(r_1 r_2 \ldots) + \sum_i \left(m_{ud} + \frac{\mathbf{p}_i^2}{2m_{ud}} \right) \tag{17.41}$$

which gives the degenerate SU(6) supermultiplet's mass. This yields a four-parameter mass formula. The contributions proportional to $\Delta m_i \equiv m_i - m_{u,d}$ and the $\mathbf{S} . \mathbf{S}$ interaction have already been discussed in section 17.2 and gave a good description of the S-wave baryon states.

A detailed discussion of the phenomenology flowing from the four-parameter formula (17.40) and the full Hamiltonian (for $p \ldots$ states) is given in de Rujula et al. (1975).

17.4   Spin–spin splitting and meson spectroscopy

We have seen how the spin–spin splitting appears to be manifested in the $(\Lambda\Sigma\Sigma^*)_i$ baryon system and, in particular, how the splitting depended upon the mass scale of the flavour index $i = s, c \ldots$. We now look at the effect of this spin–spin interaction in the meson system where it will separate the masses of quark–antiquark states with spin $S = 0$ or 1. The most immediate application therefore will be to the 0^- and 1^- states which are $L = 0$; $S = 0$ or 1.

In a meson made of quark–antiquark flavours $q_i\bar{q}_i$ the spin–spin interaction is proportional to the magnetic moments of the two quarks. Hence the 0^-1^- splittings for two different flavour combinations will be

$$\frac{(V - P)_{ij}}{(V - P)_{kl}} = \frac{\mu_i\mu_j}{\mu_k\mu_l} \tag{17.42}$$

where V and P refer to vector or pseudoscalar and the subscripts denote the flavours contained in them. Particular examples are

$$(K^* - K)_{su} = \frac{\mu_s}{\mu_u}(\rho - \pi)_{uu} \tag{17.43}$$

$$(D^* - D)_{cu} = \frac{\mu_c}{\mu_s}(K^* - K)_{su} \tag{17.44}$$

$$(F^* - F)_{cs} = \frac{\mu_s}{\mu_u}(D^* - D)_{cu} \tag{17.45}$$

Hence, since $\mu_u < \mu_s < \mu_c$, we expect

$$(F^* - F) < (D^* - D) < (K^* - K) < (\rho - \pi) \tag{17.46}$$

which is qualitatively true:

$$? < (2{\cdot}01 - 1{\cdot}86) < (0{\cdot}89 - 0{\cdot}49) < (0{\cdot}77 - 0{\cdot}14) \tag{17.47}$$

Quantitatively the mass splittings are again in remarkable agreement with the data since the ratios $\mu_s \sim \frac{2}{3}\mu_u$ and $\mu_c \sim \frac{1}{3}\mu_s$ yield

$$(K^* - K)_{us} \sim \tfrac{2}{3}(\rho - \pi)_{uu} \tag{17.48}$$
$$\text{(400 MeV)}\quad\text{(630 MeV)}$$

and

$$(D^* - D)_{cu} \sim \tfrac{1}{3}(K^* - K)_{su} \qquad (17.49)$$
$$(150 \text{ MeV}) \quad (400 \text{ MeV})$$

The $F^* - F$ splitting is therefore predicted to be

$$(F^* - F)_{cs} \sim \tfrac{2}{3}(D^* - D)_{cu} \sim 100 \text{ MeV} \qquad (17.50)$$

If this is true in Nature[1] then the dominant decay of F^* will be into $F + \gamma$ since $m_\pi > F^* - F$. Observation of this monoenergetic photon would be clear indication of this pattern of mass splittings.

17.4.1 A PROBLEM FOR MESON MASS SPLITTINGS

The spin–spin splittings so far discussed have been for hadrons carrying manifest quantum numbers such as isospin, strangeness or charm or, in the baryon case, baryon number. The empirical success might lead us to expect that the same pattern of splittings should emerge for mesons which do not carry the above manifest quantum numbers, e.g. $\eta\eta'\omega\phi\eta_c\psi \ldots$. However one should be cautious because one can now anticipate a new class of contributions, namely the annihilation of the $q_i\bar{q}_i$ into gluons, which will destroy the simple pattern in equation (17.46). Indeed, if the η_c is confirmed to be at $2 \cdot 85$ GeV (Wiik, 1975) then

$$\psi - \eta_c \sim 350 \text{ MeV} \qquad (17.51)$$

which is of the order of $K^* - K$ or $D^* - D$ in sharp contrast to the gluon exchange prediction

$$(\psi - \eta_c)_{c\bar{c}} \sim \tfrac{1}{5}(D^* - D)_{cu} \sim 30 \text{ MeV} \qquad (17.52)$$

which predicts $\eta_c(3 \cdot 07 \text{ GeV})$.

There are several excuses that one can invent here. Not only is there gluon annihilation as already mentioned but the precise $u\bar{u}$, $d\bar{d}$, $s\bar{s}$, $c\bar{c}$ content of the mesons concerned can be played with. This is currently an area of active research. Some questions related to this are discussed in section 17.6.

[1] Brandelik *et al.* (1977a) claim evidence for $F(2 \cdot 03 \pm 0 \cdot 06$ GeV) and $F^*(2 \cdot 14 \pm 0 \cdot 06$ GeV).

17.5   Electromagnetic mass shifts

17.5.1   MESONS

The electromagnetic contributions to the hadron mass (energy of the quark system) can be thought of as being of three kinds: (i) electromagnetic contributions to quark masses, e.g. $m_u \neq m_d$, (ii) Coulomb interaction between any pair of quarks $\sim e_i e_j$, (iii) magnetic interaction between any pair of quarks $\sim e_i e_j / m_i m_j$, where $e_{i,j}$, $m_{i,j}$ are the charges and masses of the i and j quarks and the total contribution to the system's energy will arise after summing over i and j.[1] The role of each of these can be well illustrated by studying

$$\pi^+ - \pi^0 = 4 \cdot 60 \text{ MeV}$$

$$K^+ - K^0 = -4 \cdot 0 \pm 0 \cdot 13 \text{ MeV} \qquad (17.53)$$

$$K^{*+} - K^{0*} = -4 \cdot 1 \pm 0 \cdot 6 \text{ MeV}$$

Notice that for kaons the charged state is lighter whereas for pions it is heavier.

The $\pi^+(u\bar{d})$ contains the same number of u and d quarks as does the $\pi^0[(u\bar{u} - d\bar{d})]/\sqrt{2}$. Consequently $\delta m(u-d)$ does not contribute to $\delta m(\pi^+ - \pi^0)$. To lift the $\pi^+ \pi^0$ degeneracy the interactions (ii) and (iii) above are necessary. These contribute to the $\pi^{+,0}$ masses amounts proportional to the product of the constituent quark charges. Since $m_u \simeq m_d$ we will lump together (ii) and (iii) and if $\langle 1/R \rangle$ is the mean separation in space of the q and $\bar{q}$ in the pion, we then have for the $\pi^+(u\bar{d})$ a contribution proportional to $\frac{2}{9}\langle 1/R \rangle$ (i.e. $e_u e_{\bar{d}}\langle 1/R \rangle$), whereas for $\pi^0[(u\bar{u} - d\bar{d})]/\sqrt{2}$ one has $-\frac{5}{18}\langle 1/R \rangle$. Hence

$$\pi^+ - \pi^0 = \tfrac{1}{2}\langle 1/R \rangle \qquad (17.54)$$

which is satisfyingly of the right sign and fitting the observed mass difference yields

$$\langle 1/R \rangle_\pi = 9 \cdot 2 \text{ MeV} \qquad (17.55)$$

For the kaon system the mass difference becomes

$$-K^+(u\bar{s}) + K^0(d\bar{s}) = -m_u + m_d - \langle 1/R \rangle_K (\tfrac{2}{9} + \tfrac{1}{9})$$

$$\equiv \Delta m_{du} - \tfrac{1}{3}\langle 1/R \rangle_K \qquad (17.56)$$

[1] Some results in section 17.5 follow from the weaker assumption that two-body forces contribute and depend on the quark flavours *independent* of e_i and e_j (Rubinstein, 1966).

where we have again lumped together the Coulombic and magnetic contributions. Since $m_s > m_{u,d}$ then $\mu_s < \mu_{u,d}$ and so the magnetic contribution to the kaon system differs from the pion case. This can be dealt with by regarding it as effectively causing $\langle 1/R \rangle_K \neq \langle 1/R \rangle_\pi$ in equations (17.54) and (17.56).

The magnetic interaction being spin dependent will in general give different magnitudes for the electromagnetic splittings in pseudoscalar and analogue vector isospin multiplets. The data in equation (17.53) are consistent with the magnetic interaction being negligible. However recent data from Aguiler-Benitez *et al.* (1977) claim that $K^{*0} - K^{*+}$ have a mass difference of $7 \cdot 7 \pm 1 \cdot 7$ MeV, rather larger than the pseudoscalar case. This suggests that the magnetic interaction has an important role to play (a conclusion which is supported by recent results on charmed meson mass splittings). For simplicity of presentation we shall ignore it and always include (ii) and (iii) together. However, you should perform the exercise of checking the role of the magnetic term in equation (17.56) explicitly to clarify the extent to which it can manifest itself in the data. In particular for charmed mesons where $\mu_c \ll \mu_{d,s}$ it can play an important role.

If we supposed that

$$\langle 1/R \rangle_\pi = \langle 1/R \rangle_K = 9 \cdot 2 \text{ MeV} \qquad (17.57)$$

then from the observed excess of 4 MeV in K^0 over K^+ we have

$$\Delta m_{ud} \simeq -7 \text{ MeV} \qquad (17.58)$$

and so the d quark is heavier than the u. This is qualitatively in agreement with the fact that the neutron has an excess of d quarks over the proton and is heavier.

For the pseudoscalar charmed mesons we have

$$D^+(c\bar{d}) - D^0(c\bar{u}) = m_d - m_u + \langle 1/R \rangle_D (\tfrac{2}{9} + \tfrac{4}{9}) \qquad (17.59)$$

Both the Δm_{ud} and the interaction contributions work towards increasing the D^+ relative to the D^0. If $\langle 1/R \rangle_D = \langle 1/R \rangle_{\pi,K}$ then an excess of 13 MeV in favour of D^+ is expected (de Rujula *et al.*, 1976; Ono, 1976a,b; Lichtenberg, 1975, 1976). However, since $\mu_c < \mu_{s,d}$ the magnetic contribution from the charmed quark is suppressed so that in effect we have

$$\langle 1/R \rangle_D < \langle 1/R \rangle_{\pi,K} \qquad (17.60)$$

and hence the $D^+ - D^0$ separation is probably less than 13 MeV. Also one may expect $D^{*+} - D^{+0}$ to differ from $D^+ - D^0$ (Ono, 1976; Lichtenberg, 1975). The data suggest that D^+ is heavier than D^0 but the absolute magnitude is not yet settled.[1]

It is important to note that there are two parameters in this problem, Δm_{ud} and $\langle 1/R \rangle$, reflecting the two $0(\alpha)$ mechanisms. From the $\pi^+ - \pi^0$ we deduced $\langle 1/R \rangle_\pi$. The Δm only arose when we *assumed* that $\langle 1/R \rangle_K \equiv \langle 1/R \rangle_\pi$. In the light of the anomalously small pion mass it may well be that this is not a valid assumption. All that we can safely conclude from $K^+ - K^0$ is therefore that

$$\Delta m_{du} \equiv m_d - m_u > 0 \tag{17.61}$$

since $\langle 1/R \rangle$ is necessarily positive. Hence the conclusion that D^+ is heavier than D^0 is unavoidable in this approach.

17.5.2 BARYONS

The nucleon system has neutron heavier than the proton by 1·5 MeV. In the quark model one has

$$n(ddu) - p(uud) = \Delta m_{du} - \tfrac{1}{3}\langle 1/R \rangle_N \tag{17.62}$$

This again fits with the $\Delta m_{du} > 0$ but would disagree with the value -7 MeV (equation 17.58) and $\langle 1/R \rangle_N = \langle 1/R \rangle_\pi = 9·2$ MeV (equation 17.57).

To obtain a value for $\langle 1/R \rangle_{\text{baryon}}$ we can study a system that receives no contribution from Δm_{du}, namely

$$\tfrac{1}{2}(\Sigma^+ + \Sigma^-) - \Sigma^0 \equiv \tfrac{1}{2}(1190 + 1197·3) - 1192·5 = 1·15 \text{ MeV} \tag{17.63}$$

The quark content and consequent energy shifts are

$$\tfrac{1}{2}(uus + dds) - uds = \tfrac{1}{2}\langle 1/R \rangle_\Sigma \tag{17.64}$$

with the result that

$$\langle 1/R \rangle_\Sigma = 2·3 \text{ MeV} \tag{17.65}$$

---

[1] Perruzi *et al.* (1977) report $\Delta m(D^+ - D^0) = 5·0 \pm 0·8$ MeV, $\Delta m(D^{*+} - D^{*0}) = 2·6 \pm 1·8$ MeV. This suggests that equation (17.67) gives a better estimate of $\Delta m(ud)$ than equation (17.58).

The $\Sigma^- - \Sigma^+$ mass difference is $+7\cdot3$ MeV. In the quark model this becomes

$$-\Sigma^+(uus) + \Sigma^-(dds) = +2\Delta m_{du} + \langle 1/R \rangle (-\tfrac{2}{3}[NS] + \tfrac{1}{3}[SS]) \tag{17.66}$$

where [NS] and [SS] are the photon exchange contributions between N (nonstrange) and S (strange) quarks. The magnetic interaction being proportional to m_{ud}^{-1} or m_s^{-1} may cause these to differ, and $\langle 1/R \rangle$ may do so also. If for the moment we take them equal we have that

$$\Delta m_{du} \simeq 4\cdot5 \text{ MeV} \tag{17.67}$$

These values for $\langle 1/R \rangle$ and Δm_{du} also give a good fit to the kaon for which

$$-K^+ + K^0 = +\Delta m_{du} - \tfrac{1}{3}\langle 1/R \rangle = +3\cdot7 \text{ MeV} \tag{17.68}$$

to be compared with 4 MeV. The neutron and proton are also predicted to have this mass difference whereas the neutron is only $1\cdot5$ MeV heavier than the proton empirically.

One other piece of information on electromagnetic mass splittings within the baryon **8** comes from Ξ^- and Ξ^0 where Ξ^- is about $6\cdot4$ MeV heavier than Ξ^0. In the quark model

$$\Xi^-(dss) - \Xi^0(uss) = \Delta m_{du} + \langle 1/R \rangle (\tfrac{2}{3}[NS]) \tag{17.69}$$

with $\Delta m_{du} = 4\cdot5$ MeV and $\langle 1/R \rangle = 2\cdot3$ MeV the predicted 6 MeV is in excellent agreement with the observed $6\cdot4 \pm 0\cdot6$ MeV.

To test how well this picture of the electromagnetic mass splittings is working it is useful to form combinations where Δm, $\langle 1/R \rangle_{[NN]}$, $\langle 1/R \rangle_{[NS]}$ and $\langle 1/R \rangle_{[SS]}$ all drop out. First, if the Δm_{du} were the *only* contribution, then from equations (17.62), (17.66) and (17.69) one would have

$$(n-p) = (\Xi^- - \Xi^0) = \tfrac{1}{2}(\Sigma^- - \Sigma^+) \tag{17.70}$$
$$(1\cdot3) \quad (6\cdot4 \pm 0\cdot6) \quad (4 \pm 0\cdot4)$$

which do not fit well with the data shown beneath. Including the $\langle 1/R \rangle_{[NS],[NN] \text{ and } [SS]}$ then there is one combination of masses that is independent of the magnitudes of these parameters. This relation (Coleman and Glashow, 1961) is

$$(n-p) + (\Xi^- - \Xi^0) = (\Sigma^- - \Sigma^+) \tag{17.71}$$
$$(7\cdot7 \pm 0\cdot6) \quad (8 \pm 0\cdot8)$$

and fits perfectly with data. This success, with the failure at equation (17.70), clearly shows the importance of the photon exchange.

One can also obtain relations between electromagnetic mass splittings in the decuplet and octet if the magnetic term is negligible. Examples include

$$-\Delta^{++} + \Delta^{-} = 3(n-p) \qquad (17.72)$$

$$(7 \cdot 9 \pm 6 \cdot 8) = (3 \cdot 9)$$

and

$$\Delta^{++} - \Delta^{+} = (p-n) + (\Sigma^{+} + \Sigma^{-} - 2\Sigma^{0}) \qquad (17.73)$$

The splitting in the decuplet between Ξ^{-*} and Ξ^{0*} has the same form as that in the octet (equation 17.69). Similar remarks apply to Σ^{+*} and Σ^{-*} (equation 17.66). The value of $\langle 1/R \rangle$ might differ in the decuplet and octet since for the magnetic term the $\langle \mathbf{S} . \mathbf{S} \rangle$ differs and is reflected in $\langle 1/R \rangle$ in our notation. The values of the electromagnetic mass splittings in **8** and **10** are (in MeV)

	8	10
$\Xi^{-} - \Xi^{0}$	6	3
$\Sigma^{-} - \Sigma^{+}$	8	4

$$(17.74)$$

so the absolute magnitude in **10** appears smaller, the ratio of **8** and **10** being the same.

As a final testing of this picture we can make estimates for the splittings of charmed baryons to supplement our earlier discussion of charmed mesons. It is straightforward to obtain

$$-\Sigma_{c}^{++} + \Sigma_{c}^{+} = \Delta m_{du} - \tfrac{4}{3}\langle 1/R \rangle \qquad (17.75)$$

and

$$-\Sigma_{c}^{+} + \Sigma_{c}^{0} = -S^{+} + S^{0} = -A^{+} + A^{0} = \Delta m_{du} - \tfrac{1}{3}\langle 1/R \rangle \quad (17.76)$$

The interesting feature common to these two results is that they involve the difference of two positive quantities and so the result is very sensitive to the actual magnitudes chosen for Δm and $\langle 1/R \rangle$. For instance if $\Delta m_{du} = 7$ MeV and $\langle 1/R \rangle = 9 \cdot 2$ MeV (cf. discussion of equations 17.57 and 17.58) then $\Sigma_{c}^{++} < \Sigma_{c}^{+}$. However, if instead we chose $\Delta m_{du} \simeq 4 \cdot 5$ MeV and $\langle 1/R \rangle = 2 \cdot 3$ MeV (cf. equations 17.65 and 17.67), then $\Sigma_c^{++} < \Sigma_c^+$. Clearly the charmed baryon masses will give us a good indication as to the relative importance of the photon exchange and quark mass contributions.

17.6 Gluon contributions, pseudoscalar and vector meson masses

The hyperfine splittings between 0^- and 1^- mesons seem to fit well with QCD calculations (section 17.4) except for the case of ψ and η_c (section 17.4a). A possible reason for this failure is that a new topology can contribute here, namely $c\bar{c} \to$ gluons $\to c\bar{c}$. This was not possible in all other examples, e.g. $\rho\pi$ have $I = 1$, K^*K have strangeness, D^*D charm and only the $SU(3)$ singlet states can be affected. Another consequence of this is that light quarks can be mixed into the ψ or η_c by

$$c\bar{c} \leftrightarrow \text{gluons} \leftrightarrow q\bar{q} \tag{17.77}$$

and so the ψ and η_c might not be pure $c\bar{c}$.

For the case of the 1^- mesons this mixing seems to be negligible (ω and ϕ, section 4.4). For the 0^- mesons the question is still under debate.

We shall not discuss the 0^-1^- splittings further, but will study the effects of the annihilation contributions to the 0^- and 1^- nonet masses individually.

The quark mass matrix with three flavours may be written

$$\hat{M} = \begin{pmatrix} 2u & 0 & 0 \\ 0 & 2u & 0 \\ 0 & 0 & 2s \end{pmatrix} \tag{17.78}$$

acting on the $(u\bar{u}, d\bar{d}, s\bar{s})$ basis and we are approximating $m_u = m_d$. If $u\bar{u}$, $d\bar{d}$, $s\bar{s}$ can annihilate with an amplitude A, which we assume to be $SU(3)$ invariant, then there will be an additional contribution to the mass matrix

$$H_I = \begin{pmatrix} A & A & A \\ A & A & A \\ A & A & A \end{pmatrix} \tag{17.79}$$

The unitary singlet

$$1 = \frac{1}{\sqrt{3}}(u\bar{u} + d\bar{d} + s\bar{s}) \tag{17.80}$$

will have a mass

$$M(\mathbf{1}, \mathbf{1}) = 3A + \tfrac{1}{3}(4u + 2s) \tag{17.81}$$

where A does not contribute to the **8** masses nor to the singlet–octet mixing. If $A \to \infty$ then the singlet is sent up to infinite mass and the octet will satisfy the Gell-Mann–Okubo formula (section 4.4).

If $m_u \neq m_s$ then there is a mixing between singlet and octet

$$\langle \mathbf{1},\mathbf{1} | M | \mathbf{8},\mathbf{1} \rangle \equiv \frac{1}{\sqrt{3}}(u\bar{u} + d\bar{d} + s\bar{s} | \hat{M} | u\bar{u} + d\bar{d} - 2s\bar{s}) \frac{1}{\sqrt{6}}$$

$$= \frac{2\sqrt{2}}{3}(u - s) \tag{17.82}$$

and so the physical states will be singlet–octet mixtures. The mass matrix for the $I = 0$ states in an $(|\mathbf{8},\mathbf{1}\rangle, |\mathbf{1},\mathbf{1}\rangle)$ basis becomes

$$\hat{M}_{I=0} = \begin{pmatrix} \dfrac{1}{3}(4s + 2u) & \dfrac{2\sqrt{2}}{3}(u - s) \\ \dfrac{2\sqrt{2}}{3}(u - s) & 3A + \dfrac{1}{3}(4u + 2s) \end{pmatrix} \tag{17.83}$$

Hence for the pseudoscalars

$$\hat{M}_{I=0} = \begin{pmatrix} \dfrac{1}{3}(4K - \pi) & \dfrac{2\sqrt{2}}{3}(\pi - K) \\ \dfrac{2\sqrt{2}}{3}(\pi - K) & 3A + \dfrac{1}{3}(2K + \pi) \end{pmatrix} \tag{17.84}$$

and the eigenvalues $\lambda_{1,2}$ will be the physical η, η' masses. Similarly for the vector mesons $\pi \to \rho$, $K \to K^*$ and $\lambda_{1,2}$ are the ω, ϕ masses.

Solving the algebra for the eigenvalues yields

$$\lambda_1 + \lambda_2 = 3A + 2K \tag{17.85}$$

(from the sum of the diagonal elements) and

$$\lambda_1 \lambda_2 = 2K\pi - \pi^2 + A(4K - \pi) \tag{17.86}$$

from the determinant.

If $A = 0$ then we find $\lambda_1 \lambda_2 = 2K$ which is well satisfied for the vector states (section 4.4)

$$\phi + \omega = 2K^* \tag{17.87}$$

This suggests that the annihilation is negligible in the vector nonet and this is supported by the $\phi(\bar{s}s)$ and its Zweig forbidden decays being small (i.e. $\bar{s}s$ mixes very little with $u\bar{u}$, $d\bar{d}$). For the pseudoscalar mesons the prediction $\eta + \eta' = 2K$ is not satisfied (equation 4.100) and hence $A \neq 0$ there.

Since we have four states $(\pi K \eta \eta')$ and three free parameters m_u, m_s, A then one relation can be found independent of the unknowns. This is obtained by eliminating A in equations (17.85) and (17.86) and yields the sum rule (Schwinger, 1964)

$$(\eta + \eta')(4K - \pi) - 3\eta\eta' = 8K^2 - 8K\pi + 3\pi^2 \qquad (17.88)$$

Inserting the pseudoscalar masses yields $1 \cdot 22 \text{ GeV}^2$ on the left-hand side and $1 \cdot 95 \text{ GeV}^2$ on the right. Hence the pseudoscalar masses do not fit well for any value of the $SU(3)$ singlet A and three quark flavours.

The analogous sum rule for the vectors is of course well satisfied because we have already seen that $A = 0$ gives an excellent result (equation 17.87).

One suggestion (de Rujula *et al.*, 1975) has been that A might be a mass-dependent quantity. In asymptotically free gauge theories (which emerge naturally in the QCD theory) $A \to 0$ as $(\log M)^{-N}$ with $N = 2$ or 3 for $J = 0$ or 1 where M is the mass. In fitting the eigenvalues of equations (17.85) and (17.86) to $\eta(550)$ and $\eta'(960)$ de Rujula *et al.* find

$$A(550) = 630 \text{ MeV}$$

$$A(960) = 83 \text{ MeV} \qquad (17.89)$$

Hence A is large and decreasing with energy. A similar analysis for the vectors yields

$$A(\omega) = 7 \cdot 2 \text{ MeV}$$

$$A(\phi) = 5 \cdot 4 \text{ MeV} \qquad (17.90)$$

Hence for vector mesons A is seen to be small (as expected—section 4.4 and equation 17.87). These results are in accord with the colour gluon QCD theory since the 1^- annihilation term involves three gluons and hence is expected to be smaller than the pseudoscalars. Moreover the decrease of A with increasing mass in the 0^- (and even the 1^-?) is anticipated from the asymptotic freedom of the QCD theory.

Since we now know that a fourth flavour of quark exists then we should allow for the possibility that the $c\bar{c}$ may mix into the mesons η, η' via the gluon annihilation. Hence consider in general the consequence of including an extra SU(3) singlet component $|R\rangle$ in the wavefunctions (R being defined as containing no $u\bar{u}$, $d\bar{d}$ or $s\bar{s}$ pairs). Then the SU(3) singlet state which mixes with the **8**, $I = 0$ will be

$$|\alpha\rangle = \cos\alpha|1\rangle + \sin\alpha|R\rangle \equiv \frac{\cos\alpha}{\sqrt{3}}(u\bar{u} + d\bar{d} + s\bar{s}) + \sin\alpha|R\rangle$$

(17.91)

The resulting change compared to the previous calculation is that:
1. The singlet mass (equation 17.81) becomes

$$4A + \frac{\cos^2\alpha}{3}(4u + 2s) + (\sin^2\alpha)R$$

(17.92)

2. The singlet–octet mixing is now $(2\sqrt{2}/3)(u-s)\cos\alpha$ (constrast equation 17.83).
Consequently

$$\hat{M}_{I=0} = \begin{pmatrix} \frac{1}{3}(4K - \pi) & \frac{2\sqrt{2}}{3}(\pi - K)\cos\alpha \\ \frac{2\sqrt{2}}{3}(\pi - K)\cos\alpha & \eta + \eta' - \frac{1}{3}(4K - \pi) \end{pmatrix}$$

(17.93)

where in the element corresponding to the singlet mass we have introduced the $\eta\eta'$ eigenvalues by exploiting the fact that the sum of the eigenvalues is equal to the sum of the diagonal elements. The determinant is given by the product $\eta\eta'$ and the resulting sum rule is

$$(\eta + \eta')(4K - \pi) - 3\eta\eta' = \tfrac{1}{3}(4K - \pi)^2 + \tfrac{8}{3}(\pi - K)^2 \cos^2\alpha$$

(17.94)

In the limit $\alpha \to 0$ we recover the previous discussion where R was not included. Indeed the sum rule in equation (17.94) collapses to the old equation (17.88).

Inserting the observed masses yields $\cos^2\alpha \simeq \tfrac{1}{5}$ (or $\tfrac{1}{4}$ if quadratic masses are used). The latter is interesting since if $|R\rangle \equiv |c\bar{c}\rangle$ then

$$|1\rangle \equiv \frac{1}{\sqrt{12}}(u\bar{u} + d\bar{d} + s\bar{s} - 3c\bar{c})$$

(17.95)

leads naturally to this magnitude of $\cos \alpha$. We have of course assumed that there is no mixing of $(\mathbf{8}, \mathbf{1})$ with the orthogonal state

$$|\alpha'\rangle = -\sin \alpha |1\rangle + \cos \alpha |R\rangle \overset{(\cos = \sqrt{\frac{3}{2}})}{=\!=\!=\!=} \tfrac{1}{2}(u\bar{u} + d\bar{d} + s\bar{s} + c\bar{c})$$

$$(17.96)$$

The conclusion appears to be that η, η' may contain significant $c\bar{c}$ contributions. This would have interesting consequences in ψ and χ decays since the η, η' can be produced through their $c\bar{c}$ content without violating Zweig's rule (e.g. $\psi' \to \psi\eta(c\bar{c})$). This is surveyed by Harari (1976a). A problem may arise with decays of the χ states which may produce η and η' pairs without inhibition and hence the narrow χ width may be problematic. All of these problems would be avoided if the solution to the $\eta\eta'$ mass problem is the mass dependence of A (equation 17.89; de Rujula et al., 1975).

For the $c\bar{c}$ solution an interesting relation has been noted by Karl (1976). The radiative decays $\psi \to \eta(\eta')\gamma$ will be proportional to the amount of $c\bar{c}$ in these pseudoscalars. Apart from phase space

$$\frac{\Gamma(\psi \to \eta'\gamma)}{\Gamma(\psi \to \eta\gamma)} = (\cot^2 \theta) = \frac{3\eta' - 4K + \pi}{4K - \pi - 3\eta} \qquad (17.97)$$

Experimentally the radiative ratio is about 3 (Braunschweig et al., 1977) and this is compatible with the right-hand side.

18　Quarks Confined to a Sphere: The MIT Bag Model

We have seen that there is good reason to think that strong interactions might be described by a non-Abelian gauge field theory (quantum chromodynamics). At very short distances the quark–gluon coupling tends asymptotically to zero and hence gives some possible justification for the phenomenological successes of the parton model. At these distances the dominant interquark force will arise from single gluon exchange and the resulting patterns of hyperfine splittings in hadron spectroscopy are not inconsistent with this.

It is plausible that as the interquark separation grows, then the interquark force will also grow in such a theory. We have seen in section 15.2.1 how this might cause quark confinement and yield only colour singlet states with finite masses. In a hypothetical world of one space and one time dimension explicit calculations in QCD have shown that the interquark potential rises linearly with the separation r, and hence the desired confinement ensues ('t Hooft, 1974).

This result is easily understood intuitively. In a world with three space dimensions the lines of "electric" force emanating from a charge spread out over the surface of a Gaussian sphere covering $4\pi r^2$ at distance r from the source. Hence $E(r) \sim 1/r^2$ and so $V(r) \sim 1/r$, the familiar Coulomb potential. In a world with only one space dimension, however, the field lines are constrained to a line and so $E(r) =$ constant. In turn therefore $V(r) \sim r$.

Consequently this result is an artefact of the particular dimensionality chosen. A proof that the confinement arises in the real world is still awaited. In the absence of such a proof, and detailed knowledge of the behaviour of the interquark forces at large distances, it is not yet possible to do complete calculations of hadronic phenomenology in QCD. What has been done so far is to assume some form for the confining potential (λr, ωr^2 ...) and to proceed from there.

If one restricts attention to massive quarks, such as the charmed quark, then nonrelativistic potential models like these might have some justification. For light quarks the situation is less clear. One approach in the latter case is to consider a system of light quarks (or even massless quarks such as are manifested in deep inelastic scattering) and demand that they be confined to a sphere of radius R. As $R \to \infty$ an unconfined free quark model will emerge. For finite R one will see what phenomenological consequences may flow from the act of confining the free quarks. These consequences are in many cases very interesting and encouraging.

18.1 Free quarks confined to a sphere

If spectroscopy had not been studied and all that we had to support the quark idea was deep inelastic scattering then we would believe that the quarks were free pointlike spin $\frac{1}{2}$ particles. We would also know that the quarks do not appear outside the proton, so therefore we take the Dirac equation for a free fermion of mass m

$$\not{\partial}\psi(x) = m\psi(x) \tag{18.1}$$

and solve it in a region of space bounded by a sphere of radius R (section 11 of Akhiezer and Berestetski, op. cit.).

There are two classes of solutions with $j = \frac{1}{2}$ corresponding to the two parity states $S_{1/2}$ and $P_{1/2}$ characterised by the values of Dirac's quantum number $\kappa = \pm(j + \frac{1}{2}) = \pm 1$. These solutions are

$$\psi_{-1} = N \begin{pmatrix} \sqrt{\left(\dfrac{\omega + m}{\omega}\right)} ij_0\left(\dfrac{rx}{R}\right) U_m \\ -\sqrt{\left(\dfrac{\omega - m}{\omega}\right)} j_1\left(\dfrac{rx}{R}\right) \dfrac{\boldsymbol{\sigma} \cdot \hat{\mathbf{r}}}{r} U_m \end{pmatrix} \tag{18.2}$$

and

$$\psi_{+1} = N' \begin{pmatrix} \sqrt{\left(\dfrac{\omega + m}{\omega}\right)}_{ij_1}\left(\dfrac{rx}{R}\right)\dfrac{\boldsymbol{\sigma}\cdot\hat{\mathbf{r}}}{r}U_m \\ \sqrt{\left(\dfrac{\omega - m}{\omega}\right)}_{j_0}\left(\dfrac{rx}{R}\right)U_m \end{pmatrix} \qquad (18.3)$$

where N, N' are normalisation coefficients, U_m is a two-component spinor for angular momentum m projected along the z-axis of quantisation and ω is the energy given by

$$\omega = \left(m^2 + \frac{x^2}{R^2}\right)^{1/2} \qquad (18.4)$$

since x is the quark momentum in units of $1/R$.

If one studies the nonrelativistic limit $\omega - m \to 0$ then only the upper two component spinors survive. Under a parity transformation

$$\psi_{\pm 1} \xrightarrow{\ \mathbf{r} \to -\mathbf{r}\ } \pm \psi_{\pm 1} \qquad (18.5)$$

in the upper components and so these are the positive and negative parity solutions (S and P states) with $j = L + \frac{1}{2} = \frac{1}{2}$. On the other hand notice that in the relativistic case only j, not L, is a good quantum number and the upper and lower components have opposite intrinsic parities.

It is reasonable to suppose that the ground state hadrons are spherically symmetric and that the quarks will consequently be confined to a sphere of radius R. Demanding that no current flows across the surface of such a sphere constrains the possible values of ω, hence quantising the system's energy levels. We can easily see what the ω values can be.

If n_μ is the outward normal to the sphere then no current crossing the boundary implies

$$n^\mu \bar{\psi} \gamma_\mu \psi = 0 \qquad (18.6)$$

This equation can be satisfied if

$$-i\gamma_\mu n^\mu \psi = \psi \qquad (18.7)$$

since this requires

$$i\bar{\psi}\gamma \cdot n = \bar{\psi} \qquad (18.8)$$

and hence

$$\bar{\psi}\psi = (i\bar{\psi}\gamma . n)\psi \equiv \bar{\psi}(i\gamma . n\psi) = -\bar{\psi}\psi \tag{18.9}$$

Clearly then $\bar{\psi}\psi = 0$ and hence $n . \bar{\psi}\gamma\psi = 0$.

Having formulated the boundary conditions as in equation (18.7) we can substitute into it the solutions $\psi_{\pm 1}$ (equations 18.2 and 18.3) and see what constraints are forced upon ω. Writing out equation (18.7) explicitly at $r = R$ yields

$$\begin{pmatrix} 0 & -i\boldsymbol{\sigma} . \hat{\mathbf{r}} \\ i\boldsymbol{\sigma} . \hat{\mathbf{r}} & 0 \end{pmatrix} \begin{pmatrix} \sqrt{\left(\dfrac{\omega + m}{\omega}\right)} ij_0(x) \\ -\sqrt{\left(\dfrac{\omega - m}{\omega}\right)} j_1(x) \end{pmatrix} = \begin{pmatrix} \sqrt{\left(\dfrac{\omega + m}{\omega}\right)} ij_0(x) \\ -\sqrt{\left(\dfrac{\omega - m}{\omega}\right)} j_1(x) \end{pmatrix} \tag{18.10}$$

leading to the constraint

$$j_1(x) = \sqrt{\left(\frac{\omega + m}{\omega - m}\right)} j_0(x) \tag{18.11}$$

Now if $j_1(z) = Aj_0(z)$ then $\tan z = z/(1 - z)$ and so in our case

$$\tan x = x/(1 - mR - m^2R^2 + x^2) \tag{18.12}$$

where we used $\omega = (m^2 + x^2/R^2)^{1/2}$ (equation 18.4).

The analogous solution for the ψ_{+1} mode is

$$\tan x = x/(1 - mR + m^2R^2 + x^2) \tag{18.13}$$

When $m \to \infty$ (nonrelativistic limit) one has

$$\tan x = \frac{x}{1 - 2mR} \to 0 \quad \text{or} \quad \tan x = x \tag{18.14}$$

respectively having solutions $x = \pi$ or $x = 0$. This leaves only the ψ_{-1} configuration as physical (the $S_{1/2}$ solution) and the $x = \pi$ value is the familiar result from the Schrödinger equation.

The ultrarelativistic extreme ($m \to 0$) yields the transcendental equation

$$\tan x = x/(1 - x) \tag{18.15}$$

and for the lowest modes the solutions are

$$\psi_{-1}(x = 2\cdot04); \qquad \psi_{+1}(x = 3\cdot81) \tag{18.16}$$

For arbitrary m we exhibit in Fig. 18.1 the solutions x as a function of the mass for the solution ψ_{-1}.

A single quark of mass m confined to a sphere of radius R therefore has energy

$$\omega = (m^2 + x^2/R^2)^{1/2}$$

When in the lowest mode $x(mR)$ is given by the curve in Fig. 18.1 and one can think of ω as the effective mass of the confined quark since as

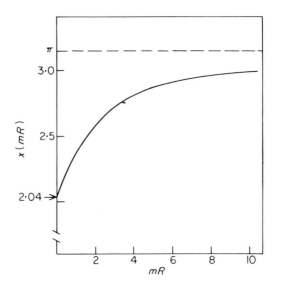

FIG. 18.1. $x(mR)$ eigenvalue of equations (18.10) and (18.12) yields the momentum (in units R^{-1}) of a fermion mass m confined to a sphere radius R.

$R \to \infty$ then $\omega \to m$. This quantity ω therefore plays the same role as the ε in our phenomenological formula equation (15.32) $\simeq 340$ MeV for u, d quarks and around 550 MeV for strange quarks. For a system of N_{ij} quarks with flavours i, j and masses $m_{i,j}$ the total quark energy will be (compare with equation 15.31)

$$E = \sum_i N_i (m_i^2 + x^2/R^2)^{1/2}$$

18.1.1 THE MIT BAG AND HADRON MASSES

As it stands this system is clearly unstable since increasing R decreases the energy monotonically until $R = \infty$. Hence there is no automatic confinement in the model. To prevent this expansion and confine the quarks *ad hoc* one could introduce a "pressure", B, which stabilises the system. This is the essential feature of the MIT bag model (Chodos *et al.*, 1974a,b; de Grand *et al.*, 1975; Johnson, 1975).

The total energy will become

$$E(R) = \sum_i N_i (m_i^2 + x^2/R^2)^{1/2} + B\frac{4\pi R^3}{3} \qquad (18.17)$$

and equilibrium will result when $E(R)$ is minimised, hence $\partial E/\partial R = 0$.

The scaling phenomenon in deep inelastic scattering suggests that the mass of the u, d quarks must be small. So to simplify the formulae let us go to the extreme and set the masses to zero.[1] The energy of the N quark system becomes

$$E = \frac{N \times 2\cdot04}{R} + \frac{4\pi R^3}{3}B \qquad (18.18)$$

since $x(0) = 2\cdot04$ and so

$$\frac{\partial E}{\partial R} = 0 = 4\pi R^2 B - \frac{N \times 2\cdot04}{R^2} \qquad (18.19)$$

with the consequence that

$$R = \frac{(N \times 2\cdot04)^{1/4}}{(4\pi B)^{1/4}} \qquad (18.20)$$

It is obvious from dimensional arguments that the size will decrease as the fourth root of the inverse pressure. The equation (18.20) has quantified this in terms of the number of massless quarks in the system. We can substitute into equation (18.18) and relate the mass (energy) of the system to the pressure

$$M_N = \tfrac{4}{3}(4\pi B)^{1/4}(N \times 2\cdot04)^{3/4} \qquad (18.21)$$

---

[1] See also Leutwyler (1974) who proposes that quarks are (nearly) massless. His motivation is rather different from that here.

or to its size

$$R = \frac{4}{3} \frac{2 \cdot 04 N}{E} \qquad (18.22)$$

Equation (18.21) implies that feeding in more quarks[1] increases the mass in such a way that

$$\frac{\text{Meson}}{\text{Baryon}} = \frac{M_2}{M_3} = \left(\frac{2}{3}\right)^{3/4} \simeq \frac{3}{4} \qquad (18.23)$$

If $N = 3$ and $M = M_{\text{proton}}$ then from equation (18.22) we find

$$R \simeq 1 \cdot 6 \, \text{fm} \qquad (18.24)$$

is the size required to generate a proton mass from the massless quarks. This mass has come from the quark's kinetic energy *and from the confining pressure*. These have generated an effective mass of some 340 MeV per u, d quark even though their mass as $R \to \infty$ may even be zero. Similarly a strange quark with mass $\simeq 100$ MeV unconfined obtains an effective mass $\varepsilon(\omega) \simeq 550$ MeV when confined to a sphere of $1 \cdot 5$ fm radius.

18.1.2   MAGNETIC MOMENTS

We saw in the nonrelativistic example that the quark and the proton have the same magnetic moments (scaled only by their overall electrical charges). To fit the $\mu_P = 2 \cdot 8$ Bohr magnetons with a nonrelativistic Dirac quark (for which $\mu_q = e/2m_q$) requires $m_q \simeq 340$ MeV. However one cannot justify this for a *physical* proton, i.e. one where the quarks are confined to a region R instead of being genuine *free* quarks. The magnetic moment of a quark with wavefunction ψ in a region $r < R$ is given in general by

$$\mu = \frac{1}{2} \int_{|r| < R} \mathrm{d}^3 \mathbf{r} \mathbf{r} \times \mathbf{j} = \frac{1}{2} \int_{|r| < R} \mathrm{d}^3 \mathbf{r} \mathbf{r} \times (\bar{\psi} \gamma \psi) e \qquad (18.25)$$

The result $\mu = e/2m_q$ is obtained only when $m_q \gg 1/R$ ($1/R$ being the characteristic momentum of the particle confined to R). To satisfy this

---

[1] Whereas fixed *pressure* yields $M_N \propto N^{3/4}$, fixed *radius* yields $M_N \propto N$. This latter gives correspondence with the approach of Chapter 17.

one must either let $R \to \infty$, which is the unconfined free limit, or let m_q be very large in which case the quark must have a large anomalous moment and contact is lost with the deep inelastic scattering phenomena. Therefore there appears to be an inconsistency in treatments like those described in Chapter 7.

For quarks with mass m confined to the sphere radius R we can explicitly calculate μ from equation (18.25). Substituting the wavefunctions in equation (18.2) into equation (18.25) yields (Allen, 1975)

$$\mu = \frac{e}{2m_Q} f(mR) \tag{18.26}$$

with

$$f(mR) \equiv \frac{1}{3} mR \frac{4\omega R + 2mR - 3}{2(\omega R)^2 - 2\omega R + mR} \tag{18.27}$$

(recall that $\omega = (m^2 + x^2/R^2)^{1/2}$ with x/R the momentum of the confined quark). We can check that as $mR \to \infty$ then $f \to 1$ and the nonrelativistic result emerges. In the other limit $mR \to 0$ we have $\omega R \equiv x(0) = 2 \cdot 04$, and so

$$f \to \tfrac{1}{3} mR \times 1 \cdot 22 \tag{18.28}$$

Hence

$$\mu \sim e \times (0 \cdot 2) R \tag{18.29}$$

is the magnetic moment of a confined massless quark.

Note that as $R \to \infty$ so $\mu \to \infty$ as $1/m$ in line with one's expectation for a free particle. What is important is that a massless *confined* Dirac particle has *finite* magnetic moment (essentially because its effective confined energy ω is non-zero). The magnetic moment as a function of mR is shown in Fig. 18.2.

To appreciate the significance of this result it is instructive to compare it with the phenomenological QCD approach (Chapter 17) which assumed that $\mu = e/2\varepsilon$ with $\varepsilon \simeq 340$ MeV (our ω).

Confining a massless quark in a sphere has yielded for us

$$\mu = \frac{e}{2\omega} \frac{1}{6} \left\{ \frac{4\omega R - 3}{\omega R - 1} \right\} \simeq \frac{5}{6} \frac{e}{2\omega} = \frac{5}{6} \mu_{\text{QCD}} \tag{18.30}$$

For a quark with a large mass $m \to \omega$ and so $\mu_{\text{bag}} = \mu_{\text{QCD}}$. Hence the phenomenological recipe $\mu \sim e/2\omega$ agrees with the spherical confinement always to better than 20 per cent. Since we expect that quantum fluctuations might affect the bag results by this order of magnitude anyway, then for spectroscopic comparisons it is simpler, and equivalent, to use the phenomenological recipes of section 17. This is illustrated again in section 19 for the spectroscopy of $q^2\bar{q}^2$ states where the recipe approach agrees well with explicit bag calculations (Jaffe, 1977a,b; Jaffe and Johnson, 1976).

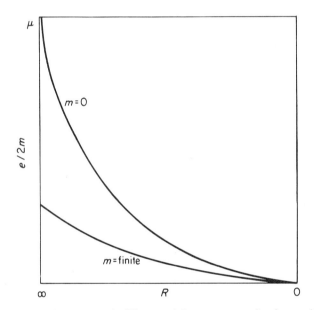

FIG. 18.2. Magnetic moment of a Dirac particle, mass m, confined to a sphere radius R.

In the explicit bag model, we can take the result of equation (18.29) and, equating the quark magnetic moment with that of the proton yields, for massless quarks

$$0\cdot 2R \simeq \frac{1}{660 \text{ MeV}} \qquad (18.31)$$

and hence $R \simeq 1\cdot 5$ fm. This shows beautiful consistency with our previous result in equation (18.25) that flowed from the proton mass. If we set $m_q > 0$ then μ_q decreases and $R > 1\cdot 5$ fm. Hence a sensible result for the size of the system is only consistent with the proton magnetic moment if the quarks are light.

18.1.3 g_A/g_V

If the operator for g_A/g_V is

$$\frac{\langle P^\uparrow|\sum_i \tau_i^+ S_z^i|N^\uparrow\rangle}{\langle P^\uparrow|\sum_i \tau_i^+|N^\uparrow\rangle} \tag{18.32}$$

then the SU(6) wavefunctions for the nucleon (section 4.2) yield a value $\frac{5}{3}$ for this quantity. In view of the empirical value being $\approx 1\cdot 1$, several excuses have been offered over the years. Originally one idea was that the strong interactions renormalized the intrinsic g_A for the quark so that at quark level one had $\gamma_\mu + R\gamma_\mu\gamma_5$ as the current with $R \approx 0\cdot 7$. Another approach has been to retain the $\gamma_\mu + \gamma_\mu\gamma_5$ structure at quark level but to take relativistic corrections into account (Bogobliubov, 1968).

As an example consider a quark moving independently of the other quarks in a scalar radially symmetric potential. The wavefunction is then

$$\psi(r) = \begin{pmatrix} f(r)U_m \\ ig(r)\boldsymbol{\sigma}\cdot\mathbf{r}U_m \end{pmatrix} \tag{18.33}$$

where U_m is the two-component quark spinor and f, g are functions of r that depend upon the specific form of the potential. The upper and lower two components of the four-component spinor have $L = 0$ and $L = 1$ respectively. Consequently the matrix element of quark spin

$$\langle q|\sigma_z|q\rangle \tag{18.34}$$

(which is unity for a quark with spin up in the nonrelativistic case) is no longer unity for the relativistic vase since the spin can be down in the lower component. Hence the result becomes

$$g_A/g_V = \tfrac{5}{3}\langle\sigma_z\rangle \tag{18.35}$$

where $\langle\sigma_z\rangle$ is the expectation value for the spin z—projection of the relativistic quark. Since the total angular momentum for each quark satisfies

$$j_z = \tfrac{1}{2}\sigma_z + l_z \tag{18.36}$$

then for a quark with $j_z = \tfrac{1}{2}$ we have

$$\langle\sigma_z\rangle = 1 - 2\delta \tag{18.37}$$

with

$$\delta = \frac{\int \psi_0^* l_z \psi_0 \, d\mathbf{r}}{\int \psi_0^* \psi_0 \, d\mathbf{r}}\bigg|_{j_z=1/2} = \frac{\frac{2}{3}\int |g(r)|^2 r^4 \, dr}{\int \{|f(r)|^2 + r^2 |g(r)|^2\} r^2 \, dr} \tag{18.38}$$

and in the last step we have supposed all the quarks to be in S-states.

Bogoliubov put the quarks in an infinite square well potential and calculated $g_A/g_V = 1\cdot 1$. The MIT bag is a modern version of this and has considerable similarity to the Bogoliubov model in this calculation.

For massless free quarks confined to the cavity and in the lowest mode, the wavefunction is given in equation (18.2). Comparison with the wavefunction for the general spherically symmetric case yields

$$f(r) = N\frac{\omega + m}{\omega} i j_0\left(\frac{rx}{R}\right) \tag{18.39}$$

$$g(r) = N\frac{\omega - m}{\omega} i j_1\left(\frac{rx}{R}\right) r^{-2} \tag{18.40}$$

with N the normalisation coefficient such that

$$\int dr \, r^2 (|f(r)|^2 + |g(r)|^2) = 1 \tag{18.41}$$

For massless quarks this yields

$$\delta = \frac{\frac{2}{3}\int_0^R |j_1(rx/R)|^2 \, dr}{\int_0^R dr \, r^2\{|j_0(rx/R)|^2 + 1/r^2 |j_1(rx/R)|^2\}} = \frac{2x - 3}{6(x - 1)} \tag{18.42}$$

where $x = 2\cdot 04$ for the lowest cavity mode. Hence

$$\frac{g_A}{g_V} = \frac{5}{3}\left[1 - \frac{2x - 3}{3(x - 1)}\right] \approx 1\cdot 1 \tag{18.43}$$

in excellent agreement with the data.

The 30 per cent deviation from the result $g_A/g_V = \frac{5}{3}$ is due entirely to the lower components of ψ_0 being sizeable for the cavity-confined massless quarks. As a function of quark mass g_A/g_V ranges from $1\cdot 1$ ($m = 0$) to $\frac{5}{3}$ ($m = \infty$). As an exercise one can calculate g_A/g_V as a function of m by using the full wavefunction at equations (18.39) and (18.40) and find the best value of m to fit the data ($g_A/g_V = 1\cdot 25$). This has been done by Golowich who found $mR \approx 1$ (Golowich, 1975).

We can check the related phenomenology of the $mR \simeq 1$ case. This yields $x = 2\cdot4$ and hence $\omega R = (x^2 + m^2 R^2)^{1/2} = 2\cdot6$. If $\omega_{u,d} \simeq 330$ MeV (as required for reasonable mass or proton magnetic moment phenomenology) then $1/R \simeq 130$ MeV $= 1\cdot6$ fm. Hence the proton radius is quite able to accept a value $m_{u,d} \simeq 100$ MeV in place of zero.[1] In turn the strange quark mass will have to be about 300 MeV (i.e. some 150 to 200 MeV heavier than the u, d quarks) and hence $m_s R \simeq 2\cdot5$ and $x \simeq 2\cdot7$. This yields $\omega_s \simeq \sqrt{2m} = 520$ MeV as the phenomenological effective mass of a strange quark.

18.2   Gluons in the bag and hyperfine splitting

In the QCD model we showed how the exchange of coloured gluons generated a splitting of mass between nucleon and delta and between 0^- and 1^- mesons. In the MIT bag qualitatively similar results will be found. The main difference is that the overall scale of the splitting (the quantity ΔE_M in equation 17.4) is arbitrary *a priori* in the earlier approach: here it can be related to the quark–gluon coupling constant α_c which can be calculated from other phenomenology in the bag model. Furthermore the treatment is now relativistic.

The resulting expressions for the flavour dependence of the spin-dependent mass splittings are algebraically complicated as a result of the integration over the spherical cavity; however they are the same (modulo 20 per cent effects) as the naive prescription of section 17 and hence the resulting systematics of the splittings are almost identical in the two approaches. We will see this later and first describe the cavity calculation.

We will work to lowest order in α_c and hence no self-gluon couplings will contribute since all such diagrams are of higher order. This has the consequence that the gluons act as if they were eight independent Abelian fields and hence the problem is analogous to one of conventional electromagnetism in a cavity.

In the gluon exchange process the quarks remain in the lowest cavity mode and so the vertex current is time independent. Hence we are dealing with conventional electro- or magnetostatics.

---

[1] This magnitude has also been found by Jaffe and Llewellyn Smith (1973) using quite a different approach. However, see also Leutwyler (1974).

The colour magnetic interaction energy between a pair of quarks is

$$\Delta E_M = -4\pi\gamma_c \sum_{a=1,8} \sum_{i>j} \int_{\text{bag}} d^3r \, \mathbf{B}_i^a(r) \cdot \mathbf{B}_j^a(r) \qquad (18.44)$$

where B_i^a is the colour magnetic field generated by the i-th quark in the system and the colour index a runs from 1 to 8. The colour magnetic field in the bag satisfies

$$\left. \begin{array}{l} \nabla \times \mathbf{B}_i^a = \mathbf{j}^a \\ \nabla \cdot \mathbf{B}_i^a = 0 \end{array} \right\} r < R \qquad (18.45)$$

where j_i^a is the i-the quark's colour current. This is

$$\mathbf{j}_i^a = q_i^+ \alpha \lambda^a q_i$$

$$= \frac{-3}{4\pi} \hat{\mathbf{r}} \times \sigma_i \lambda_i^a \mu_i'(r)/r^3 \qquad (18.46)$$

where $\mu_i'(r)$ is the i-th quark's scalar magnetisation density when in the lowest cavity mode. When integrated over the bag this yields the quark mode's magnetic moment $\mu(m_i, R)$ given by

$$\int_0^R \mu_i'(r) \, dr = \mu(m_i, R) = \frac{1}{2\omega} \frac{1}{3} \left\{ \frac{4\omega R + 2mR - 3}{2(\omega R - 1) + m/\omega} \right\}$$

$$\approx \left(\frac{5}{6} \text{ to } 1 \right) \frac{1}{2\omega} \qquad (18.47)$$

The equations (18.45) and (18.46) can be integrated to determine $B_i^a(r)$ and is a standard exercise in electromagnetic theory. However we must demand that no colour gluon crosses the boundary of the bag and this necessitates inclusion of a boundary condition

$$\hat{\mathbf{r}} \times \sum_i \mathbf{B}_i^a = 0, \qquad r = R \qquad (18.48)$$

The resulting $B_i(r)$ that also satisfies this condition is

$$B_i^a(r) = \frac{\lambda_i^a}{4\pi} \left(\frac{\mu(m_i, r)}{r^3} \right) \{ 3\hat{r}(\sigma_i \cdot \hat{r}) - \sigma_i \}$$

$$+ \frac{\lambda_i \sigma_i}{4\pi} \left(\frac{\mu(m_i R)}{R^3} + 2 \int_r^R \frac{dr'}{r'^3} \mu'(r') \right) \qquad (18.49)$$

where $\mu(m_i, r)$ is the integral of $\mu'_i(r')$ to radius r and

$$M_i(r) = \int_r^R \frac{dr'}{r'^3} \mu'_i(r') \tag{18.50}$$

This result is familar from electromagnetic theory, apart perhaps from the presence of terms in the final brackets whose origin is the boundary condition in equation (18.48).

Having found the colour magnetic field at position r originating from the i-th quark we have for the interaction energy with the j-th quark:

$$\Delta E_M^{ij}(r) = \frac{\lambda_j^a \lambda_i^a}{4\pi} \frac{\mu'_i(r)\mu(m_i, r)}{r^3} \{3\sigma_j \cdot \hat{\mathbf{r}}\sigma_i \cdot \hat{\mathbf{r}} - \sigma_j\sigma_i\}$$

$$+ \frac{\lambda_j^a \lambda_i^a}{4\pi} \sigma_i \cdot \sigma_j \mu'_i(r') \left\{ \frac{\mu(m_i R)}{R^3} + 2\int_r^R \frac{dr'}{r'^3} \mu'(r') \right\} \tag{18.51}$$

The structure of this equation is already familiar from the two-body Fermi–Breit interaction (compare equation 17.35). For the S-wave states the first line gives no contribution. Integrating over the bag we obtain finally the total magnetic interaction energy

$$\Delta E_M = -3\alpha_c \sum_a \sum_{i>j} (\lambda_i^a \sigma_i) \cdot (\lambda_j^a \sigma_j) \left\{ \frac{\mu(m_j, R)\mu(m_i, R)}{R^3} \right.$$

$$\left. + 2\int_r^R \frac{dr}{r^4} \mu(m_i, r)\mu(m_j, r) \right\} \tag{18.52}$$

If the second term in parentheses were absent, since $\mu(m_i R) \simeq 1/2\omega_i$ the form for ΔE_M in equation (18.52) would be identical to that of the QCD development in equation (17.35). However the second term introduces a multiplicative correction itself dependent upon $(\omega_i\omega_j)^{-1}$.

For baryons $\sum_a^1 \lambda \cdot \lambda = -\frac{8}{3}$ and for mesons it is $-\frac{16}{3}$, so with $N = 1, 2$ for mesons, baryons we have

$$\Delta E_M = N 8\alpha_c \sum_{i>j} \sigma_i \cdot \sigma_j \frac{\mu(m_i, R)\mu(m_j, R)}{R^3} I(m_i R, m_j R) \tag{18.53}$$

where I is the parenthesis in equation (18.52).

To obtain a feeling for the order of magnitudes here let's take massless quarks where $\mu \simeq 0.2R$. Then $I(0, 0) \simeq 1.5$ and so the $\Delta - N$

---

[1] Note that $\lambda \equiv 2F$; compare equations (15.38) and (15.39).

splitting will be[1] (recall $\langle \boldsymbol{\sigma} . \boldsymbol{\sigma} \rangle = +3$ for Δ and -3 for nucleon)

$$\Delta E_M = 48\alpha_c \frac{0\cdot04}{R} \times 1\cdot5 \qquad (18.54)$$

Hence if $R \simeq 1$ fm $\simeq (200 \text{ MeV})^{-1}$ then

$$\Delta E_M \sim 570\alpha_c \text{ MeV} \qquad (18.55)$$

The observed separation of nucleon and Δ then requires $\alpha_c \simeq 0\cdot55$.

For a state composed of quarks with arbitrary flavours and masses we will rewrite equation (18.53) to read

$$\Delta E_N = \sum_{ij} (\boldsymbol{\sigma}_i . \boldsymbol{\sigma}_j) M_{ij} N \qquad (18.56)$$

where

$$M_{ij} \equiv \frac{8\alpha_c}{R} \left[\frac{\mu(m_i R)}{R} \frac{\mu(m_j R)}{R} \right] I(m_i R, m_j R) \qquad (18.57)$$

and $\mu(m_i R)/R \simeq (2\omega_i R)^{-1}$, the exact expression being given by equation (18.47). The quantity $R M_{ij}/8\alpha_c$ is exhibited in Fig. 18.3.

To illustrate the use of the Figure, and to compare the results with the QCD we shall calculate $(\Sigma^* - \Sigma)/(\Delta - N)$ mass splittings. This ratio essentially probes the ratio of strange–nonstrange to nonstrange–non-strange couplings. In QCD we took $\omega_s \simeq 500$ MeV and $\omega_{u,d} \simeq 330$ MeV; hence

$$\left(\frac{\Sigma^* - \Sigma}{\Delta - N} \right)_{\text{QCD}} = \frac{\omega_{u,d}}{\text{fi}\omega_s} \simeq \frac{2}{3} \qquad (18.58)$$

If, in the bag, we take $m_{u,d} \simeq 0$ then

$$\frac{M_{uu}R}{8\alpha_c} \simeq \frac{0\cdot18}{3} \qquad (18.59)$$

with $m_s \sim 300$ MeV and $m_s R \sim 2\cdot5$ so that the interaction with massless u, d quarks at $m_s R = 2\cdot5$ yields

$$\frac{M_{us}R}{8\alpha_c} \simeq \frac{0\cdot12}{3} \qquad (18.60)$$

and the ratio of $2:3$ is again obtained.

---

[1] That $I(0, 0) \simeq 1\cdot5$ can be seen by comparing equation (18.57) with Fig. 18.3.

If instead we take $m_{u,d} \simeq 100$ MeV so that $mR \simeq 1$ then $M_{uu}R/8\alpha_c$ is reduced to about $0 \cdot 13/3$ (lower curve in Fig. 18.3). The interaction of u and s will have to be interpolated from these curves but will be nearly the same as assuming both u, d are of the same mass and $mR \simeq 2 \cdot 5$. The ratio of $2:3$ still arises at this qualitative level.

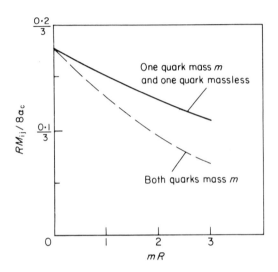

FIG. 18.3. $RM_{ij}/8\alpha_c$, the magnetic interaction between confined quarks with masses m_i and m_j. When one quark is massless the upper curve is used, m denoting the other quark's mass. The lower curve is for both quarks having the same mass, m. The curves for unequal massive quarks may be estimated by interpolation.

Hence phenomenologically the same results approximately arise for the magnetic separations in this explicit model as in the QCD model with phenomenological masses ε employed for the magnetic moments. The reason is that the flavour dependence in equation (18.53) is dominantly in the $\mu(m_iR)\mu(m_jR)$ ($\simeq 1/\omega_i\omega_j$ as in QCD) and only weakly in the $I(m_iR, m_jR)$. Consequently a similar phenomenology emerges.

Detailed calculations have been reported in the literature and there are some finer details such as the R dependence of different hadrons, volume energy etc. that cause contributions to energies (Johnson, 1975, 1977; de Grand et al., 1975).

18.3   Act of confinement

The quarks in a bag carry both electric and colour charge. Gauge
invariance necessarily requires that electromagnetic and colour gauge
fields (photons and gluons) are also present. Consider first the familiar
case of electromagnetism.

Gauss' theorem applied to the surface implies that

$$4\pi Q = \int \mathbf{\nabla} . \mathbf{E} \, dV = \int \mathbf{E} . d\mathbf{S}$$

where Q is the total electrical charge of the system. Hence if the system
of quarks has net charge non-zero then lines of electric field necessarily
cross the normal to the sphere. Conversely, if the photon field were also
confined to the sphere the total charge would necessarily be zero.

Consider the chromodynamics case which is analogous to the above.
If the demand that no quark current crosses the boundary is supplemented by the demand that colour gluons are also confined then
application of Gauss' theorem implies that the system have zero colour
charge. Hence the introduction of a pressure B that counterbalances the
flow of colour flux automatically requires the system to be colour
neutral. If colour symmetry is exact then the system must be a colour
singlet.

Notice that the quark confinement arises as a result of colour
confinement. The imposition *ad hoc* of a boundary condition that
confines the coloured gluons has, by Gauss, confined the coloured
quarks. A dynamical origin for this boundary condition has not been
presented.

18.4   Glueballs

The boundary condition that no colour current flow across the bag
surface allows only colour singlet states to exist with finite masses. In
addition to $q\bar{q}$, qqq states one expects that colour singlet states containing only gluons will exist (Jaffe and Johnson, 1976). Such states are
often referred to as "gluonium" or "glueballs".

The possible modes for a massless vector field confined to a cavity are
discussed in many books on electromagnetism. For any value of the
total angular momentum $J \geqslant 1$ either transverse electric (TE) or trans-

verse magnetic (TM) modes exist with parities $(-)^{J+1}$ and $(-)^J$ respectively.

The boundary condition $n_\mu F_{\mu\nu} = 0$ $(F_{\mu\nu} \equiv \partial_\mu A_\nu - \partial_\nu A_\mu)$ leads to the equation of constraint for the TE modes

$$\tan x = \frac{x}{1 - x^2}$$

(x being the momentum in units of $1/R$). The lowest value of x satisfying this is 2.74 (contrast 2·04 for the spin $\frac{1}{2}$ case in equation 18.16).

If we neglect any gluon self-coupling then the mass of an n gluon state with all the gluons in the lowest mode, and with the same radius as an n quark state, will be

$$\frac{M(n \text{ gluons})}{M(n \text{ quarks})} = \frac{2·74}{2·04} \simeq \frac{4}{3}$$

since $M \propto (nx)$ in equation (18.22). Hence the colour singlet state $(\text{TE})^2$ will be about $\frac{4}{3}$ times as massive as the mean $(q\bar{q})$ ground state; thus

$$M(\text{TE})^2 \simeq 960 \text{ MeV}$$

In the lowest mode two $J^P = 1^+$ gluons will couple to a total $J^{\pi c} = 0^{++}$ or 2^{++}.

For the TM modes the transcendental equation has $x = 4·49$ as its lowest solution. The two-gluon state with the same radius as $(\text{TE})^2$ will therefore have a mass

$$(\text{TM})^2 = \frac{4·49}{2·74}(\text{TE})^2 \simeq 1600 \text{ MeV}$$

The possible $J^{\pi c}$ are again 0^{++} and 2^{++}.

A two-gluon state $(\text{TE})(\text{TM})$ can also exist with $J^{\pi c} = 0^{-+}, 1^{-+}, 2^{-+}$ and masses 1290 MeV (before any spin-dependent splitting from gluon exchanges).

The two-gluon colour singlet states have charge conjugation $C = +1$. Colour singlet states with three gluons have $C = +1$ (d_{ijk}) or $C = -1$ (f_{ijk}). The lowest mass states are $(\text{TE})^3$ with $J^{\pi c} = 0^{+-}, 1^{++}, 2^{+-}, 3^{++}$ and mass

$$(\text{TE})^3 = \frac{2·74}{2·04} \times 1100 \text{ MeV} = 1400 \text{ MeV}$$

(where we compared with the mean mass of qqq, N and Δ, taking account of the different mode frequency $x = 2{\cdot}74$ as against $2{\cdot}04$ in the spin-$\frac{1}{2}$ case).

If the ideas of QCD are correct then colour singlets of pure glue must exist in addition to the $q\bar{q}$, qqq states. *A priori* they could be just a continuum and not show any peaks. In the bag model however there are eigenmodes of the glue in the cavity which generate a rich structure in the 1 to 2 GeV region. Some of these states have $J^{\pi c} = 1^{--}$ and may be seen in e^+e^- annihilation. Since the glue is electrically neutral its coupling to e^+e^- must be via a $q\bar{q}$ loop (Fig. 18.4(b)). The decay to hadrons consisting entirely of $q\bar{q}$ combinations will be diagrammatically as in Fig. 18.4(a). Hence as an estimate

$$\frac{\Gamma(\text{glue} \to e^+e^-)}{\Gamma(\text{glue} \to \text{all})} \sim 0(\alpha^2)$$

which suggests that J/ψ ($3{\cdot}1$ GeV) is not a glue ball.

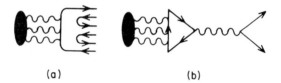

(a)   (b)

FIG. 18.4. Gluonium decay to (a) hadrons, (b) e^+e^-.

The phenomenology of glue balls is in its infancy and seems likely to be much discussed in the immediate future. The mixing of glue with, in particular, massive quark states (e.g. ψ, ψ') has been discussed by Jaffe and Kiskis (1976).

19 Multiquark Hadrons

19.1 $q\bar{q}$ and qqq states in QCD

In section 15.2.7 we saw that a system of coloured quarks interacting by exchanging octets of vector gluons would have an overall energy or mass $M(n) = nM_q + V(\lambda^2_{tot} - \frac{4}{3}n)$ for n quarks and antiquarks with λ^2_{tot} the colour SU(3) Casimir operator for the representation of the n quark system. If

$$V = \frac{3}{4}(m_q - \varepsilon)$$

then

$$M(n) = \frac{3}{4}m_q\lambda^2_{tot} + \frac{3}{4}\varepsilon\left(\frac{4}{3}n - \lambda^2_{tot}\right) \tag{19.1}$$

The values for λ^2_{tot} for various SU(3) representations have been given in Table 2.2. Since $\lambda^2_{tot} = 0$ only for colour singlets, then in the $m_q \to \infty$ limit only colour singlets will have finite masses. The finite mass colour singlet states will be

$$M(n) = n\varepsilon \tag{19.2}$$

Phenomenologically $\varepsilon \approx 350$ MeV for u, d quarks, ≈ 500 MeV for s and ≈ 1500 MeV for c quarks and approximately 400 MeV is to be added for each unit of L excitation. Hence typical masses of colour **1** states are

$$M(q\bar{q})_{L=0} \sim 700 \text{ MeV}(\pi\rho); \quad 800 \text{ MeV}(KK^*); \quad 1000 \text{ MeV}(\eta\eta'\phi) \text{ etc.} \tag{19.3}$$

while

$$M(qqq)_{L=0} \sim 1100 \text{ MeV } (N\Delta)$$
$$M(qqq)_{L=1} \sim 1500 \text{ MeV } (N^*) \text{ etc.} \tag{19.4}$$

These are qualitatively all right. The $\mathbf{F} \cdot \mathbf{F} \, \mathbf{s} \cdot \mathbf{s}$ coupling in the gluon exchange then shifts the energy levels of different spin states in the same supermultiplet.

We recall that $\langle \mathbf{F}_1 \cdot \mathbf{F}_2 \rangle_1 = -\frac{4}{3}$ and $\langle \mathbf{F}_1 \cdot \mathbf{F}_2 \rangle_{\bar{3}} = -\frac{2}{3}$. The former is relevant to the gluon exchange between $q\bar{q}$ in a colour singlet meson and the latter between qq in a colour singlet baryon (since the remaining q is necessarily $\mathbf{3}$ of colour and so the qq must be in $\bar{\mathbf{3}}$ if an overall qqq singlet is to be formed). The fact that the signs are both negative caused 0^- to be lower in mass than 1^- and $\frac{1}{2}^+$ lower than $\frac{3}{2}^+$ while the respective magnitudes of $\lambda \cdot \lambda$ and $\mathbf{s} \cdot \mathbf{s}$ caused the shifts relative to the mean multiplet energy to be

$$0^- - \Delta; \quad \tfrac{1}{2}^+ - \tfrac{1}{2}\Delta$$
$$1^- + \tfrac{1}{3}\Delta; \quad \tfrac{3}{2}^+ + \tfrac{1}{2}\Delta \tag{19.5}$$

Phenomenologically therefore Δ is approximately 300 MeV and hence

$$M(u\bar{u}, d\bar{d})_{0^-} \sim 400 \text{ MeV}$$
$$M(u\bar{u}, d\bar{d})_{1^-} \sim 800 \text{ MeV} \tag{19.6}$$
$$M(uud)_{\frac{1}{2}^+} \sim 950 \text{ MeV}$$
$$M(uud)_{\frac{3}{2}^+} \sim 1250 \text{ MeV} \tag{19.7}$$

which are in good first-order agreement with the real world.

19.2 Colour singlet $q^2\bar{q}^2$ mesons

After $q\bar{q}$, the simplest meson system that can have colour $\mathbf{1}$ quantum numbers is $qq\bar{q}\bar{q}$ (denoted $q^2\bar{q}^2$). Given the above mass estimates then the $q^2\bar{q}^2$ S-wave masses will be of order 1400 ($u^2\bar{u}^2$) to 2000 MeV($s^2\bar{s}^2$) before the spin-dependent splittings are taken into account. The $q^2\bar{q}^2$ in S-wave can couple to $J^P = 0^+1^+2^+$. These values of J^P can also be attained by $q\bar{q}$ in P-wave where the above rules will lead one to expect masses of order 1100 MeV before spin-dependent splittings.

After spin–spin splitting is included, the state of lowest J tends to be pushed down in mass. Jaffe (1977a,b) has shown that in the $q^2\bar{q}^2$ the 0^+ state is lowered by a greater amount than is the 0^+ in $q\bar{q}$. Consequently the $0^+ q\bar{q}$ and $q^2\bar{q}^2$ states may be of the same order of mass. Indeed it has even been suggested that the 0^+ states $\varepsilon(700)$ $S^*(993)$ $\delta(976)$ may be $q^2\bar{q}^2$ and lower in mass than the $q\bar{q}$ states (Jaffe, 1977a,b).

19.2.1   CHARACTERISTICS OF $q^2\bar{q}^2$ DECAYS

We have met the idea of decays which are OZI forbidden (requiring several gluons to enable the decay to arise (Fig. 19.1(a)) and OZI allowed (where a single $q\bar{q}$ is created and only one gluon is needed (Fig. 19.1(b)). In the case of $q^2\bar{q}^2$ mesons there can exist a class of diagrams

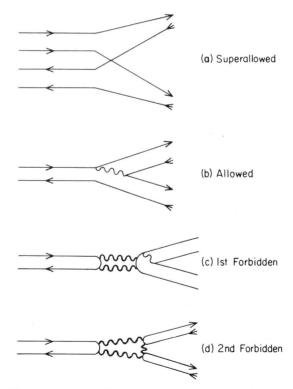

FIG. 19.1. Hierarchy of decay diagrams in quark–gluon models.

which we may denote "OZI superallowed". If phase space allows it the $q^2\bar{q}^2$ can simply collapse into $q\bar{q} + q\bar{q}$ as in Fig. 19.1(c),(d).

One can imagine a limit where the quark–gluon coupling is set equal to zero. In this case diagrams of OZI forbidden or allowed topologies will have zero amplitude and hence all $q\bar{q}$ states will be stable and hence of zero width. The OZI superallowed diagrams will still exist in this limit and so the $q^2\bar{q}^2$ states will have finite width.

In the real world the $q\bar{q}$ states have finite widths. The OZI super-allowed decays may be expected to give rise to extremely broad states which may be difficult to identify as true resonant states. However if $q^2\bar{q}^2$ states are low enough in mass then the superallowed decays may be kinematically forbidden or suppressed and these states will have widths typical of $q\bar{q}$ states. A particular example of this is in the case of the 0^+ $q^2\bar{q}^2$ nonet.

Jaffe (1977) has shown that a 0^+ nonet $q^2\bar{q}^2$ is expected to exist with rather low mass due to the large spin–spin splitting. In a specific model (the MIT bag)[1] the masses of these states are predicted as follows, and compared with candidate states:

		M	Γ
$u\bar{u}d\bar{d}$ (650)	$\ldots \varepsilon(700)$	660 ± 100	640 ± 140
$\dfrac{1}{\sqrt{2}}s\bar{s}(u\bar{u}+d\bar{d})$ (1100)	$\ldots S^*(993)$	$993 \cdot 2 \pm 4 \cdot 4$	$40 \cdot 0 \pm 7 \cdot 4$
$ud\bar{s}\bar{s}$ etc. (1100)	$\ldots \delta(976)$	$976 \cdot 4 \pm 5 \cdot 4$	$46 \cdot 9 \pm 11 \cdot 2$
$u\bar{s}\,d\bar{d}$ etc. (900)	$\ldots \kappa(?)$	~ 1300 broad	

The $\varepsilon(700) = u\bar{u}\,d\bar{d}$ can fall apart into $\pi\pi$ without inhibition. The very broad width of this state is therefore natural. The $S^*(993)$ and $\delta(976)$ mass degeneracy is natural as both contain $s\bar{s}$. Furthermore the

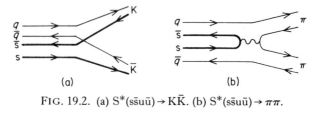

FIG. 19.2. (a) $S^*(s\bar{s}u\bar{u}) \to K\bar{K}$. (b) $S^*(s\bar{s}u\bar{u}) \to \pi\pi$.

---

[1] Our mass formula equation (19.2) corresponds to assuming a fixed bag *radius* (equation 18.22).

excess of some 300 MeV in mass over ε is natural due to the extra pair of strange quarks. The S^* can fall apart into $K\bar{K}$ whereas $\pi\pi$ requires a topology like Fig. 19.2(b). Hence the dominant decay mode is to $K\bar{K}$ as observed but the $K\bar{K}$ threshold is so near that the S^* width is consequently narrow. Similar remarks apply to the $\delta(976)$ whose width is similar to that of S^*.

19.2.2 WHAT IS THE FLAVOUR CONTENT OF THESE MULTIPLETS?

A qq system can be in $\mathbf{3} \otimes \mathbf{3} = \mathbf{6} \oplus \bar{\mathbf{3}}$ representations of SU(3) of flavour. The explicit content and charges of these multiplets are

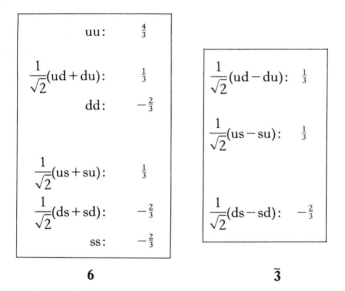

Notice the charges $\frac{1}{3}, \frac{1}{3}, -\frac{2}{3}$ characteristic of $\bar{\mathbf{3}}$ (like a single $\bar{q}$) as against $\mathbf{3}$. The $\mathbf{6}$ is symmetric and $\bar{\mathbf{3}}$ antisymmetric under interchange of any pair. Just as $\mathbf{1}$ and $\mathbf{8}$ states with similar isospin and charge can mix in producing the physical states (e.g. ω and ϕ) so can the $\bar{\mathbf{3}}$ mix with the three analogous states in the $\mathbf{6}$. For ease of notation we shall imagine that they are ideally mixed qq, $\bar{\mathbf{3}}$, and $\bar{q}\bar{q}$, $\mathbf{3}$, are shown in hypercharge $-I_3$ space in Fig. 19.3(a),(b). The resulting nonet of $q^2\bar{q}^2$ is then in Fig. 19.3(c).

Hence a nonet contains at most two strangeness carrying q or $\bar{q}$. An **18** contains at most three and the $\bar{s}\bar{s}ss$ occurs only in the 36plet.

The Pauli principle limits the possible number of allowed J^P, flavour combinations. In the next section we will see that a $J^P = 0^+$ nonet can

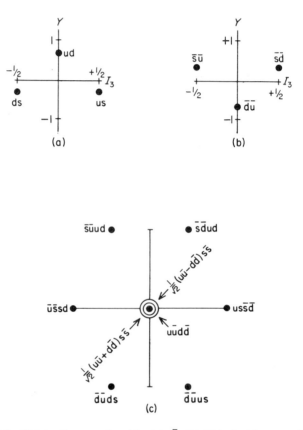

FIG. 19.3. Weight diagrams for: (a) qq in $\bar{3}$; (b) $\bar{q}\bar{q}$ in 3; (c) nonet of $q^2\bar{q}^2$.

exist. The three neutral members are $(u\bar{u}d\bar{d}) \equiv \varepsilon$; $\dfrac{1}{\sqrt{2}}s\bar{s}(u\bar{u} + d\bar{d}) \equiv S^*$;

$\dfrac{1}{\sqrt{2}}s\bar{s}(u\bar{u} - d\bar{d}) \equiv \delta$; $ud\bar{s}\bar{d} \equiv \kappa$ whose phenomenology has already been discussed.

19.2.3   WHAT STATES OF $q^2\bar{q}^2$ ARE ALLOWED BY THE PAULI PRINCIPLE?

A pair of quarks can have

$$\text{Spin} = 0 \text{ or } 1 \tag{19.9}$$

$$SU(3)_{\text{flavour}} = \bar{3} \text{ or } 6 \tag{19.10}$$

$$SU(3)_{\text{colour}} = \bar{3} \text{ or } 6 \tag{19.11}$$

In each of these separate cases the first option is antisymmetric and the second symmetric under interchange of the labels. The Pauli exclusion principle allows only antisymmetric combinations under interchange of all labels. Hence under $\text{Spin} \otimes SU(3) \otimes \widetilde{SU(3)}$ the qq can be as in equation (19.12).

ASS	$S = 0, \mathbf{6}, \tilde{\mathbf{6}}$	$S = 0, \bar{\mathbf{6}}, \tilde{\bar{\mathbf{6}}}$
SAS	$S = 1, \bar{\mathbf{3}}, \tilde{\mathbf{6}}$	$S = 1, \mathbf{3}, \tilde{\bar{\mathbf{6}}}$
SSA	$S = 1, \mathbf{6}, \tilde{\bar{\mathbf{3}}}$	$S = 1, \bar{\mathbf{6}}, \tilde{\mathbf{3}}$
AAA	$S = 0, \bar{\mathbf{3}}, \tilde{\bar{\mathbf{3}}}$	$S = 0, \mathbf{3}, \tilde{\mathbf{3}}$

$$\tag{19.12}$$

$$qq \qquad\qquad \bar{q}\bar{q}$$

All possible pairings can now take place between qq and $\bar{q}\bar{q}$ consistent with overall colour $\mathbf{1}$. This leads to the following states

$$
\begin{aligned}
(0^+, \mathbf{6}) \otimes (0^+, \bar{\mathbf{6}}) & \qquad \tilde{\mathbf{6}} \times \tilde{\bar{\mathbf{6}}} \to 0^+, \mathbf{36} \\
(0^+, \mathbf{6}) \otimes (1^+, \mathbf{3}) & \qquad \tilde{\mathbf{6}} \times \tilde{\bar{\mathbf{6}}} \to 1^+, \mathbf{18} \\
(1^+, \bar{\mathbf{3}}) \otimes (0^+, \bar{\mathbf{6}}) & \qquad \tilde{\mathbf{6}} \times \tilde{\bar{\mathbf{6}}} \to 1^+, \mathbf{18} \\
(1^+, \bar{\mathbf{3}}) \otimes (1^+, \mathbf{3}) & \qquad \tilde{\mathbf{6}} \times \tilde{\bar{\mathbf{6}}} \to 0^+, 1^+, 2^+, \mathbf{9} \\
(1^+, \mathbf{6}) \otimes (1^+, \bar{\mathbf{6}}) & \qquad \tilde{\bar{\mathbf{3}}} \times \tilde{\mathbf{3}} \to 0^+, 1^+, 2^+, \mathbf{36} \\
(1^+, \mathbf{6}) \otimes (0^+, \mathbf{3}) & \qquad \tilde{\bar{\mathbf{3}}} \times \tilde{\mathbf{3}} \to 1^+, \mathbf{18} \\
(0^+, \bar{\mathbf{3}}) \otimes (1^+, \bar{\mathbf{6}}) & \qquad \tilde{\bar{\mathbf{3}}} \times \tilde{\mathbf{3}} \to 1^+, \overline{\mathbf{18}} \\
(0^+, \bar{\mathbf{3}}) \otimes (0^+, \mathbf{3}) & \qquad \tilde{\bar{\mathbf{3}}} \times \tilde{\mathbf{3}} \to 0^+, \mathbf{9}
\end{aligned}
\tag{19.13}
$$

where the total spin-parity S-wave states are shown in the final column.

Hence

$$J^P = 2^+: \quad \mathbf{9}, \mathbf{36}$$

$$J^P = 1^+: \quad \mathbf{9}, \mathbf{18}, \mathbf{18}^*, \overline{\mathbf{18}}, \overline{\mathbf{18}}^*, \mathbf{36} \qquad (19.14)$$

$$J^P = 0^+: \quad \mathbf{9}, \mathbf{9}^*, \mathbf{36}, \mathbf{36}^*$$

is the sum total of states allowed in S-wave.

19.3 The spin-dependent splitting of $q^2 \bar{q}^2$ states

As in previous examples (sections 15.2 and 17.2) we wish to compute the expectation values of the gluon exchange interaction

$$\mathcal{H} = \Delta \sum_{i \neq j} \mathbf{F}_i \cdot \mathbf{F}_j \mathbf{S}_i \cdot \mathbf{S}_j \equiv -\Delta \tfrac{1}{4} \sum_{i \neq j} \lambda_i \lambda_j \mathbf{S}_i \cdot \mathbf{S}_j \qquad (19.15)$$

between any pair of q or $\bar{q}$ in $q_1 q_2 \bar{q}_3 \bar{q}_4$ where Δ has dimensions of mass to set the scale. We will illustrate this for the case of the $\mathbf{9}$ with 2^+, 1^+ and $\mathbf{36}$ with 2^+, 1^+. These states are

$$\mathbf{9} \equiv (1^+, \mathbf{6}_c)_{3,f} \otimes (1^+, \overline{\mathbf{6}}_c)_{\bar{3},f} \quad \text{and} \quad \mathbf{36} \equiv (1^+, \mathbf{3}_c)_{6,f} \otimes (1^+, \overline{\mathbf{3}}_c)_{6,f}$$

First let us separate $\mathcal{H}$ into two pieces, one involving qq or $\bar{q}\bar{q}$ and the other involving $q\bar{q}$. Then

$$\frac{4H}{\Delta} = - [\lambda_1 \cdot \lambda_2 \mathbf{S}_1 \cdot \mathbf{S}_2 + \lambda_3 \cdot \lambda_4 \mathbf{S}_3 \cdot \mathbf{S}_4] - [\lambda_1 \cdot \lambda_3 \mathbf{S}_1 \cdot \mathbf{S}_3 + \lambda_2 \cdot \lambda_4 \mathbf{S}_3 \cdot \mathbf{S}_4$$

$$+ \lambda_1 \cdot \lambda_4 \mathbf{S}_1 \cdot \mathbf{S}_4 + \lambda_2 \cdot \lambda_3 \mathbf{S}_2 \cdot \mathbf{S}_3] \qquad (19.16)$$

The expectation values of the first set of bracketed terms are immediate. Since $q_1 q_2$ and $\bar{q}_3 \bar{q}_4$ are each in spin 1 then

$$\langle \mathbf{S}_1 \cdot \mathbf{S}_2 \rangle = \langle \mathbf{S}_3 \cdot \mathbf{S}_4 \rangle = \tfrac{1}{4} \qquad (19.17)$$

Similarly, since the dimensionality of their colour group representations are known, then

$$\langle \lambda_1 \cdot \lambda_2 \rangle = \langle \lambda_3 \cdot \lambda_4 \rangle = \begin{cases} \tfrac{4}{3} \text{ in } \mathbf{6}_c \\ -\tfrac{8}{3} \text{ in } \mathbf{3}_c \end{cases} \qquad (19.18)$$

In order to compute the $\langle \mathbf{S}_i \cdot \mathbf{S}_j \lambda_i \cdot \lambda_j \rangle$ in the second set of bracketed terms we need to know the total spin and colour group representation of the $q\bar{q}$ pairs. *A priori* we do not know these but we can find out what they are by recoupling the spins (colours) to answer the question.

If q_1q_2 couple to j_A, q_3q_4 to j_B and j_A and j_B to J

$$[(j_1 \otimes j_2)_{j,A} \otimes (j_3 \otimes j_4)_{j,B}]_{j,A\otimes j,B=J}$$

$$= \sum_{j,C,j,D} k_{CD}[(j_1 \otimes j_3)_{j,C} \otimes (j_2 \otimes j_4)_{j,D}]_{j,C\otimes j,D=J} \quad (19.19)$$

then what are $j_{C,D}$ and coefficients k_{CD} such that $q_1\bar{q}_3$ couple to j_C, $q_2\bar{q}_4$ to j_D and j_C with j_D couple to J? We will answer this in detail in a moment. First we shall look at a very simple example, namely the 2^+ states.

$$\text{If} \begin{Bmatrix} q_1q_2 \rightarrow 1^+ \\ \bar{q}_3\bar{q}_4 \rightarrow 1^+ \end{Bmatrix} \quad \text{and} \quad 1^+ \otimes 1^+ \rightarrow 2^+ \quad (19.20)$$

then clearly when all q and $\bar{q}$ are in relative S-waves

$$\begin{Bmatrix} q_1\bar{q}_3 \rightarrow 1^- \\ q_2\bar{q}_4 \rightarrow 1^- \end{Bmatrix} \quad \text{and} \quad 1^- \otimes 1^- \rightarrow 2^+ \quad (19.21)$$

is the only possibility for the $q\bar{q}$ spins, i.e. the $q\bar{q}$ are in vector (V) spin states.

For this particular example, therefore, *all* pairs qq, $\bar{q}\bar{q}$ or $q\bar{q}$ are in spin 1 and hence $\langle S_i S_j \rangle = \frac{1}{4}$ for each. Hence

$$\frac{4H}{\Delta} = -\frac{1}{4}\left\langle \sum_{i\neq j} \lambda_i \cdot \lambda_j \right\rangle \equiv -\frac{1}{4}\langle \lambda_1 \cdot \lambda_3 + \lambda_2 \cdot \lambda_4 + (\lambda_1 + \lambda_3) \cdot (\lambda_2 + \lambda_4)\rangle \quad (19.22)$$

$$\equiv -\frac{1}{4}\langle \lambda_1 \cdot \lambda_3 + \lambda_2 \cdot \lambda_4 + \lambda_A^{q\bar{q}} \cdot \lambda_B^{q\bar{q}}\rangle \quad (19.23)$$

where A, B refer to the $q\bar{q}$ systems 13 and 24. Now for any pair i, j we have

$$2\lambda_i \cdot \lambda_j \equiv \lambda_{i+j}^2 - \lambda_i^2 - \lambda_j^2 \quad (19.24)$$

and so

$$\left. \begin{array}{l} \lambda_1 \cdot \lambda_3 \equiv \frac{1}{2}(\lambda_A^2 - 2\lambda_q^2) \\ \lambda_2 \cdot \lambda_4 \equiv \frac{1}{2}(\lambda_B^2 - 2\lambda_q^2) \\ \lambda_A \cdot \lambda_B \equiv \frac{1}{2}(\lambda_{tot}^2 - \lambda_A^2 - \lambda_B^2) \end{array} \right\} \quad (19.25)$$

(since $\lambda_q^2 \equiv \lambda_{\bar{q}}^2 = +\frac{16}{3}$ for 3_c or $\bar{3}_c$). In the sum we have finally

$$\langle H \rangle = -\frac{1}{16} \cdot \frac{1}{2}(\lambda_{tot}^2 - 4\lambda_q^2)\Delta = +\frac{2}{3}\Delta \quad (19.26)$$

since $\lambda_{tot}^2 = 0$ for a colour singlet system. This result is independent of $\lambda_{A,B}^2$ individually, these having cancelled (equation 19.25) as a result of the factorisation which in turn was due to the spin couplings all being maximally stretched to 2^+. Therefore one necessarily predicts that the **9($6_c \otimes \bar{6}_c$)** and the nine-dimensional analogous subgroup of the **36($3_c \otimes \bar{3}_c$)** will be degenerate in mass. This will be hard to test, however, because the $\langle H \rangle$ is positive which means that these states have been shifted up in mass relative to the mean mass of the supermultiplet and so we may anticipate that they will be very broad, decaying by Zweig superallowed modes.

From the guide in section 17 we expect $\Delta \simeq 300$ MeV and hence these 2^+ states will be shifted up by some 200 MeV relative to the mean mass of 1400 ($u^2\bar{u}^2$). Hence we anticipate

$$\left. \begin{aligned} 2^+(u\bar{u}\,d\bar{d}\text{ etc.}) &\sim 1600 \text{ MeV} \\ 2^+(u\bar{u}\,s\bar{d}\text{ etc.}) &\sim 1750 - 1800 \text{ MeV} \\ 2^+(u\bar{u}\,s\bar{s}\ldots) &\sim 1900 - 1950 \text{ MeV} \end{aligned} \right\} \tag{19.27}$$

for the **9** states. In the **36** representation 2^+ states also arise with masses

$$\left. \begin{aligned} 2^+(u\bar{s}\,s\bar{s}\ldots) &\sim 2050 - 2100 \text{ MeV} \\ 2^+(s\bar{s}\,s\bar{s}) &\sim 2200 - 2250 \text{ MeV} \end{aligned} \right\} \tag{19.28}$$

These orientations all agree, to better than 50 MeV, with explicit calculations in the MIT bag model (Jaffe, 1977a,b).

19.3.1   SPIN RECOUPLING IN $q^2\bar{q}^2 \to (q\bar{q})(q\bar{q})$

The spin and colour recoupling are straightforward to calculate. We will give explicit examples for one case and quote the results for the others which should be checked as an exercise by the reader.

We label $(q_1q_2)_{J,A}(\bar{q}_3\bar{q}_4)_{J,B} \leftrightarrow (q_1\bar{q}_3)_{J,C}(q_2\bar{q}_4)_{J,D}$. $J_{A,B}$ will be 0^+ or 1^+ while $J_{C,D}$ are 0^- (P) or 1^- (V). Then

$$[J_A \otimes J_B]_J = [\alpha PP + \beta PV + \gamma VP + \delta VV]_J \tag{19.29}$$

and the object is to find $\alpha\beta\gamma\delta$. The results are

$$(0^+ \otimes 0^+)_{0^+} = \tfrac{1}{2}PP + \tfrac{3}{2}VV \tag{19.30}$$

$$(1^+ \otimes 1^+)_{2^+} = VV \tag{19.31}$$

$$(1^+ \otimes 1^+)_{1^+} = \tfrac{1}{2}(PV + VP) \qquad (19.32)$$

$$(1^+ \otimes 1^+)_{0^+} = \tfrac{3}{2}PP - \tfrac{1}{2}VV \qquad (19.33)$$

$$(1^+ \otimes 0^+)_{1^+} = -\tfrac{1}{2}(VP - PV) + \tfrac{1}{2}(VV) \qquad (19.34)$$

$$(0^+ \otimes 1^+)_{1^+} = \tfrac{1}{2}(VP - PV) + \tfrac{1}{2}VV \qquad (19.35)$$

The result (equation 19.31) is obvious and was already used in the derivation of equations (19.27) and (19.28). We will exhibit the derivation of equation (19.30).

The spin 0 state of two spin $\tfrac{1}{2}$ objects is given in Chapter 3. Coupling a spin 0 pair with another spin 0 pair yields a state

$$\tfrac{1}{2}(u_1 d_2 - d_1 u_2)(u_3 d_4 - d_3 u_4) \qquad (19.36)$$

which necessarily has total $J^P = 0^+$ when in S-wave. Coupling $(13) \otimes (24)$ to yield $J^P = 0^+$ means that (13) and (24) can both be $0^-(P)$ or both $1^-(V)$. In the latter case the three combinations of $JM = (11, 1-1), (1-1, 11), (10, 10)$ are possible. We first need to know the particular combination that corresponds to 0^+. From a table of Clebsch–Gordan coefficients for $1 \otimes 1 \to 2 \oplus 1 \oplus 0$ each with $J_z = 0$ we can solve to obtain

$$(0^+, J_z = 0) = \frac{1}{\sqrt{3}}\{(1, -1) + (-1, 1) - (0, 0)\} \qquad (19.37)$$

$$(1^+, J_z = 0) = \frac{1}{\sqrt{2}}\{(1, -1) - (-1, 1)\} \qquad (19.38)$$

$$(2^+, J_z = 0) = \frac{1}{\sqrt{6}}\{(1, -1) + (-1, 1) + 2(0, 0)\} \qquad (19.39)$$

where only the J_z values are quoted in parentheses in an obvious notation. From equation (19.37) and Chapter 3 we can write the VV wavefunction that corresponds to 0^+. We have for $\alpha PP + \delta VV$ that

$$[(13) \otimes (24)]_{0^+} = \frac{\alpha}{2}(u_1 d_3 - d_1 u_3)(u_2 d_4 - d_2 u_4)$$

$$+ \frac{\delta}{\sqrt{3}}[u_1 u_3 d_2 d_4 + d_1 d_3 u_2 u_4 - \tfrac{1}{2}(u_1 d_3 + d_1 u_3)(u_2 d_4 + d_2 u_4)]$$

$$(19.40)$$

Comparing this with the $[(12) \otimes (34)]_{0^+}$ state (equation 19.36) yields

$$\alpha = \frac{1}{2} \quad \text{and} \quad \delta = \frac{\sqrt{3}}{2} \tag{19.41}$$

and hence

$$[0^+ \otimes 0^+]_{0^+} = \frac{1}{2}PP + \frac{\sqrt{3}}{2}VV \tag{19.42}$$

The other entries in the list can be obtained by a similar procedure using the equations (19.38) and (19.39) as necessary.

19.3.2 COLOUR RECOUPLING $q^2\bar{q}^2 \to (q\bar{q})(q\bar{q})$

We can label the SU(3) colour group by UDS (ultraviolet, deep purple and silver) and the $\mathbf{6}, \bar{\mathbf{3}}$ representations can be written by analogy with the SU(3) flavours of UDS, Table 3.6. The coupling $(\bar{\mathbf{3}}_c \otimes \mathbf{3}_c)_1$ is then

$$\frac{1}{\sqrt{3}}[\tfrac{1}{2}(U_1D_2 - D_1U_2)(\bar{U}_3\bar{D}_4 - \bar{D}_3\bar{U}_4) + \tfrac{1}{2}(D_1S_2 - S_1D_2)(\bar{D}_3\bar{S}_4 - \bar{S}_3\bar{D}_4)$$

$$+ \tfrac{1}{2}(S_1U_2 - U_1S_2)(\bar{S}_3\bar{U}_4 - \bar{U}_3\bar{S}_4)] \tag{19.43}$$

In the $(q\bar{q})(q\bar{q})$ this will be a linear combination of $(\mathbf{1}_c \otimes \mathbf{1}_c)_{1,c}$ and $(\mathbf{8}_c \otimes \mathbf{8}_c)_{1,c}$. Hence it must equal

$$\frac{A}{3}(U_1\bar{U}_3 + D_1\bar{D}_3 + S_1\bar{S}_3)(U_2\bar{U}_4 + D_2\bar{D}_4 + S_2\bar{S}_4)$$

$$+ \frac{B}{\sqrt{8}}\left\{ \begin{array}{l} U\bar{D}D\bar{U} + U\bar{S}S\bar{U} + D\bar{S}S\bar{D} + D\bar{U}U\bar{D} + S\bar{U}U\bar{S} + S\bar{D}D\bar{S} \\ \\ + \tfrac{1}{2}(U\bar{U} - D\bar{D})(U\bar{U} - D\bar{D}) \\ \quad + \tfrac{1}{6}(U\bar{U} + D\bar{D} - 2S\bar{S})(U\bar{U} + D\bar{D} - 2S\bar{S}) \end{array} \right\}$$

$$\tag{19.44}$$

Equating terms yields $A = 1/\sqrt{3}$ and $B = -\sqrt{\dfrac{2}{3}}$.

It is simple to perform the analogous computation for $(6_c \otimes 6_c)_{1,c}$. The resulting recoupling Table for overall colour **1** is

	$1_c \otimes 1_c$	$8_c \otimes \dot{8}_c$
$\bar{3}_c \otimes 3_c$	$1/\sqrt{3}$	$-\sqrt{\left(\frac{2}{3}\right)}$
$6_c \otimes 6_c$	$\sqrt{\left(\frac{2}{3}\right)}$	$1/\sqrt{3}$

$$(19.45)$$

Knowing these colour recouplings and the spin recouplings enables the calculation of the $\lambda \cdot \lambda\, \mathbf{s} \cdot \mathbf{s}$ involving the gluon exchange between q and $\bar{q}$ (the second set of terms in equation 19.16). Hence the mass shifts relative to the mean supermultiplet mass can now be calculated for each of the states in equation (19.13) and compared with the results of Jaffe (1977a,b).

Some care is needed when discussing the **9** and **36** 0^+ states and the **18** and **18** states since each of these can be formed in more than one way. For example the **9**, 0^+ can have qq in 6_{col} or in 3_{col} (equation 19.13). The physical eigenstates will be linear combinations of these. This is discussed in detail by Jaffe (1977b).

A general discussion of $qq\bar{q}\bar{q}$ states with both charm and strange quarks has been made by Lipkin (1977).

20   Quark–Lepton Unification

When I began writing this book four flavours of lepton and three flavours of quark were well established and theorists expected that a fourth flavour also existed. The first clear evidence for this fourth quark emerged in November 1974 and its existence was confirmed by the discovery of charmed hadrons in 1976. The properties of this quark appear to be those required if a renormalisable non-Abelian gauge theory of weak and electromagnetic interactions is realised in Nature where the left-handed quarks and leptons are in doublets of $SU(2) \times U(1)$. In a sense this discovery has closed a chapter in high-energy physics. In the future the charm discovery may well be seen as a pivotal one in that it has added weight to the suspicion that non-Abelian gauge theories may be relevant to weak, electromagnetic *and* strong interactions and this in turn raises the question as to whether a grand unification of all three of these may be achieved. Secondly it seems possible that it might signal the start of a new era where hadron spectroscopy is superseded by quark and lepton spectroscopy and profound questions are now being raised as to the possible connection between quarks and leptons.

20.1   Quarks and leptons

The leptons appear to have many properties similar to the quarks manifested in deep inelastic phenomena (they are structureless spin $\frac{1}{2}$ fields with $V - A$ currents). One suspects that there may be some deeper connection between them but a tantalising puzzle exists: Why do quarks have fractional electric charges and come in three colours (hence confined?) while leptons have integer charges and are colour singlets?

A necessary condition for a pure $V - A$ theory to be renormalisable is that (Bouchiat *et al.*, 1972)

$$\sum_i Q_i = \sum_{\text{quarks}} Q_i + \sum_{\text{leptons}} Q_i = 0 \qquad (20.1)$$

For a $(\nu_e e^-)$ lepton doublet $\sum Q_i = -1$. The constraint in equation (20.1) is therefore satisfied by the quark doublet (u, d) for which, in three colours

$$\sum Q_i = 3(\tfrac{2}{3} - \tfrac{1}{3}) = 1 \qquad (20.2)$$

If the constraint (equation 20.1) is fundamental then we see that three colours of quark *require* the quark to be fractionally charged (or vice versa) if leptons and quarks form doublets.

As far as we know a satisfactory universe could have been built from just these four building blocks of fermions and our everyday experiences would have been the same as in the real world (apart from esoteric phenomena like the pion lifetime and *CP* violation). Yet for some reason Nature is not satisfied with $\nu_e e^-$ but repeated itself by invoking $\nu_\mu \mu^-$. The constraint in equation (20.1) is satisfied by Nature also providing a second generation of quarks in the left-handed doublet (c, s).

The discovery of a heavy charged lepton (τ^-) (Perl *et al.*, 1977) now brings us to the possibility that there exists a third generation of leptons (and quarks?). If the previous pattern is repeated and this new lepton has a weak coupling

$$\begin{pmatrix} \nu_\tau \\ \tau^- \end{pmatrix}_L$$

then the constraint (equation 20.1) will no longer be satisfied. It can be restored if we add a further doublet of quarks

$$\begin{pmatrix} t \\ b \end{pmatrix}_L$$

occurring in three colours and with the familiar $(\tfrac{2}{3}, -\tfrac{1}{3})$ electrical charges.

This gives potentially great significance to the discovery of Y (9·5 GeV) in the summer of 1977 (Herb *et al.*, 1977) which suggests that the b quark probably exists with an effective mass around 4·5 to 5 GeV (see, e.g. Ellis *et al.*, 1977). Searches are now planned for the first evidence that a t quark exists and there is a firm belief that its associated spectroscopy will emerge in experiments at the next generation of accelerators (PEP and PETRA e^+e^- colliding beams at SLAC and DESY respectively). One attractive feature of six quark models is that *CP* violation can be accommodated rather naturally. If the weak charged current of the quarks is written

$$
J = (\bar{u}, \bar{c}, \bar{t} \ldots) \gamma_\mu (1 - \gamma_5) U \begin{pmatrix} d \\ s \\ b \\ \cdot \\ \cdot \end{pmatrix}
\tag{20.3}
$$

where U is a unitary matrix, then for a 2×2 matrix (two quark doublets) all elements of the matrix can be real

$$
U \equiv \begin{pmatrix} \cos \theta & \sin \theta \\ -\sin \theta & \cos \theta \end{pmatrix}
\tag{20.4}
$$

where θ is the Cabibbo angle. For 3×3 one can no longer absorb all phases into the phase convention for the quark fields and the matrix may be written

$$
U \equiv \begin{pmatrix} c_1 & s_1 c_3 & s_1 s_3 \\ -s_1 c_2 & c_1 c_2 c_3 - s_2 s_3 \, e^{i\delta} & c_1 c_2 s_3 + s_2 c_3 \, e^{i\delta} \\ -s_1 s_2 & c_1 s_2 c_3 + c_2 s_3 \, e^{i\delta} & c_1 s_2 s_3 - c_2 c_3 \, e^{i\delta} \end{pmatrix}
\tag{20.5}
$$

where $c_i(s_i) \equiv \cos \theta_i (\sin \theta_i)$, $i = 1, 2, 3$. The three Euler angles $\theta_{1,2,3}$ generalise the conventional Cabibbo angle θ giving all possible mixings between the (u, d): (c, s) and (t, b) doublets. The phase δ leads to *CP*-violating effects in the $K^0 - \overline{K^0}$ system. A quantitative discussion of this and empirical limits on the angles $\theta_{1,2,3}$, δ are given in Ellis *et al.* (1977). The original observation that *CP* could be violated in this way was made by Kobayashi and Maskawa (1972) two years before even the charmed quark was discovered.

20.2   Grand unification

Is there a large multiplet of fundamental fermions which contains both quarks and leptons? Are the non-Abelian vector gauge theories of weak electromagnetism $(SU(2) \times U(1))$ and the $SU(3)(QCD)$ of strong interactions both embedded in one large gauge group?

As a first example we will look at just the $SU(2) \times U(1)$ weak electromagnetism where the gauge fields are the weak isotriplet $W^{+,-,0}$ and the singlet B^0. The coupling of the electromagnetic current and the isotriplet are related by $e = g \sin \theta_W$ (equations 15.10 and 15.11) and hence

$$\sin \theta_W = \frac{\langle q | J^{e.m} | q \rangle}{\langle q | J_3 | q \rangle} \qquad (20.6)$$

In $SU(2) \times U(1)$ this angle is arbitrary. If the $SU(2) \times U(1)$ is embedded in a larger group G ($SU(3)$ is an obvious example but the following remarks are true in more generality) then the ratio in equation (20.6) is fixed. It is (Fritsch and Minkowski, 1975)

$$\sin^2 \theta_W = \sum I_{3i}^2 / \sum Q_i^2 \qquad (20.7)$$

where the sum is over all the fermions in one irreducible representation of $G \supset SU(2) \times U(1)$.

As an example consider a $V - A$ theory with the $(\nu_e e)$ and (u, d) doublets of left-handed fermions and the right-handed fermions in singlets of the $SU(2) \times U(1)$ group. If these are all in a single irreducible representation of G then

$$\sum I_{3i}^2 = (I_3^2)_{u+d+e+\nu} = 2$$
$$\sum Q_i^2 = 2\{3([\tfrac{2}{3}]^2 + [-\tfrac{1}{3}]^2) + (-1)^2\} = \tfrac{16}{3} \qquad (20.8)$$

(only the left-handed fermions contributed to $\sum I_3^2$ but both left- and right-handed appear in $\sum Q_i^2$, hence the overall factor of 2 there). Hence

$$\sin^2 \theta_W = \tfrac{3}{8} \qquad (20.9)$$

If the right-handed fermions were also in doublets (vector theory) then

$$(\sum I_{3i}^2)_{L+R} = 2(\sum I_{3i}^2)_L \qquad (20.10)$$

and so

$$\sin^2 \theta_W = \tfrac{3}{4} \qquad (20.11)$$

This result presumably only has meaning in some symmetry limit. In the real world this symmetry is badly broken. Within the framework of a particular grand unification scheme based on the group $G = SU(5) \supset SU(2) \times U(1) \times SU(3)_{col}$ (Georgi and Glashow, 1974), Buras *et al.* (1978) have estimated the amount by which this result should be renormalised in the real world. They claim that the theory yields $\sin^2 \theta_W \simeq 0.2$ which is in fair agreement with data (see Fig. 11.19).

In conclusion we give a few examples in the SU(5) superunification scheme to illustrate the sort of ideas and patterns that emerge in such models.

The fundamental representation is a quintet which has the following decomposition into $SU(3)_{colour}$ and $SU(2)_{flavour}$:

$$\mathbf{5} = (\mathbf{3, 1}) + (\mathbf{1, 2})$$

The product of two fundamental representations is immediately given by the Young diagrams of Chapter 3

$$\mathbf{5 \times 5 = 15 + 10}$$

$$\mathbf{15} = (\mathbf{6, 1}) + (\mathbf{1, 3}) + (\mathbf{3, 2})$$

$$\mathbf{10} = (\bar{\mathbf{3}}, \mathbf{1}) + (\mathbf{1, 1}) + (\mathbf{3, 2})$$

(20.12)

and the $SU(3) \times SU(2)$ decompositions are also shown.

How are the fermions classified? There are fifteen left-handed states, namely the three leptons $\nu_e e^- e^+$ and twelve quarks (u, d, $\bar{u}$, $\bar{d}$ in three colours apiece). These can be accommodated in a **10** and **5** of SU(5) as follows

$$\bar{\mathbf{5}} = (\bar{\mathbf{3}}, \mathbf{1}) + (\mathbf{1, 2}) = (\bar{d})_L + (\nu_e, e^-)_L \qquad (20.13)$$

$$\mathbf{10} = (\bar{\mathbf{3}}, \mathbf{1}) + (\mathbf{1, 1}) + (\mathbf{3, 2}) = (\bar{u})_L + (e^+)_L + (u, d)_L \qquad (20.14)$$

The gauge bosons belong to the 24-dimensional regular representation of $\mathbf{5 \times \bar{5} = 1 + 24}$. The $SU(3) \times SU(2)$ colour–flavour decomposition is

$$\mathbf{24} = \underbrace{(\mathbf{8, 1})}_{\text{gluons}} + \underbrace{(\mathbf{1, 3}) + (\mathbf{1, 1})}_{W^{+,-,0}, \gamma} + (\mathbf{3, 2}) + (\bar{\mathbf{3}}, \mathbf{2}) \qquad (20.15)$$

The familiar flavour singlet colour octet of gluons (QCD) is seen as are the colour singlet bosons of the weak electromagnetic theory. In addition there are twelve bosons predicted which carry colour and flavour. These can cause transition between leptons and quarks, e.g.

$$(u, d)_L[3, 2] \rightarrow (e^+)_L[1, 1] + X[3, 2] \qquad (20.16)$$

Hence lepton and baryon number are not separately conserved and the proton can decay $p \rightarrow e^+ + \pi$ for example. This can be suppressed if these "leptoquark" bosons are very massive, say 10^{-10} gm!

A final theory must go far beyond this. There is no need for further generations of quarks and leptons and even if one puts them in *ad hoc* there is no reason why they should be Cabibbo mixed. One extension of the SU(5) scheme is to the group SO(10). This has a sixteen-dimensional representation which contains both **10** and **5̄** in its SU(5) subgroup. Hence all the left-handed fermions and antifermions in a given generation can be accommodated in a single representation.

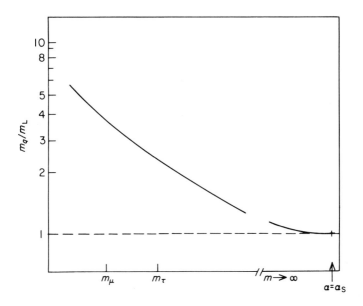

FIG. 20.1. Possible pattern of mass ratios of $Q = -1$ leptons and the associated $Q = -\frac{1}{3}$ quarks as a function of lepton mass. At grand unification mass where $\alpha = \alpha_s$ ($\approx\frac{1}{50}$) the lepton and quark masses are equal. At low masses α_s ($\approx 0\cdot 2) > \alpha (=\frac{1}{137})$ and the quark mass is bigger.

As a final comment on the SU(5) scheme we return to the paper of Buras *et al.* (1978). At the grand unification mass one has

$$m_e \simeq m_d, \qquad m_\mu = m_s, \qquad m_\tau = m_b \qquad (20.17)$$

Buras *et al.* study how these predictions are affected by renormalising to physical energies. With the muon mass and $m_\tau \simeq 1\cdot9$ GeV as input and using $\alpha_s(Q^2 \simeq 10 \text{ GeV}^2) \simeq 0\cdot2$–$0\cdot3$ (compare Chapter 16) then the results are

$$m_s \simeq (0\cdot4 - 0\cdot5) \text{ GeV}, \qquad m_b \simeq 5 \text{ GeV}$$

if there are six quark flavours. This is in remarkable agreement with the observed strange mass scale and the observation of Υ and Υ' at around 9 to 10 GeV. This may be a hint that a deep symmetry between quarks and leptons is indeed present in Nature. If heavier quarks and leptons exist then it is possible that lepton–quark mass equality will become realised (Fig. 20.1).

Only twenty years ago the search was on for the correct group structure of hadrons. The multitude of hadrons that were found had properties consistent with there being a deeper layer of matter containing quarks, originally proposed as a *raison d'être* for the approximate SU(3) flavour symmetry observed in the hadron spectrum. Today we believe that the quarks, and leptons, are fundamental and that the weak electromagnetic and strong interactions are described by some group which contains $SU(2) \times U(1)$ and $SU(3)_{\text{colour}}$ as subgroups. The search is now on for the correct group structure. At least five leptons and fifteen quarks (five flavours in three colours) appear to exist. This compares with the number of hadrons known at the time SU(3) was first proposed. Are we now entering an era of quark and lepton spectroscopy which will ultimately take us beyond the quark?

Bibliography

There are many gaps in this book. Some readers will want more background offered to them while others will wish for more details about various topics. To help remedy these deficiencies I give some suggestions. The works quoted are primarily those from which I have learned something at some time. There are numerous other similar works which the reader can find in any good library.

A general introduction to the field in which the quark model fits is S. Gasiorowicz, "Elementary Particle Physics" (John Wiley, New York, 1967). Also there is "Photon–Hadron Interactions" by R. P. Feynman (Benjamin, New York, 1972). This book also gives a detailed exposition of the role of the parton model in high-energy physics.

One of the few works specifically addressed to quark models is J. J. J. Kokkedee, "The Quark Model" (Benjamin, New York, 1969). My aim has been to extend the base on which these books were written and to bring out the developments of the subsequent decade. Kokkedee's book covers some topics in strong interactions that we have not discussed here, e.g. quark additivity and total cross-sections, high-energy two-body processes at small momentum transfers and the question of nonrelativistic motion for massive deeply bound quarks. It also contains some reprints of early papers.

A recent summer school addressed specifically to quarks is reported in "Fundamentals of Quark Models" (Proceedings of the 17th Scottish Universities Summer School in Physics (Eds I. M. Barbour and A. T. Davies), University of Glasgow, Scotland, 1976). This includes a detailed description of the Pati Salam model and attempts that are being made to investigate quark confinement using gauge theories quantised on a lattice. There is also an introduction to dual string models of hadrons. Some of the problems concerned with quark confinement, in particular the role of colour, are described in "Colour Symmetry and Quark Confinement" (Proceedings of XII Rencontre de Moriond (Ed. J. Tran Thanh Van), CNRS, France, 1977).

The background to unitary symmetries and group theory is described in many places. Some examples include:

P. Carruthers, "Introduction to Unitary Symmetry" (Interscience, New York, 1966).
M. Hammermesh, "Group Theory" (Addison-Wesley, Reading, Massachusetts, 1963).
M. Gell-Mann and Y. Ne'eman, "The Eightfold Way" (Benjamin, New York, 1964).
H. Lipkin, "Lie Groups for Pedestrians" (North-Holland, Amsterdam, 1966).

As an introduction to gauge theories see:

J. Iliopoulos, "An Introduction to Gauge Theories" (Lectures on the CERN Academic training programme, 1975–1976, CERN report 76-11, CERN, Geneva).

I apologise to my colleagues who have been inadequately referenced. To give a full and just bibliography would require a companion volume and it would be out of date as soon as it was completed. To remedy this in part I refer the reader to some articles in the review series "Physics Reports" (section C of *Physics Letters*, published by North-Holland, Amsterdam). These articles have a very extensive list of references to their particular corner of the field. I also hope that this list will aid the reader who has been stimulated to pursue some item in more detail than it was covered here.

Part I

LIPKIN, H. J. (1973). Quark models for pedestrians. **18**, 175.
ROSNER, J. (1974). Classification and decays of resonant particles. **11**, 193.
GREENBERG, O. W. AND NELSON, C. A. (1977). Colour models of hadrons. **32**, 1.

Part II

WEST, G. B. (1975). Electron scattering from atoms, nuclei and nucleons. **18**, 264.

KOGUT, J. AND SUSSKIND, L. (1973). Parton models of elementary particles. **8**, 77.

FRISHMAN, Y. (1974). Light cone and short distances. **13**, 1.

LANDSHOFF, P. V. AND POLKINGHORNE, J. C. (1972). Models for hadronic and leptonic processes at high energies. **5**, 1.

GILMAN, F. J. (1972). Photoproduction and electroproduction. **4**, 95.

LLEWELLYN SMITH, C. H. (1972). Neutrino interactions at accelerators. **3**, 261.

Part III

ABERS, E. AND LEE, B. W. (1973). Gauge theories. **9**, 1.

POLITZER, H. D. (1974). Asymptotic freedom: An approach to strong interactions. **14**, 130.

FELDMAN, G. AND PERL, M. (1975). Electron positron annihilation above 2 GeV, and the new particles. **19**, 234; and also: Recent results in electron–positron annihilation above 2 GeV, **33**, 285.

MANDELSTAM, S., WILSON, K., KOGUT, J. AND SUSSKIND, L. (1976). Contributions in: Extended systems in field theory. **23**, 240.

References

ABRAMOWITCH, M. AND STEGUN, I. (1965). "Handbook of Mathematical Functions". Dover Publications, New York.

ADLER, S. L. (1965a). *Phys. Rev. Lett.* **14**, 1051.

ADLER, S. L. (1965b). *Phys. Rev.* **140**, B736.

ADLER, S. L. (1966). *Phys. Rev.* **143**, 1144.

ADLER, S. L. (1969). *Phys. Rev.* **177**, 2426.

ADLER, S. L. AND DASHEN, R. (1968). "Current Algebras". Benjamin, New York.

AGUILAR-BENITEZ, M. *et al.* (1977). *Nucl. Phys.* **B124**, 189.

AKHIEZER, A. AND BERETSTETSKI, V. B. (1963). "Quantum Electrodynamics". John Wiley, New York.

ALEKSEEV, A. I. (1958). *Sov. Phys. JETP.* **7**, 826.

ALGUARD, M. J. *et al.* (1976). *Phys. Rev. Lett.* **37**, 1261.

ALLEN, E. (1975). *Phys. Lett.* **57B**, 263.

ALTARELLI, G. AND PARISI, G. (1977). *Nucl. Phys.* **B126**, 298.

ALTARELLI, G. *et al.* (1974). *Nucl. Phys.* **B69**, 531.

AMATI, D. *et al.* (1964). *Nuovo Cimento*, **34**, 1732.

ANDERSON, H. L. *et al.* (1976). *Phys. Rev. Lett.* **37**, 4.

ANDERSON, H. L. *et al.* (1977). *Phys. Rev. Lett.* **38**, 1450.

ANDREADIS, P. *et al.* (1974). *Ann. Phys. (N.Y.)*, **88**, 242.

ANTREASYAN, D. *et al.* (1977). *Phys. Rev. Lett.* **39**, 906.

APPELQUIST, T. AND GEORGI, H. (1973). *Phys. Rev.* **D8**, 4000.

APPELQUIST, T. AND POLITZER, H. D. (1975a). *Phys. Rev.* **D12**, 1404.

APPELQUIST, T. AND POLITZER, H. D. (1975b). *Phys. Rev. Lett.* **34**, 43.

ASCOLI, G. *et al.* (1968). *Phys. Rev. Lett.* **21**, 1411.

ASCOLI, G. *et al.* (1970). *In* Proceedings of 15th International Conference on High Energy Physics, Kiev, USSR, p. 221.

ATWOOD, W. *et al.* (1976). *Phys. Lett.* **64B**, 479.

AUBERT, J. J. *et al.* (1974). *Phys. Rev. Lett.* **33**, 1404.

AUGUSTIN, J. E. *et al.* (1974). *Phys. Rev. Lett.* **33**, 1406.

BARBIERI, R. *et al.* (1975). *Phys. Lett.* **56B**, 477.

BARBIERI, R., GATTO, R. AND KOGERLER, R. (1976a). *Phys. Lett.* **60B**, 183.

BARBIERI, R., GATTO, R. AND REMIDDI, E. (1976b). *Phys. Lett.* **61B**, 465.

BARBIERI, R. *et al.* (1976c). *Nucl. Phys.* **B105**, 125.

BARGER, V. AND PHILLIPS, R. J. N. (1974). *Nucl. Phys.* **B73**, 269.

BARISH, B. C. *et al.* (1977). *In* Proceedings of International Neutrino Conference, Aachen, 1976, p. 289. Vieweg, Braunschweig.

BARNES, K. J., CARRUTHERS, P. AND VON HIPPEL, F. (1965). *Phys. Rev. Lett.* **14**, 82.

BEBEK, C. J. *et al.* (1973). *Phys. Rev. Lett.* **30**, 624.

BECCHI, C. AND MORPURGO, G. (1965a). *Phys. Lett.* **17**, 352.

BECCHI, C. AND MORPURGO, G. (1965b). *Phys. Rev.* **140B**, 687.

BEG, M. A., LEE, B. W. AND PAIS, A. (1964). *Phys. Rev. Lett.* **13**, 514.

BELL, J. S. (1974). *Acta Physica Austriaca Suppl.* **13**, 395.

BELL, J. S. AND HEY, A. J. G. (1974). *Phys. Lett.* **51B**, 365.

BELL, J. S. and JACKIW, R. (1969). *Nuovo Cimento,* **51**, 47.

BENVENUTI, A. *et al.* (1974). *Phys. Rev. Lett.* **32**, 800.

BENVENUTI, A. *et al.* (1977). *In* Proceedings of International Neutrino Conference, Aachen, 1976, p. 296. Vieweg, Braunschweig.

BERGE, J. P. *et al.* (1975). Fermilab report 75/84 (unpublished).

BERMAN, S. M., BJORKEN, J. D. AND KOGUT, J. (1971). *Phys. Rev.* **D4**, 3388.

BETHE, H. A. AND SALPETER, E. E. (1951). *Phys. Rev.* **84**, 1232.

BHARADWAJ, V. K. (1977). Ph.D. thesis, University of Oxford.

BJORKEN, J. D. (1966). *Phys. Rev.* **148**, 1467.

BJORKEN, J. D. (1967). *In* Proceedings of 3rd International Symposium on Electron and Photon Interactions, Stanford, California.

BJORKEN, J. D. (1969). *Phys. Rev.* **179**, 1547.

BJORKEN, J. D. (1971). *In* "Hadronic Interactions of Electrons and Photons" (Eds J. Cumming and H. Osborn), p. 475. Academic Press, London and New York.

BJORKEN, J. D. AND DRELL, S. D. (1964). "Relativistic Quantum Mechanics". McGraw-Hill, New York.

BJORKEN, J. D. AND GLASHOW, S. L. (1964). *Phys. Lett.* **11**, 255.

BJORKEN, J. D. AND PASCHOS, E. A. (1969). *Phys. Rev.* **185**, 1975.

BJORKEN, J. D. AND WALECKA, J. D. (1966). *Ann. Phys.* (*N.Y.*), **38**, 35.

BLANKENBECLER, R. AND BRODSKY, S. J. (1974). *Phys. Rev.* **D10**, 2973.

BLANKENBECLER, R., BRODSKY, S. J. AND GUNION, J. F. (1972). *Phys. Lett.* **39B**, 649; and *Phys. Rev.* **D6**, 2652.

BLIETSCHAU, J. *et al.* (1976). *Nucl. Phys.* **B114**, 169.

BLIETSCHAU, J. *et al.* (1977). *Nucl. Phys.* **B118**, 218.

BLOOM, E. (1973). *In* Proceedings of 6th International Symposium on Electron and Photon Interactions, Bonn, 1973.

BLOOM, E. (1975). *In* Proceedings of the Summer Institute on Particle Physics, Stanford University (SLAC red report 191), California.

BODEK, A. *et al.* (1974). *Phys. Lett.* **51B**, 417.

BOGOLIUBOV, P. N. (1968). *Ann. Inst. Henri Poincare*, **8**, 163.

BOGOLIUBOV, N. N. AND SHIRKOV, D. (1959). "Introduction to the Theory of Quantised Fields". Interscience, New York.

BOHM, M., JOOS, H. AND KRAMMER, M. (1972). *Nuovo Cimento*, **7A**, 21.

BOHM, M., JOOS, H. AND KRAMMER, M. (1973). *Nucl. Phys.* **B51**, 397.

BOHM, M., JOOS, H. AND KRAMMER, M. (1974). *Nucl. Phys.* **B69**, 349.

BOUCHIAT, C. *et al.* (1972). *Phys. Lett.* **38B**, 519.

BOWLER, K. C. (1970). *Phys. Rev.* **D1**, 929.

BRANDELIK, R. *et al.* (1977a). *Phys. Lett.* **70B**, 132.

BRANDELIK, R. *et al.* (1977b). *Phys. Lett.* **70B**, 387.

BRAUNSCHWEIG, W. *et al.* (1977). *Phys. Lett.* **67B**, 243.

BRODSKY, S. J. AND FARRAR, G. (1973). *Phys. Rev. Lett.* **31**, 1153.

BRODSKY, S. J. AND FARRAR, G. (1975). *Phys. Rev.* **D11**, 1309.

BRODSKY, S. J., CLOSE, F. E. AND GUNION, J. F. (1973). *Phys. Rev.* **D8**, 3678.

BRODSKY, S. J., ROSKIES, R. AND SUAYA, R. (1973). *Phys. Rev.* **D8**, 4574.

BROUT, R. AND ENGLERT, F. (1964). *Phys. Rev. Lett.* **13**, 321.

BUCELLA, F. *et al.* (1970). *Nuovo Cimento*, **69A**, 133.

BURAS, A. AND GAEMERS, K. (1977). *Phys. Lett.* **71B**, 106.

BURAS, A. *et al.* (1978). *Nucl. Phys.* **B135**, 66.

CABIBBO, N. (1963). *Phys. Rev. Lett.* **10**, 531.

CABIBBO, N., PARISI, G. AND TESTA, M. (1970). *Lett. al Nuovo Cimento*, **4**, 35.

CALLAN, C. AND GROSS, D. (1969). *Phys. Rev. Lett.* **22**, 156.

CALLAN, C., COOTE, N. AND GROSS, D. (1976). *Phys. Rev.* **D13**, 1649.

CAPPS, R. H. (1974). *Phys. Rev. Lett.* **33**, 1637.

CAPPS, R. H. (1975). *Phys. Rev.* **D12**, 3606.

CARLITZ, R. (1975). *Phys. Lett.* **58B**, 345.

CARLITZ, R. AND KISLINGER, M. (1970). *Phys. Rev.* **D2**, 336.

CAZZOLI, E. *et al.* (1976). *Phys. Rev. Lett.* **34**, 1125.

CASHMORE, R. J., HEY, A. J. G. AND LITCHFIELD, P. (1975a). *Nucl. Phys.* **B98**, 237.

CASHMORE, R. J., HEY, A. J. G. AND LITCHFIELD, P. (1975b). *Nucl. Phys.* **B95**, 516.

CHAN, H. M. AND TSOU, S. T. (1977). *In* Proceedings of Bielefeld Summer Institute (1976); *Acta Physica Austriaca*.

CHANOWITZ, M. AND GILMAN, F. J. (1976). *Phys. Lett.* **63B**, 178.

CHODOS, A. *et al.* (1974a). *Phys. Rev.* **D9**, 3471.

CHODOS, A. *et al.* (1974b). *Phys. Rev.* **D10**, 2599.

CLEYMANS, J. AND RODENBERG, R. (1974). *Phys. Rev.* **D9**, 155.

CLINE, D. *et al.* (1976). *Phys. Rev. Lett.* **37**, 252. ibid. 648.

CLOSE, F. E. (1974a). *Nucl. Phys.* **B80**, 269.

CLOSE, F. E. (1974b). *In* Proceedings of 17th International Conference on High Energy Physics, London, 1974, pp. II–157. Rutherford Laboratory, England.

CLOSE, F. E. (1975). *Acta Physica Polonica*, **B6**, 785.

CLOSE, F. E. AND GILMAN, F. J. (1972). *Phys. Lett.* **38B**, 541.

CLOSE, F. E., OSBORN, H. AND THOMSON, A. M. (1974). *Nucl. Phys.* **B77**, 281.

COLEMAN, S. AND GLASHOW, S. L. (1961). *Phys. Rev. Lett.* **6**, 423.

COLGLAZIER, E. W. AND ROSNER, J. (1971). *Nucl. Phys.* **B27**, 349.

COMBE, J. *et al.* (1966). *Z. Naturforsche*, **21A**, 1757.

COMBRIDGE, B. L.. KRIPFGANZ, J. AND RANFT, J. (1977). *Phys. Lett.* **70B**, 234.

COPLEY, L. A., KARL, G. AND OBRYK, E. (1969a). *Phys. Lett.* **29B**, 117.

COPLEY, L. A., KARL, G. AND OBRYK, E. (1969b). *Nucl. Phys.* **B13**, 303.

CUNDY, D. (1974). *In* Proceedings of 17th International Conference on High Energy Physics, London, 1974, pp. iv–145. Rutherford Laboratory, England.

CUTLER, R. AND SIVERS, D. (1977). *Phys. Rev.* **D17**, 196.

DAKIN, J. AND FELDMAN, G. (1973). *Phys. Rev.* **D8**, 2862.

DALITZ, R. H. (1965). *In* "High Energy Physics" (Eds C. de Witt and M. Jacob). Gordon and Breach, New York.

DALITZ, R. H. (1967). *In* Proceedings of Second Hawaii Topical Conference in Particle Physics (Eds S. Pakvasa and S. Tuan), p. 398. University of Hawaii Press, Honolulu.

DALITZ, R. H. (1976). *In* "Fundamentals of Quark Models" (Eds I. Barbour and A. Davies). Scottish Universities Summer School in Physics.

DALITZ, R. H. (1977). *In* "Quarks and Hadronic Substructure". Plenum Press, New York.

DALITZ, R. H. AND SUTHERLAND, D. (1966). *Phys. Rev.* **146**, 1180.

DASHEN, R. AND GELL-MANN, M. (1965). *Phys. Lett.* **17**, 142.

DASHEN, R. AND GELL-MANN, M. (1966a). *Phys. Rev. Lett.* **17**, 340.

DASHEN, R. AND GELL-MANN, M. (1966b). *In* "Symmetry Principles at High Energies", Proceedings of Coral Gables Conferences. Freeman and Company, San Francisco.

DE GRAND, T. *et al.* (1975). *Phys. Rev.* **D12**, 2060.

DE RUJULA, A., GEORGI, H. AND GLASHOW, S. L. (1975). *Phys. Rev.* **D12**, 147.

DE RUJULA, A., GEORGI, A. AND GLASHOW, S. L. (1976). *Phys. Rev. Lett.* **37**, 398.

DE SWART, J. J. (1963). *Rev. Mod. Phys.* **35**, 916.

DEVENISH, R. AND LYTH, D. (1975). *Nucl. Phys.* **B93**, 109.

DOMBEY, N. (1969). *Rev. Mod. Phys.* **41**, 236.

DOMBEY, N. (1971). *In* "Hadronic Interactions of Electrons and Photons" (Eds J. Cumming and H. Osborn). Academic Press, London and New York.

DRELL, S. D. (1970). *In* "Subnuclear Phenomena" (Ed. A. Zichichi). Academic Press, New York and London.

DRELL, S. D. AND WALECKA, J. D. (1964). *Ann. Phys.* (*N.Y.*), **28**, 18.

DRELL, S. D. AND YAN, T. M. (1970). *Phys. Rev. Lett.* **24**, 181.

DRELL, S. D. AND YAN, T. M. (1971). *Ann. Phys.* (*N.Y.*), **66**, 595.

DRELL, S. D., LEVY, D. J. AND YAN, T. M. (1970). *Phys. Rev.* **D1**, 1035.

EDWARDS, B. J. AND KAMAL, A. N. (1976). *Phys. Rev. Lett.* **36**, 241.

EICHTEN, E., WILLEMSEN, J. AND FEINBERG, F. (1973). *Phys. Rev.* **D8**, 1204.

EICHTEN, E. *et al.* (1975). *Phys. Rev. Lett.* **34**, 369.

EICHTEN, E. *et al.* (1976). *Phys. Rev. Lett.* **36**, 500.

EINHORN, M. (1976). *Phys. Rev.* **D14**, 3451.

EINHORN, M. AND ELLIS, S. D. (1975). *Phys. Rev.* **D12**, 2007.

ELLIS, J. (1977). Lectures at 1977 Cracow Summer School in Physics. *Acta Physica Polonica.*

ELLIS, J. AND JAFFE, R. L. (1974). *Phys. Rev.* **D9**, 1444; erratum **D10**, 1669.

ELLIS, J. *et al.* (1977). *Nucl. Phys.* **B131**, 285.

EVANGELIDES, E. *et al.* (1974). *Nucl. Phys.* **B71**, 381.

FAIMAN, D. AND HENDRY, A. W. (1968). *Phys. Rev.* **173**, 1720.

FAIMAN, D. AND HENDRY, A. W. (1969). *Phys. Rev.* **180**, 1572.

FAISSNER, H. *et al.* (1976). *In* Proceedings of International Neutrino Conference, Aachen, 1976, p. 223. Vieweg Braunschweig.

FARRAR, G. (1974). *Nucl. Phys.* **B77**, 429.

FARRAR, G. AND JACKSON, D. (1975). *Phys. Rev. Lett.* **35**, 1416.

FEDERMAN, P., RUBINSTEIN, H. R. AND TALMI, I. (1966). *Phys. Lett.* **22**, 208.

FELDMAN, G. J. (1976). *In* Proceedings of Summer Institute on Particle Physics (Ed. M. Zipf), p. 81. SLAC Report No. 198 (Nov. 1976). SLAC, Stanford, California.

FELDMAN, G. J. (1977). *In* Proceedings of Summer Institute on Particle Physics (Ed. M. Zipf), p. 241. SLAC Report No. 204 (Nov. 1977). SLAC, Stanford, California.

FELDMAN, G. J. AND PERL, M. (1975). *Phys. Reports,* **19C**, 234.

FELDMAN, G. J. AND PERL, M. (1977). *Phys. Reports.* **33C**, 286.

FELDMAN, G. J. *et al.* (1975). *Phys. Rev. Lett.* **35**, 821.

FEYNMAN, R. P. (1969). *Phys. Rev. Lett.* **23**, 1415.

FEYNMAN, R. P. (1972). "Photon Hadron Interactions". Benjamin, New York.

FEYNMAN, R. P. AND GELL-MANN, M. (1958). *Phys. Rev.* **109**, 193.

FEYNMAN, R. P., KISLINGER, M. AND RAVNDAL, F. (1971). *Phys. Rev.* **D3**, 2706.

FEYNMAN, R. P., FIELD, R. AND FOX, G. (1977). *Nucl. Phys.* **B128**, 1.

FIELD, R. AND FEYNMAN, R. P. (1977). *Phys. Rev.* **D15**, 2590.

FRAUTSCHI, S. (1963). "Regge Poles and S-Matrix Theory". Benjamin, New York.

FREGAU, J. H. AND HOFSTADTER, R. (1955). *Phys. Rev.* **99**, 1503.

FREUND, P. G. O. AND LEE, B. W. (1964). *Phys. Rev. Lett.* **13**, 592.

FRITZSCH, H. AND GELL-MANN, M. (1972). *In* Proceedings of 16th International Conference on High Energy Physics, Chicago-Batavia, 1972, vol. 2. Fermilab, Batavia.

FRITZSCH, H., GELL-MANN, M. AND LEUTWYLER, H. (1974). *Phys. Lett.* **B47**, 365.

FRITZSCH, H., GELL-MANN, M. AND MINKOWSKI, P. (1975). *Phys. Lett.* **59B**, 256.

FUBINI, S. AND FURLAN, G. (1965). *Physics*, **1**, 229.

FUJIMURA, K., KOBAYASHI, T. AND NAMIKI, M. (1970). *Prog. Theor. Phys. (Kyoto)*, **43**, 73.

GAILLARD, M. AND LEE, B. W. (1974). *Phys. Rev.* **D10**, 897.

GAILLARD, M., LEE, B. W. AND ROSNER, J. L. (1975). *Rev. Mod. Phys.* **47**, 277.

GASIOROWICZ, S. (1967). "Elementary Particle Physics". John Wiley, New York.

GELL-MANN, M. (1953). *Phys. Rev.* **92**, 833.

GELL-MANN, M. (1961). California Institute of Technology Report CTSL-20 (1961). (Reprinted on p. 11 of "The Eightfold Way", Gell-Mann and Ne'eman, op. cit.)

GELL-MANN, M. (1962). *Phys. Rev.* **125**, 1067.

GELL-MANN, M. (1964). *Phys. Lett.* **8**, 214.

GELL-MANN, M. (1972). *In* "Elementary Particle Physics" (Ed. P. Urban), p. 733. Springer-Verlag, Vienna.

GELL-MANN, M. AND LOW, F. (1954). *Phys. Rev.* **95**, 1300.

GELL-MANN, M. AND NE'EMAN, Y. (1964). "The Eightfold Way". Benjamin, New York.

GEORGI, H. AND POLITZER, H. D. (1974). *Phys. Rev.* **D9**, 416.

GILMAN, F. J. (1968). *Phys. Rev.* **167**, 1365.

GILMAN, F. J. (1975). *In* Proceedings of 1975 International Symposium on Lepton and Photon Interactions at High Energies, Stanford, California. SLAC, Stanford, California.

GILMAN, F. J. AND HARARI, H. (1968). *Phys. Rev.* **165**, 1803.

GILMAN, F. J. AND KARLINER, I. (1973). *Phys. Lett.* **46B**, 426.

GILMAN, F. J. AND KARLINER, I. (1974). *Phys. Rev.* **D10**, 2194.

GILMAN, F. J., KUGLER, M. AND MESHKOV, S. (1973). *Phys. Lett.* **45B**, 481.

GILMAN, F. J., KUGLER, M. AND MESHKOV, S. (1974). *Phys. Rev.* **D9**, 715.

GLASHOW, S. L. (1961). *Nucl. Phys.* **22**, 579.

GLASHOW, S. L., ILIOPOULOS, J. AND MAIANI, L. (1970). *Phys. Rev.* **D2**, 1285.

GOEBEL, C. (1956). *Phys. Rev.* **103**, 258.

GOLOWICH, E. (1975). *Phys. Rev.* **D12**, 2108.

GOURDIN, M. (1972). *Nucl. Phys.* **B38**, 418.

GREENBERG, O. W. (1964). *Phys. Rev. Lett.* **13**, 598.

GREENBERG, O. W. (1975). Proceedings of Second Orbis Scientiae, University of Miami, Jan. 1975.

GREENBERG, O. W. AND RESNIKOFF, M. (1967). *Phys. Rev.* **163**, 1844.

GREENBERG, O. W. AND ZWANZIGER, D. (1966). *Phys. Rev.* **150**, 1177.

GRONAU, M., RAVNDAL, F. AND ZARMI, Y. (1973). *Nucl. Phys.* **B51**, 611.

GROSS, D. J. AND LLEWELLYN SMITH, C. H. (1969). *Nucl. Phys.* **B14**, 337.

GROSS, D. J. AND WILCZEK, F. (1973a). *Phys. Rev. Lett.* **30**, 1343.

GROSS, D. J. AND WILCZEK, F. (1973b). *Phys. Rev.* **D8**, 3633.

GROSS, D. J. AND WILCZEK, F. (1974). *Phys. Rev.* **D9**, 980.

GROSSE, H. (1977). *Phys. Lett.* **68B**, 343.

GUNION, J. (1974). *Phys. Rev.* **D10**, 242.

GUNION, J. (1976). *Phys. Rev.* **D14**, 1400.

GUNION, J., BRODSKY, S. AND BLANKENBECLER, R. (1973). *Phys. Rev.* **D8**, 287.

GURALNIK, G., HAGEN, C. AND KIBBLE, T. (1964). *Phys. Rev. Lett.* **13**, 585.

GURSEY, F. AND RADICATI, L. A. (1964). *Phys. Rev. Lett.* **13**, 173.

HAMMERMESH, M. (1963). "Group Theory". Addison-Wesley, Reading, Massachusetts.

HAN, M. AND NAMBU, Y. (1965). *Phys. Rev.* **139B**, 1006.

HAN, M. AND NAMBU, Y. (1974). *Phys. Rev.* **D10**, 674.

HAND, L. N. (1963). *Phys. Rev.* **129**, 1834.

HANSON, G. (1975). *In* Proceedings of 1975 International Symposium on Lepton and Photon Interactions at High Energies, Stanford, California. SLAC, Stanford, California.

HANSON, G. *et al.* (1975). *Phys. Rev. Lett.* **35**, 1609.

HARA, Y. (1964). *Phys. Rev.* **136**, 507.

HARARI, H. (1968). *In* "Spectroscopic and Group Theoretic Methods in Physics". North-Holland.

HARARI, H. (1971). *In* "Hadronic Interactions of Electrons and Photons". Academic Press, London and New York.

HARARI, H. (1974). §B.6 *in* "ψchology", SLAC-PUB-1514, Stanford, California, (Unpublished.)

HARARI, H. (1976a). *Phys. Lett.* **60B**, 172.

HARARI, H. (1976b). *Phys. Lett.* **64B**, 469.

HASERT, F. J. *et al.* (1973). *Phys. Lett.* **46B**, 138.

HEIMANN, R. (1977). University of Bern preprint (Unpublished.)

HERB, S. W. *et al.* (1977). *Phys. Rev. Lett.* **39**, 252.

HEY, A. J. G. (1974a). Polarisation effects in electron–proton scattering. Daresbury Lecture Note Series, No. 13, SRC Daresbury Lab, Warrington, U.K.

HEY, A. J. G. (1947b). *In* "High Energy Lepton Interactions", Proceedings of 9th Rencontre de Moriond (Ed. J. Tran Thanh Van). CNRS, France.

HEY, A. J. G. AND WEYERS, J. (1974). *Phys. Lett.* **48B**, 69.

HEY, A. J. G., ROSNER, J. L. AND WEYERS, J. (1973). *Nucl. Phys.* **B61**, 205.

HIGGS, P. W. (1964a). *Phys. Lett.* **12**, 232.

HIGGS, P. W. (1964b). *Phys. Rev. Lett.* **13**, 508.

HINCHLIFFE, I. AND LLEWELLYN SMITH, C. H. (1977a). *Phys. Lett.* **66B**, 281.

HINCHLIFFE, I. AND LLEWELLYN SMITH, C. H. (1977b). *Nucl. Phys.* **B128**, 93.

HORGAN, R. R. (1974). *Nucl. Phys.* **B71**, 514.

HWA, R. C., MATSUDA, S. AND ROBERTS, R. G. (1977). Rutherford Laboratory Report RL 77/117 A. (Unpublished.)

IIZUKA, J. (1966). Prog. Theor. Phys. Supplement No. 37–38, p. 21.

INNES, W. R. *et al.* (1977). *Phys. Rev. Lett.* **39**, 1240.

JACKSON, J. D. (1976A). *Rev. Mod. Phys.* **48**, 417.

JACKSON, J. D. (1976b). *In* Proceedings of Summer Institute on Particle Physics, p. 147. Stanford University, (Ed. M. Zipf), SLAC Report No. 198 (November 1976).

JACKSON, J. D. (1977). "New Particle Spectroscopy" *in* Proceedings of the European Conference on Particle Physics, Budapest.

JACOB, M. AND WICK, G. C. (1959). *Ann. Phys.* **7**, 404.

JAFFE, R. L. (1975). *Phys. Rev.* **D11**, 1953.

JAFFE, R. L. (1977a). *Phys. Rev.* **D15**, 267.

JAFFE, R. L. (1977b). *Phys. Rev.* **D15**, 281.

JAFFE, R. L. AND JOHNSON, K. (1976). *Phys. Lett.* **60B**, 201.

JAFFE, R. L. AND KISKIS, J. (1976). *Phys. Rev.* **D13**, 1355.

JAFFE, R. L. AND LLEWELLYN SMITH, C. H. (1973). *Phys. Rev.* **D7**, 2506.

JAFFE, R. L. AND PATRASCIOIU, A. (1975). *Phys. Rev.* **D12**, 1314.

JAUCH, J. AND ROHRLICH, F. (1955). "Theory of Photons and Electrons". Addison-Wesley, Reading Massachusetts.

JOHNSON, K. (1975). *Acta Physica Polonica*, **B6**, 865.

JOHNSON, K. (1977). *In* "Fundamentals of Quark Models" (Eds I. Barbour and A. Davies). Scottish University Summer School in Physics.

KANE, G. (1977). *In* "Colour Symmetry and Quark Confinement", Proceedings of XII Rencontre de Moriond, p. 9. CNRS, France, 1977.

KARL, G. (1977). *Nuovo Cimento*, **38**, 315.

KARL, G. AND OBRYK, E. (1968). *Nucl. Phys.* **B8**, 609.

KELLETT, B. H. (1974a). *Ann. Phys.* **87**, 60.

KELLETT, B. H. (1974b). *Phys. Rev.* **D10**, 2269.

KENNY, B. G., PEASLEE, D. C. AND TASSIE, L. J. (1975). *Phys. Rev. Lett.* **34**, 429 and 1482.

KINOSHITA, T. (1962). *J. Math. Phys.* **3**, 650.

KNAPP, B. *et al.* (1976). *Phys. Rev. Lett.* **37**, 882.

KNIES, G., MOORHOUSE, R. G. AND OBERLACK, H. (1974). *Phys. Rev.* **D9**, 2680.

KOBAYASHI, M. AND MASKAWA, T. (1972). *Prog. Theor. Phys.* **49**, 282.

KOGUT, J. AND SUSSKIND, L. (1974a). *Phys. Rev.* **D9**, 697.

KOGUT, J. AND SUSSKIND, L. (1974b). *Phys. Rev.* **D9**, 3391.

KRAMMER, M., SCHILDKNECHT, D. AND STEINER, F. (1974). Hamburg Report DESY 74/64. (Unpublished.)

KOBUTA, T. AND OHTA, K. (1976). *Phys. Lett.* **65B**, 374.

KUTI, J. AND WEISSKOPF, V. F. (1971). *Phys. Rev.* **D4**, 3418.

LANDAU, L. F. (1948). *Dokl. Akad. Nauk USSR*, **60**, 207.

LANDSHOFF, P. V. (1976). *Phys. Lett.* **66B**, 452.

LANDSHOFF, P. V. AND POLKINGHORNE, J. C. (1971). *Nucl. Phys.* **B28**, 240.

LANDSHOFF, P. V. AND POLKINGHORNE, J. C. (1972). *Phys. Reports*, **5C**, 1.

LANDSHOFF, P. V. AND POLKINGHORNE, J. C. (1973a). *Phys. Lett.* **45B**, 361.

LANDSHOFF, P. V. AND POLKINGHORNE, J. C. (1973b). *Phys. Rev.* **D8**, 927.

LANDSHOFF, P. V. AND POLKINGHORNE, J. C. (1973c). *Phys. Rev.* **D8**, 4157.

LANDSHOFF, P. V. AND POLKINGHORNE, J. C. (1974). *Phys. Rev.* **D10**, 891.

LANDSHOFF, P. V., POLKINGHORNE, J. C. AND SHORT, R. (1971). *Nucl. Phys.* **B28**, 225.

LEDERMAN, L. M. AND POPE, B. G. (1977). *Phys. Lett.* **66B**, 486.

LEE, T. D. AND YANG, C. N. (1956). *Nuovo Cimento.* **3**, 749.

LEE, W. *et al.* (1977). *Phys. Rev. Lett.* **38**, 202.

LEUTWYLER, H. (1974). *Phys. Lett.* **48B**, 431.

LE YOUANC, A. *et al.* (1973). *Phys. Rev.* **D8**, 2223.

LE YOUANC, A. *et al.* (1974a). *Phys. Rev.* **D9**, 1415.

LE YOUANC, A. *et al.* (1974b). *Phys. Rev.* **D9**, 2636.

LE YOUANC, A. *et al.* (1975a). *Phys. Rev.* **D11**, 1272.

LICHT, A. L. AND PAGNAMENTA, A. (1970). *Phys. Rev.* **D2**, 1150.

LICHTENBERG, D. B. (1968). *Phys. Rev.* **178**, 2197.

LICHTENBERG, D. B. (1975). *Phys. Rev.* **D12**, 3760.

LICHTENBERG, D. B. (1976). *Phys. Rev.* **D14**, 1412.

LIPES, R. G. (1972). *Phys. Rev.* **D8**, 2849.

LIPKIN, H. J. (1973). *Phys. Lett.* **45B**, 267.

LIPKIN, H. J. (1974). *Phys. Rev.* **D9**, 1579.

LIPKIN, H. J. (1977). *Phys. Lett.* **70B**, 113.

LIPKIN, H. J. AND MESHKOV, S. (1965). *Phys. Rev. Lett.* **14**, 670.

LIPKIN, H. J. AND MESHKOV, S. (1966). *Phys. Rev.* **143**, 1269.

LLEWELLYN SMITH, C. H. (1967). *Phys. Lett.* **28B**, 335.

LLEWELLYN SMITH, C. H. (1969). *Ann. Phys.* **53**, 521.

LLEWELLYN SMITH, C. H. (1974). *Phys. Reports.* **3C**, 264.

LLEWELLYN SMITH, C. H. (1975). *In* Proceedings of International Symposium on Lepton and Photon Interactions At High Energies, Stanford, 1975 (Ed. W. T. Kirk), p. 709. SLAC, Stanford, California.

LLEWELLYN SMITH, C. H. (1976). *In* Proceedings of 1976 Les Houches Summer School in Physics.

MARINESCU, N. AND STECH, B. (1975). University of Heidelberg preprint. (Unpublished.)

MARTIN, A. (1977). *Phys. Lett.* **67B**, 330.

MATHEWS, P. T. (1975). *In* Proceedings of European Physical Society International Conference on High Energy Physics, Palermo (Ed. A. Zichichi). Editrice Compositori, Bologna.

MATSUDA, Y. (1976). *Prog. Theor. Phys.* **55**, 777.

MATVEEV, V. A., MURADYAN, R. M. AND TAVKHELIDZE, A. N. (1973). *Lett al Nuovo Cimento*, **7**, 719.

MCELHANEY, R. AND TUAN, S. F. (1973). *Phys. Rev.* **D8**, 2267.

MELOSH, H. J. (1974). *Phys. Rev.* **D9**, 1095.

METCALF, W. J. AND WALKER, R. L. (1974). *Nucl. Phys.* **B76**, 253.

MICU, L. (1969). *Nucl. Phys.* **B10**, 521.

MO, L. (1975). *In* Proceedings of 1975 International Symposium on Lepton and Photon Interactions at High Energies, Stanford, California SLAC, Stanford, California.

MOORHOUSE, R. G. (1966). *Phys. Rev. Lett.* **16**, 772.

MOORHOUSE, R. G. AND OBERLACK, H. (1973). *Phys. Lett.* **43B**, 44.

MOORHOUSE, R. G., OBERLACK, H. AND ROSENFELD, A. H. (1974). *Phys. Rev.* **D9**, 1.

MORPURGO, G. (1965). *Physics*, **2**, 95.

MORPURGO, G. (1974). *In* Proceedings of Erice School in Subnuclear Physics, 1974, (Ed. A. Zichichi). Academic Press, London and New York.

MORPURGO, G. (1977). "Quarks and Hadronic Substructure". Plenum Press, New York.

MOTT, N. F. (1929). *Proc. R. Soc.* **A124**, 425.

NACHTMANN, O. (1972). *Nucl. Phys.* **B38**, 397.

NGUYEN, H. *et al.* (1977). *Phys. Rev. Lett.* **39**, 262.

NISHIJIMA, K. AND NAKANO, T. (1953). *Prog. Theoret. Phys.* **10**, 581.

NOVIKOV, V. A. *et al.* (1976). Moscow Report ITEP (112) (Unpublished).

O'DONNELL, P. J. (1976). *Phys. Rev. Lett.* **36**, 177.

OKUBO, S. (1962). *Prog. Theoret. Phys.* **27**, 949.

OKUBO, S. (1963). *Phys. Lett.* **5**, 163.

ONO, S. (1976a). *Phys. Rev. Lett.* **37**, 398.

ONO, S. (1976b). *Phys. Rev. Lett.* **37**, 655.

PANOFSKY, W. (1968). *In* Proceedings of International Symposium on High Energy Physics, Vienna, 1968.

PARTICLE DATA GROUP (T. TRIPPE *et al.*) (1976). *Rev. Mod. Phys.* **48** (2), Part II.

PATI, J. C. (1977). *In* "Fundamentals of Quark Models" (Ed. I. Barbour and A Davies). Scottish University Summer School in Physics.

PATI, J. C. AND SALAM, A. (1973a). *Phys. Rev.* **D8**, 1240.

PATI, J. C. AND SALAM, A. (1973b). *Phys. Rev. Lett.* **31**, 661.

PATI, J. C. AND SALAM, A. (1974). *Phys. Rev.* **D10**, 275.

PATI, J. C. AND SALAM, A. (1975). *Phys. Lett.* **58B**, 333.

PERL, M. *et al.* (1977). *Phys. Lett.* **70B**, 487.

PERRUZI, I. *et al.* (1977). *Phys. Rev. Lett.* **39**, 1301.

PETERSON, W. P. AND ROSNER, J. L. (1972). *Phys. Rev.* **D6**, 820.

PETERSON, W. P. AND ROSNER, J. L. (1973). *Phys. Rev.* **D7**, 747.

PERKINS, D. H. (1975). *Contemp. Phys.* **16**, 173.

PERKINS, D. H., SCHREINER, P. AND SCOTT, W. (1977). *Phys. Lett.* **67B**, 347.

POLITZER, H. D. (1973). *Phys. Rev. Lett.* **30**, 1346.

POLITZER, H. D. (1974). *Phys. Reports*, **14C**, 129.

PREPARATA, G. (1973). *Phys. Rev.* **D7**, 2973.

PREPARATA, G. (1974). *Nucl. Phys.* **B80**, 299.

PREPARATA, G. (1975). *In* "Lepton and Photon Structure" (Ed. A. Zichichi). International School of Subnuclear Physics, Erice, Sicily, 1974, Academic Press, London and New York.

RANDA, J. AND DONNACHIE, A. (1977). *Nucl. Phys.* **B125**, 303.

REINES, F., GURR, H. S. AND SOBEL, H. W. (1976). Proceedings of International Neutrino Conference, Aachen, 1976, p. 217. Vieweg, Braunschweig.

RIORDAN, E. M. *et al.* (1975). SLAC PUB. 1634. SLAC, Stanford, California. (Unpublished.)

ROSNER, J. (1969). *Phys. Rev. Lett.* **22**, 362.

RUBINSTEIN, H. R. (1966b). *Phys. Rev. Lett.* **17**, 41.

RUTHERFORD, E. (1911). *Phil. Mag.* **21**, 669.

RUTHERGLEN, J. (1969). Proceedings of IV International Symposium on Electron and Photon Interactions at High Energies (Ed. D. Braben), p. 163. Daresbury Laboratory, England.

SALAM, A. (1967). Lectures at Imperial College, London. (Unpublished.) (As reported in Salam, 1968.)

SALAM, A. (1968). *In* "Elementary Particle Theory". Proceedings of 8th Nobel Symposium (Ed. N. Svartholm), p. 367. Almqvist and Wiksell, Stockholm.

SALAM, A. AND WARD, J. C. (1964). *Phys. Lett.* **13**, 168.
SANDA, A. AND TEREZAWA, H. (1975). *Phys. Rev. Lett.* **34**, 1403.
SCHMIDT, M. (1974). *Phys. Rev.* **D9**, 408.
SCHWINGER, J. (1957). *Ann. Phys.* **2**, 407.
SCHWINGER, J. (1964). *Phys. Rev. Lett.* **12**, 237; and *Phys. Rev.* **135**, B816.
SCHWITTERS, R. F. (1975). *In* Proceedings of 1975 International Symposium on Lepton and Photon Interactions at High Energies, Stanford, California. (SLAC, Stanford, California.)
SCHWITTERS, R. F. AND STRAUCH, K. (1976). *Ann. Rev. Nucl. Sci.* **26**, 89.
SCHWITTERS, R. F. *et al.* (1975). *Phys. Rev. Lett.* **35**, 1320.
SEGRE, G. AND WEYERS, J. (1976). *Phys. Lett.* **62B**, 91.
SEHGAL, L. M. (1974). *Phys. Rev.* **D10**, 1663.
SIVERS, D., BRODSKY, S. J. AND BLANKENBECLER, R. (1976). *Phys. Reports,* **23C**, 1.
SMITH, A. J. (1976). *In* Proceedings of Summer Institute on Particle Physics, Stanford Report 198, p. 449.
STEINBERGER, J. (1977). Some recent experimental results in high energy neutrino physics. CERN Report, Geneva, 1977. To be published in Proceedings of Summer Institute, Cargese, 1977.
SUNDARESON, M. K. AND WATSON, P. J. S. (1970). *Ann. Phys.* **59**, 375.
TAYLOR, R. E. (1975). Proceedings of International Symposium on Lepton and Photon Interactions at High Energies, Stanford (1975) (Ed. W. T. Kirk), p. 679. SLAC, Stanford, California.
'T HOOFT, G. (1971a). *Nucl. Phys.* **B33**, 173.
'T HOOFT, G. (1971b). *Nucl. Phys.* **B35**, 167.
'T HOOFT, G. (1972). Unpublished remark in Proceedings of 1972 Marseilles Conference on Yang–Mills fields (as quoted in footnote 3 on p. 132 of Politzer, 1974).
'THOOFT, G. (1974). *Nucl. Phys.* **B75**, 461.
TUNG, WU-KI (1975). *Phys. Rev.* **D12**, 3613.
WALKER, R. L. (1969). *In* Proceedings of IV International Symposium on Electron and Photon Interactions at High Energies (Ed. D. Braben), p. 23. Daresbury Laboratory, England.
WEINBERG, S. (1967). *Phys. Rev. Lett.* **19**, 1264.
WEINBERG, S. (1973). *Phys. Rev. Lett.* **31**, 494.
WEINBERG, S. (1974). *Rev. Mod. Phys.* **46**, 255.
WEISBERGER, W. I. (1965). *Phys. Rev. Lett.* **14**, 1047.
WERBROUCK, A. *et al.* (1970). *Nuovo Cimento Lett.* **4**, 1267.
WEST, G. B. (1970). *Phys. Rev. Lett.* **24**, 1206.
WEST, G. B. (1975). *Phys. Reports,* **18**, 263.
WHITAKER, J. S. *et al.* (1976). *Phys. Rev. Lett.* **37**, 1596.

WIIK, B. (1975). *In* Proceedings of International Symposium on Electron and Photon Interactions at High Energies, Stanford, California (Ed. W. Kirk), p. 679. SLAC, Stanford, California.

WIIK, B. AND WOLF, G. (1977). Lectures at Les Houches Summer School 1976. DESY report 77/01. DESY, Hamburg.

YAMAGUCHI, Y. (1975). *In* Proceedings of European Physical Society International Conference on High Energy Physics, Palermo (Ed. A. Zichichi). Editrice Compositori, Bologna.

YANG, C. N. (1950). *Phys. Rev.* **77**, 242.

YANG, C. N. AND MILLS, R. C. (1954). *Phys. Rev.* **96**, 191.

ZEE, A. (1973). *Phys. Rev.* **D8**, 4038.

ZWEIG, G. (1964a). CERN Report No. 8182/TH 401 (Unpublished).

ZWEIG, V. (1964b). CERN Report No. 8419/TH 412 (Unpublished).

Subject Index

A

Active Quark, 249, 303; *see also* Counting rules

Additivity, quark electromagnetic transitions, 143

Adler sum rule, 235

Adler–Weisberger sum rule, 105

Angular distributions
in e^+e^- annihilation, 281–290
in lepton scattering, 175, 177, 180, 188–189
in neutrino interactions, 237, 240–241
ψ' decay, 366–369

Anomalous magnetic moment
light quark, 176, 186, 392, 416–418
massive quark, 126, 128, 417
proton, 176–177, 418

Antiparticles, SU(2) transformation, 28

Antiquarks, 11, 38, 54–59; *see also* Sea quarks
in SU(2), 26–28, 54

Antisymmetric representation, 40, 45, 46, 58
SU(4), 379–380, 383
SU(6), 56, 60–61

Antisymmetric wavefunctions, colour, 338–339

Asymptotic freedom, 217–221, 255, 344, 351, 357, 407, 409–410, 448, 451

Automodelity, 316; *see also* Counting rules

Axial charge, Melosh transformation, 114

B

$B(x\mathscr{F}_3/\mathscr{F}_2)$, 241–243

b quark, 444, 447–448

Bad operators, 206; *see also* Good operators

Bag, *see* MIT bag

Bar magnet, Zweig rule analogy, 74

Baryons, 5, 7; *see also* SU(6) × O(3)
colour, 338–339
decuplet, 19
qqq orbital excitations, 80–86
qqq states, 39–40, 339
SU(6), 59–62

Baryon number, of q and $\bar{q}$, 38

Bethe–Salpeter model, 157–158

Bjorken sum rule, 298–301